Intelligent Networks

The book presents the latest developments in intelligent communication networks based on applicability from various domains of artificial intelligence and machine learning, including channel modeling, model-based structure, channel prediction, and signal detection. It further explains important topics such as vehicular mobility modeling, human-centric network applications, security and privacy in social networks, and trust-based intelligent transportation systems.

This book:

- Presents a model-based approach to constructing an effective network by using state-of-the-art artificial intelligence techniques.
- Discusses the theoretical and practical applications of channel prediction and signal detection.
- Introduces the fundamental concepts and application of vehicular networks in conjunction with artificial intelligence.
- Explores wireless communication network techniques enabled by human-centric applications, and designed and developed with artificial intelligence characteristics.
- Highlights the challenges in designing and developing an effective and intelligent communication network that can be applied in different domains of human activities to find sustainable solutions.

This book is primarily written for senior undergraduate and graduate students, and academic researchers in the fields of electrical engineering, electronics and communications engineering, computer engineering, and information technology.

Wireless Communications and Networking Technologies: Classifications, Advancement and Applications

Series Editor: D.K. Lobiyal, R.S. Rao and Vishal Jain

The series addresses different algorithms, architecture, standards and protocols, tools and methodologies which could be beneficial in implementing next generation mobile network for communication. Aimed at senior undergraduate students, graduate students, academic researchers and professionals, the proposed series will focus on the fundamentals and advances of wireless communication and networking, and their such as mobile ad-hoc network (MANET), wireless sensor network (WSN), wireless mess network (WMN), vehicular ad-hoc networks (VANET), vehicular cloud network (VCN), vehicular sensor network (VSN) reliable cooperative network (RCN), mobile opportunistic network (MON), delay tolerant networks (DTN), flying ad-hoc network (FANET) and wireless body sensor network (WBSN).

Cloud Computing Enabled Big-Data Analytics in Wireless Ad-hoc Networks
Sanjoy Das, Ram Shringar Rao, Indrani Das, Vishal Jain and Nanhay Singh

Smart Cities
Concepts, Practices, and Applications
Krishna Kumar, Gaurav Saini, Duc Manh Nguyen, Narendra Kumar and Rachna Shah

Wireless Communication
Advancements and Challenges
Prashant Ranjan, Ram Shringar Rao, Krishna Kumar and Pankaj Sharma

Wireless Communication with Artificial Intelligence
Emerging Trends and Applications
Anuj Singal, Sandeep Kumar, Sajjan Singh and Ashish Kr. Luhach

Computational Intelligent Security in Wireless Communications
Suhel Ahmad Khan, Rajeev Kumar, Omprakash Kaiwartya, Raees Ahmad Khan and Mohammad Faisal

Networking Technologies in Smart Healthcare: Innovations and Analytical Approaches
Pooja Singh, Omprakash Kaiwartya, Nidhi Sindhwani, Vishal Jain and Rohit Anand

Artificial Intelligence in Cyber Physical Systems: Principles and Applications
Anil Kumar Sagar, Parma Nand, Neetesh Kumar, Sanjoy Das and Subrata Sahana

Intelligent Networks: Techniques and Applications
Vivek Kumar Singh, Anil Kumar Sagar, Parma Nand, Rani Astya and Omprakash Kaiwartya

For more information about this series, please visit: https://www.routledge.com/Wireless%20Communications%20and%20Networking%20Technologies/book-series/WCANT

Intelligent Networks

Techniques and Applications

Edited by
Vivek Kumar Singh, Anil Kumar Sagar,
Parma Nand, Rani Astya
and Omprakash Kaiwartya

CRC Press
Taylor & Francis Group
Boca Raton London New York

CRC Press is an imprint of the
Taylor & Francis Group, an **informa** business

Designed cover image: shutterstock

First edition published 2025
by CRC Press
2385 NW Executive Center Drive, Suite 320, Boca Raton FL 33431

and by CRC Press
4 Park Square, Milton Park, Abingdon, Oxon, OX14 4RN

CRC Press is an imprint of Taylor & Francis Group, LLC

ISBN: 9781032657950 (hbk)
ISBN: 9781032891392 (pbk)
ISBN: 9781003541363 (ebk)

DOI: 10.1201/ 9781003541363

Typeset in Sabon
by Deanta Global Publishing Services, Chennai, India

Contents

Preface

The intelligent network is one of the most advanced concepts, designs, and architectures of intelligent communication networks that jointly utilize artificial intelligence and machine learning concepts with computer networks to facilitate intelligent communication in terms of controlling functions of the computer networks. Artificial intelligence and machine learning improve the efficiency of network communication with appropriate security. Human social-centric applications are required for intelligent networks to improve communication and increase network capacity. Furthermore, the application-oriented architecture of the communication network is enabled with a learning mechanism. This approach presents fundamental concepts of intelligent networks with human activity and assesses application-based network solutions in detail. Wireless communication network techniques have been enabled by human-centric design and developed with artificial intelligence characteristics. This approach provides effective solutions for human-centric problems. In addition, the protection-driven framework for solving social communication issues in terms of security and privacy is enabled by intelligent techniques. Security and privacy are crucial in social communication models.

This book is a formal introduction to artificial intelligence and communication networks in human social activities. It provides more efficient and robust approaches for enhancing the thinking processes of networks to deliver packets in an appropriate manner. It also discusses the significant future challenges and issues of communication networks that could be addressed by employing artificial intelligence techniques. It provides a comprehensive study that discovers more effective interaction between artificial intelligence and communication networks. This book is a comprehensive and valuable approach to the theory, applications, and current research developments in the technology related to artificial intelligence and communication networks.

Editors

Vivek Kumar Singh is currently working as Assistant Professor in the Department of Computer Science and Engineering, School of Engineering and Technology at Sharda University, India. He received a Ph.D. in Computer Science and Engineering from the National Institute of Technology, Uttarakhand; M. Tech in Software Engineering from Motilal Nehru National Institute of Technology, Allahabad; and a B.E. in Computer Science and Engineering from Rajiv Gandhi Proudyogiki Vishwavidyalaya, Bhopal. He has more than 12 years of industry, teaching, and research experience in various reputed organizations. He has published good quality research articles in reputed international journals and conferences, including six SCI and SCI-E journals and nine conferences, from which two were awarded "Best Paper." His areas of interest are computer vision, artificial intelligence, machine learning, and visual attention modeling. He is involved in the reviewing process for various reputed journals and conferences.

Anil Kumar Sagar is currently working as Professor of Department of Computer Science and Engineering in the School of Engineering and Technology, Sharda University, India. Dr. Anil Kumar Sagar obtained his doctorate from JNU, New Delhi, in the area of ad hoc networks. He obtained his B.E. in Computer Science and Engineering from G B Pant Engineering College Pauri, Garhwal, and an M. Tech from JSSATE Noida. He formerly served as a Dean of Academics at Raj Kumar Goel Institute of Technology, Ghaziabad. He also worked as a Member of the Board of Studies in the Computer Science Department at Galgotias University, Greater Noida, and RKGIT, Ghaziabad. He is a member of the editorial board/review committee for many international and national journals and served as a program/organizing committee member for several conferences. He has guided ten M. Tech (Computer Science) students and is currently guiding five Ph.D. (Computer Science) students. He has about 20 years of teaching experience in various prestigious colleges and universities.

Parma Nand received a Ph.D. in Computer Science and Engineering from IIT Roorkee and an M. Tech and a B. Tech in Computer Science and Engineering from IIT Delhi. He has more than 27 years of experience both in industry and academia. He has published more than 150 papers in peer-reviewed international and national journals and conferences. He has also published a number of book chapters with reputed publishers, such as Springer Berlin Heidelberg. He has reviewed a number of books for publishers such as Tata McGraw-Hill and Galgotias Publications, and papers in journals of international repute. He has delivered many invited/keynote talks at international and national conferences/workshops/seminars in India and abroad. He is an active member of the advisory/technical program committee of reputed international and national conferences. He is a reviewer for a number of reputed journals such as Springer, Elsevier, and Computers & Electrical Engineering. His research interests include computer graphics, algorithms, distributed computing, and wireless and sensor networking. He has successfully completed a number of government-funded projects.

Rani Astya is an Associate Professor in the Computer Science and Engineering Department at the School of Engineering and Technology, Sharda University, India. She has a B. Tech, M. Tech (IIT Roorkee), and Ph.D. (Banasthali Vidyapith). She has 20 years of experience in academics, currently teaching computer science subjects for B. Tech, M. Tech, MCA, and BCA students. She has published 45 research papers in well-known peer-reviewed international conferences and journals, and is currently researching underwater sensor networking. She has also published two book chapters. She has guided a number of students on their projects. Under her guidance, a number of students participated in and won the Hackathon competition. She has filed over 20 patents, of which 15 have been published. Others are lined up in the future.

Omprakash Kaiwartya is currently working at the Department of Computer Science, Nottingham Trent University (NTU), UK, as a Senior Lecturer and Course Leader for MSc Engineering (Electronics). Previously, he was a Research Associate (equivalent to Senior Lecturer) in the Department of Computer and Information Science at Northumbria University, Newcastle, UK, where he was working on a European Union project. Prior to this, he was a Post-Doctoral Fellow (equivalent to Lecturer) in the Faculty of Computing, University of Technology (UTM), Malaysia. Before moving to Malaysia, he completed his BSc in Computer Science from Guru Ghansidas Central University, Bilaspur Chhattisgarh, India, and a combined master's and Ph.D. degree from the School of Computer and Systems Science, Jawaharlal Nehru University (JNU), New Delhi, India. Overall, he has authored/co-authored over 100 international publications including IEEE journal articles, conference proceedings, book chapters, and books. Dr. Omprakash's research interest focuses on cyber security in connected systems for diverse domain areas including transport, healthcare, and industrial production. His recent scientific contributions are on the Internet of Connected Vehicles (IoV), e-mobility, electronic vehicle-charging management (EV), Internet of Healthcare Things (IoHT), smart use case implementation of sensor networks, and next-generation wireless communication technologies (6G and beyond). Furthermore, Omprakash is a Fellow of the Higher Education Academy (FHEA), an IEEE senior member, and a BCS professional member. He has served as a TPC member or reviewer at more than 100 international conferences and workshops including IEEE Globecom, IEEE ICC, IEEE CCNC, IEEE ICNC, IEEE VTC, IEEE INFOCOM, ACM CoNEXT, ACM MobiHoc, and ACM SAC. Furthermore, he reviews papers for over 30 international journals including IEEE magazines Wireless Communications, Networks, Communications, IEEE Communications Letters, IEEE Sensors Letters, IEEE Transactions on Industrial Informatics, Vehicular Technologies, Intelligent Transportation Systems, Big Data, and Mobile Computing. Moreover, Omprakash has been an editorial member of various special issues of top-ranked journals of the Communication Society. He also serves as an Associate Editor of IET Intelligent Transport Systems, IEEE Internet of Things Journal, Springer, EURASIP Journal on Wireless Communication and Networking, MDPI Electronics, Ad-Hoc and Sensor Wireless Networks, and KSII Transactions on Internet & Information Systems.

Contributors

Vaibhav Agarwal
Department of Computer Engineering
NIT Kurukshetra
Haryana, India

Rishabh Aggarwal
Jagan Institute of Management Studies
New Delhi, India

Tasneem Ahmed
Advanced Computing and Research Laboratory
Department of Computer Application
Integral University
Lucknow, Uttar Pradesh, India

Shalom Akhai
Department of Mechanical Engineering
Maharishi Markandeshwar Engineering College
Maharishi Markandeshwar (Deemed to be University)
Haryana, India

Shaik Vaseem Akram
Department of Electronics and Communication Engineering
SR University
Telangana, India

M. Anandaraj
Department of Information Technology
PSNA College of Engineering and Technology
Dindigul, Tamilnadu, India

Iqra Ashraf
Department of Hospital Administration
Era University
Lucknow, Uttar Pradesh, India

Rani Astya
School of Engineering and Technology
Department of Computer Science and Engineering
Sharda University
Greater Noida, Uttar Pradesh, India

Megha Chhabra
School of Engineering and Technology
Department of Computer Science and Engineering
Sharda University
Greater Noida, Uttar Pradesh, India

Sayani Das
Department of Mass Communication
The University of Burdwan
West Bengal, India

K. Deepa
Manav Rachna University
Faridabad, Haryana, India

D. Devasena
Sri Ramakrishna Engineering College
Coimbatore, Tamil Nadu, India

Mohammad Faisal
Advanced Computing and Research Laboratory
Department of Computer Application
Integral University
Lucknow, Uttar Pradesh, India

Nafees Akhter Farooqui
Department of Computer Science Engineering
Koneru Lakshmaiah Education Foundation
Vaddeswaram, Andhra Pradesh, India

P. Ganeshkumar
College of Computer and Information Science
AI-Imam Mohammad Ibn Saud Islamic University
Saudi Arabia

Raju Singh Gaur
Department of Mathematics
D.A-V. College
C.S.J.M. University
Kanpur, Uttar Pradesh, India

Vinay Gautam
Chitkara University Institute of Engineering and Technology
Chitkara University
Punjab, India

Anita Gehlot
Division of Research and Innovation
Uttaranchal Institute of Technology
Uttaranchal University
Dehradun, Uttarakhand, India

Ramkrishna Ghosh
Department of Information Technology
Haldia Institute of Technology
Haldia, West Bengal, India

Ratnalata Gupta
Shivajirao Kadam Institute of Technology and Management (SKITM)
Indore, Madhya Pradesh, India

Vishal Jain
Sharda School of Engineering and Technology
Sharda University
Greater Noida, Uttar Pradesh, India

Rupam Jaiswal
Department of Hospital Administration
Era University
Lucknow, Uttar Pradesh, India

Nashra Javed
Advanced Computing and Research Laboratory
Department of Computer Application
Integral University
Lucknow, Uttar Pradesh, India

Meet Kapoor
Jagan Institute of Management Studies
New Delhi, India

Samta Kathuria
Parul Institute of Law
Parul University
Vadodara, Gujarat, India

Christian Kaunert
School of Law and Government
Professor of International Security
Dublin City University
Ireland

Kazi Kutubuddin Sayyad Liyakat
Department of Electronics and Telecommunication Engineering
Brahmdevdada Mane Institute of Technology
Solapur, Maharashtra, India

Suman Madan
Jagan Institute of Management Studies
New Delhi, India

Jhansi Bharathi Madavarapu
Information Technology
University of the Cumberlands
United States

Manoj Kumar Mahto
Vignan Institute of Technology and Science
Hyderabad, Telangana, India

Praveen Malik
Discipline of Electronics and Communication Engineering
Lovely Professional University
Punjab, India

Tushar Mehrotra
Department of Computer Science and Engineering
Sharda School of Engineering and Technology
Sharda University
Greater Noida, Uttar Pradesh, India

Venugopal Reddy Modhugu
Independent Researcher
Fairfax, USA

Suneeta Mohanty
School of Computer Engineering
KIIT (Deemed to be University)
Bhubaneswar, Odisha, India

Archan Mitra
School of Media Studies
Presidency University
Bangalore, India

S. Naganandhini
Department of Computer Science and Engineering
PSNA College of Engineering and Technology
Dindigul, Tamilnadu, India

Parma Nand
School of Engineering and Technology
Department of Computer Science and Engineering
Sharda University
Greater Noida, Uttar Pradesh, India

Sasmita Padhy
VIT Bhopal University
Sehore, Madhya Pradesh, India

Shweta Pandey
Law College of Dehradun
Uttaranchal University
Dehradun, India

Prasant Kumar Pattnaik
School of Computer Engineering
KIIT (Deemed to be University)
Bhubaneswar, Odisha, India

Mamta Punjabi
Shivajirao Kadam Institute of Technology and Management (SKITM)
Indore, Madhya Pradesh, India

G. Rajavikram
Vignan Institute of Technology and Science
Hyderabad, Telangana, India

Jyoti Rani
Chandigarh University
Punjab, India

Shivani G Sabharwal
Jagan Institute of Management Studies
New Delhi, India

Halima Sadia
Department of Computer Science and Engineering
Integral University
Lucknow, Uttar Pradesh, India

Anil Kumar Sagar
School of Engineering and Technology
Department of Computer Science and Engineering
Sharda University
Greater Noida, Uttar Pradesh, India

Subrata Sahana
School of Engineering and Technology
Department of Computer Science and Engineering
Sharda University
Greater Noida, Uttar Pradesh, India

Sakshi
Sharda School of Engineering and Technology
Department of Computer Science and Applications
Sharda University
Greater Noida, Uttar Pradesh, India

K. Selvaraj
Department of Information Technology
PSNA College of Engineering and Technology
Dindigul, Tamilnadu, India

Bhagwati Sharan
School of Engineering and Sciences
Department of Computer Science and Engineering
SRM University-AP
Amaravati, Andhra Pradesh, India

Mousam Sharma
School of Law
Queen Mary University of London
London, United Kingdom

Rattan Sharma
Centre for Sustainable Development
VIPS-TC & DSB
New Delhi, India

Bhupinder Singh
Sharda School of Law
Sharda University
Greater Noida, Uttar Pradesh, India

Rajesh Singh
Division of Research and Innovation
Uttaranchal Institute of Technology
Uttaranchal University
Dehradun, Uttarakhand, India

Vivek Kumar Singh
Sharda School of Engineering and Technology
Department of Computer Science and Engineering
Sharda University
Greater Noida, Uttar Pradesh, India

Amrit Suman
Sharda School of Engineering and Technology
Sharda University
Greater Noida, Uttar Pradesh, India

Preetam Suman
VIT Bhopal University
Sehore, Madhya Pradesh, India

Shashikala Tapaswi
Department of Computer Science and Engineering
ABV-IIITM
Gwalior, Madhya Pradesh, India

Priyanka Tyagi
Department of Computer Science and Engineering
Galgotias College of Engineering and Technology
Greater Noida, Uttar Pradesh, India

Kanishka Vaish
Law College of Dehradun
Uttaranchal University
Dehradun, Uttarakhand, India

Amandeep Singh Wadhwa
University Institute of Engineering and Technology
Panjab University
Chandigarh, India

Pawan Whig
Vivekananda Institute of Professional Studies – Technical Campus
New Delhi, India

Sana Zeba
Department of Computer Engineering
Jamia Millia Islamia University
New Delhi, India

Chapter 1

Fundamentals of AI and communication networks

Applications in human social activities

Manoj Kumar Mahto and G. Rajavikram

1.1 INTRODUCTION

Artificial intelligence (AI) and communication networks, two autonomous yet interrelated technologies, influence modern digital life. AI integrates machine learning, natural language processing, and computer vision to imitate human intelligence (Russell & Norvig, 2016). Since its mid-20th century birth, it has revolutionised machine learning and human intelligence.

Nevertheless, communication networks are an intricate framework that allows devices and systems to communicate with one another. Since the beginning of telecommunication networks, the World Wide Web has grown, and new wireless technologies like 5G have emerged (Thompson et al., 2014). These networks enable worldwide data transmission and communication. Where AI's data processing skills harmoniously connect with communication networks' enormous reach, new possibilities have grown. Edge computing reduces latency and improves real-time decision-making by processing AI closer to the data source (Shi et al., 2016). Additionally, AI in communication networks has enabled predictive maintenance, intelligent network management, and data route optimisation for efficiency. Communication networks using AI revolutionise consumer relationships. AI-powered chatbots and virtual assistants respond instantly and with precision to natural language enquiries. Customer service is improved, and support is always available, transcending geographical and temporal obstacles. Users feel appreciated and heard since these interactions are personalised and responsive. The association can alter knowledge distribution and education. AI algorithms may recommend instructional content based on learning styles, timing, and understanding. Communication networks distribute these resources to learners globally, democratising education and levelling the playing field for varied learners. Human social activities have changed due to this convergence. AI-powered algorithms evaluate massive communication information from networks to offer personalised social media content. Communication networks currently include AI-driven virtual assistants and chatbots for human-like conversations and assistance with customers.

DOI: 10.1201/9781003541363-1

1.2 FOUNDATIONS OF AI

1.2.1 AI

In AI, machines simulate human-like intelligence to do tasks that require human intellect. Each approach and discipline of AI addresses particular intelligence emulation features. The subdivisions add to AI systems' extensive capabilities. The field of AI has a rich history marked by significant developments and milestones that have shaped its evolution into the transformational discipline it is today. Figure 1.1 illustrates the fundamental underpinnings of AI.

1950s – The birth of AI The inception of AI can be traced back to the 1950s when researchers like Alan Turing and John McCarthy laid the foundation. Turing introduced the concept of a machine capable of intelligent behaviour, known as the Turing Test, which remains a touchstone for AI evaluation. McCarthy coined the term "Artificial Intelligence" and organised the seminal Dartmouth Workshop in 1956, marking the formal birth of AI research (Turing, 1950; McCarthy, 1955).

1956–1970 – Early exploration: The following years saw AI researchers exploring fundamental concepts such as problem-solving, theorem proving, and symbolic reasoning. The Logic Theorist, developed by Allen Newell and Herbert A. Simon, showcased machine capabilities for proving mathematical theorems. The General Problem Solver, another creation of Newell and Simon, demonstrated AI's ability to solve a wide range of problems using symbolic manipulation (Newell et al., 1960).

1980s – Expert systems and knowledge representation: The 1980s saw a rise in expert systems and AI algorithms that replicated human experts' recommendations. AI can capture and use domain-specific knowledge, which has been demonstrated by rule-based systems like MYCIN for medical diagnosis. The period emphasises knowledge representation and reasoning in AI systems.

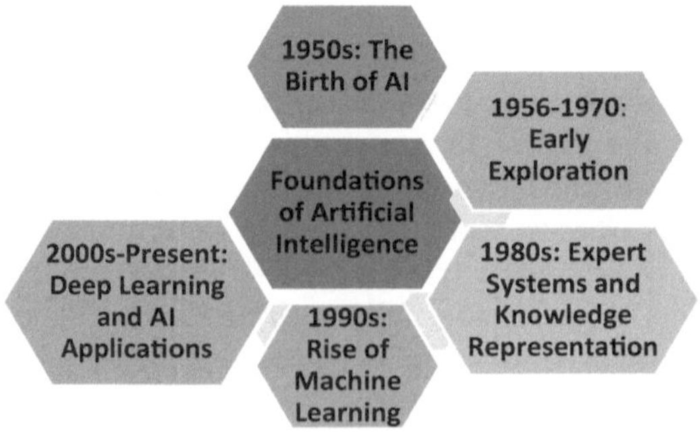

Figure 1.1 Foundation of AI.

1990s – Rise of machine learning: Advances in machine learning and neural networks reinvigorated AI in the 1990s. Machine learning techniques like SVMs and decision trees became prominent for the classification and recognition of patterns. AI defeated Garry Kasparov in challenging strategy games with IBM's Deep Blue. This era witnessed the prominence of machine learning techniques such as Support Vector Machines (SVMs) and decision trees, which played pivotal roles in the classification and recognition of intricate patterns. Notably, AI achieved a historic milestone when IBM's Deep Blue defeated Garry Kasparov in challenging strategy games, showcasing the prowess of AI in complex decision-making scenarios. This period marked a crucial chapter in the evolution of AI, setting the stage for continued innovation and exploration in the field of machine learning (Russell & Norvig, 2018).

2000s–present – Deep learning and AI applications: Massive datasets and sophisticated processing enabled breakthrough deep learning developments in the 21st century. Linguistic translation, picture and audio recognition, and many other tasks are achieved using deep neural networks. AlphaGo, which defeated human Go champions, demonstrated AI's strategic reasoning (Silver et al., 2017). Recent contributions from prominent researchers have significantly advanced our understanding of the ever-evolving landscape of AI applications. In their comprehensive review, Smith and Johnson (2022) delve into the latest advancements in deep learning, offering valuable insights into the current state of the field. Their work not only synthesises existing knowledge but also contributes new perspectives, providing a nuanced understanding of the ongoing trends in AI. Ma et al. (2023) contribute to the discourse by focusing on innovations in neural network architectures tailored for multimodal tasks. Furthermore, Gawlikowski et al. (2023) offer a detailed analysis of the practical applications of deep neural networks in real-world scenarios. Their case study-driven approach sheds light on the effectiveness and challenges of deploying AI technologies, enhancing our comprehension of AI's impact on diverse industries. These recent works collectively contribute to the evolving narrative of AI research, offering not only an update on the latest trends but also deepening our understanding of the potential and challenges within the field.

1.3 ESSENTIALS OF COMMUNICATION NETWORKS

Communication networks serve as the backbone of our connected world, enabling a seamless exchange of data and information between devices and users. At its core are the internet, wireless networks, and communication protocols, each playing a pivotal role in shaping the modern landscape of communication. Figure 1.2 depicts the essentials of network components.

The **internet** is a global network of interconnected computers and devices that operates on standardised protocols such as TCP/IP, ensuring reliable data. It has revolutionised the way we access information, communicate,

Figure 1.2 Basics of network components.

and collaborate, fostering a digital ecosystem that transcends geographical boundaries.

Wireless networks have transformed the way we communicate by eliminating the need for physical connections. Wi-Fi, based on IEEE 802.11 standards, enables devices to connect wirelessly to local networks (Tanenbaum & Wetherall, 2018). Cellular networks, including 4G and emerging 5G technologies, provide ubiquitous mobile connectivity, enabling real-time communication and powering the Internet of Things (IoT).

Communication protocols are standardised rules that govern data exchange within networks. Protocols like HTTP facilitate web communication, SMTP ensures reliable email transmission, and TCP guarantees robust data transfer (Stallings, 2016). These protocols enable devices with diverse characteristics to communicate seamlessly, fostering interoperability.

1.3.1 Introduction to network architecture and components

Network architecture defines the structure and arrangement of network components, which is crucial for designing efficient and reliable communication infrastructures. Key components include **nodes**, such as computers and routers, that facilitate data transmission; **links**, physical or wireless connections between nodes; **routers** that direct data; **switches** that manage data within local networks; and **firewalls** that safeguard against threats. Two common network architectures are **client-server architecture**, where servers provide services to clients, and **peer-to-peer architecture**, where devices communicate directly. Architectural knowledge is vital when developing networks that satisfy varied communication demands. In Figure 1.3, the structure of networks and their components are explained.

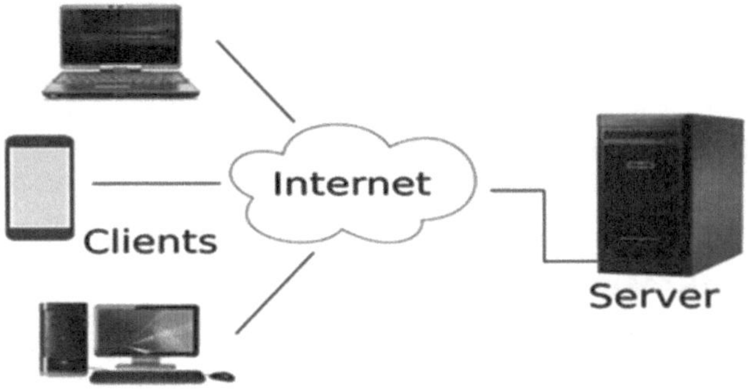

Figure 1.3 Basics of network architecture and components.

1.3.2 Evolution of communication networks: from 1G to 5G and beyond

Starting with 1G analogue cellular networks, communication networks have advanced to 5G and beyond. 2G offered digital communication and messaging through texts, whereas 1G introduced the wireless telephone. 3G provided mobile data and connection to the internet; 4G added apps and faster data.

The latest generation 5G assures transformation. It promises autonomous vehicles, remote surgery, and augmented reality with high speeds, low latency, and significant device connection (Andrews et al., 2014). Beyond 5G, research focuses on millimetre waves and quantum communication, forecasting a future where communication networks redefine a sense of connection. Recent studies by Li et al. (2022) delve into the optimisation of millimetre-wave communication for enhanced data transfer, while the work of Narottama (2023) explores the applications and security implications of quantum communication, offering insights into the evolving landscape of communication technologies. These latest contributions underscore ongoing efforts to push the boundaries of connectivity, paving the way for unprecedented applications and advancements in the field.

1.4 CONVERGENCE OF AI AND COMMUNICATION NETWORKS

AI and communication systems have been brought together to bring about a new era that is characterised by increased productivity and creativity. This synergistic link is transforming data processing, analysis, and use across fields. Communication networks have been optimised using AI's superior algorithms and capabilities, improving performance and user experiences. AI improves communication networks by processing and analysing data.

AI-based machine learning algorithms help networks find patterns, anomalies, and trends in enormous datasets. It enables network operators to quickly identify and resolve faults, maintaining uninterrupted connectivity (Haykin, 2019). AI-driven analytics that forecast detect network congestion and failures, optimising proactively.

1.4.1 Edge computing and its role in optimising AI applications over networks

The use of edge computing emerges as an essential component of this synergy, addressing the challenges of latency and bandwidth restrictions brought about by the combination of both of these variables. In this technique, information is processed closer to the source of information, which may be devices or sensors, rather than being sent to a centralised data hub. According to Shi et al. (2016), AI algorithms installed at the edge of the network enable real-time decision-making, thereby lowering latency and optimising bandwidth utilisation. The implementation of this strategy is essential for applications that require instantaneous reactions, such as autonomous vehicles and industrial IoT projects.

1.4.2 Cloud computing's contribution to AI-powered communication services

A further extremely important ingredient, cloud computing, enhances the theoretical possibilities of AI by offering scalable resources for performing complicated computations. A significant amount of processing power is required for AI models, which cloud services provide on demand. In order to provide seamless experiences, communication services that are driven by AI that are based on natural language processing and individualised recommendation engines, such as chatbots, exploit the adaptable nature of cloud computing (Marston et al., 2011).

1.4.3 Pioneering the future of communication networks

Data management, analysis, and application delivery are all being transformed because of the combination of AI and communication networks. The capacity of artificial intelligence to understand and predict future trends improves network performance, while the use of edge computing provides optimisations for AI applications that are sensitive to delay. The forthcoming wave of communication experiences will be shaped by cloud computing, which enables services that are powered by AI. As AI keeps on transforming the environment of communication networks, this synergistic journey opens doors to breakthroughs that we have not yet imagined may be possible.

1.5 AI-DRIVEN PERSONALISED CONTENT AND SOCIAL MEDIA

AI-driven customised content transforms social media through data-driven personalisation. Relevant content and adverts have been produced by machine learning algorithms depending on interests, behaviour, and interactions. This increases user engagement as they connect with the appropriate information. Confidentiality and algorithmic bias are problems. To effectively employ AI to create engaging, inclusive, and responsible social media interactions, customisation and integrity must be balanced. AI-driven individualised social media information is given in Figure 1.4.

1.5.1 AI algorithms personalise content recommendations on social media platforms

Social media suggestions for content are now personalised by AI algorithms. These algorithms capitalise on machine learning and data analysis to customise information depending on user decision-making, behaviours, and interactions. AI systems capitalise on collaborative filtering to identify trends in users' prior interactions and propose comparable material (Ekstrand et al., 2011). The customisation improves user engagement and platform retention by presenting relevant and interesting information.

1.5.2 Sentiment analysis and emotion detection for understanding user engagement

Natural language processing (NLP) sentiment analysis is important for understanding social media user involvement. AI systems analyse user-generated information like posts and comments to identify emotions and attitudes. Platforms learn users' attitudes towards subjects, goods, and events by assessing their positivity, negativity, or neutrality (Pang & Lee,

Figure 1.4 AI- driven personalised content and social media.

2008). This data helps platforms improve content suggestions and marketing according to users' emotions.

1.5.3 Social network analysis and its applications in targeting specific audiences

AI is employed in social network analysis to gather insights from the complicated connections developed by users of social media. AI technologies help platforms understand user interactions in networks through connection mapping, contributor identification, and communication pattern analysis (Wasserman & Faust, 1994). Targeted advertising, community identification, and enhancing content distribution approaches to reach certain customer segments can all benefit from this type of information. This nuanced understanding facilitates targeted advertising, community identification, and optimisation of content distribution strategies tailored to specific customer segments. Recent advancements in this domain, as highlighted by the works of Chen et al. (2023), further refine AI-driven social network analysis techniques, introducing novel algorithms and methodologies that enhance the accuracy and depth of insights derived from complex user connections. These contemporary contributions underscore the ongoing evolution of AI applications in the realm of social media analytics, paving the way for more sophisticated and effective strategies in targeted marketing and community engagement.

1.5.4 Enhancing user engagement through AI insights

Social media platforms have transformed into personalised and interconnected environments because of their incorporation of AI algorithms. The organisation uses sentiment measurement to gauge responses from consumers, consumer behaviour for personalised content suggestions, and statistical analysis of social networks to determine specific demographics and identify prominent individuals. AI and social media employed simultaneously could increase engagement among consumers and offer insightful data that helps platforms and advertisers create campaigns that are more successful.

1.6 VIRTUAL ASSISTANTS AND CHATBOTS

A substantial shift in user engagement and service delivery is now underway due to the implementation of AI-powered chatbots and artificially intelligent assistants into communication networks. New technologies have brought about this transformation. AI-powered entities are influencing how

Figure 1.5 Chatbot virtual assistant.

people and organisations interact by offering personalised aid and simplifying daily tasks. Furthermore, they encourage communication between both individuals and organisations from the two categories. Figure 1.5 shows how a chatbot may act as a virtual assistant.

1.6.1 Integration in communication networks

Chatbots and virtual assistants may be immediately incorporated into interpersonal relationships, allowing users to immediately have context-relevant conversations. They support messaging apps, websites, and real-time interactivity across channels. The incorporation enhances the user experience by increasing information and service accessibility. This is done by responding quickly, answering enquiries, and rendering transactions easy.

1.6.2 NLP for human-like interactions

AI extends NLP, which enables machines to interpret, generate, and comprehend human language. It enables the design and development of chatbots and AI-powered assistants. These AI-driven systems can read user enquiries, adapt to several languages, and interact like people. NLP makes this achievable. Mikolov et al. (2013) found that technological innovation allows fluid exchanges that are available, thereby rendering interaction more natural. This attribute is essential for technological adoption.

1.6.3 Case studies of AI-powered virtual assistants in social activities

The integration of virtual assistants is becoming ubiquitous in an assortment of interpersonal activities, consequently revolutionising the way in which human beings connect with one another and communicate. One example of a product that improves home automation is Amazon's Alexa, which makes it possible for customers to operate smart devices by using voice commands from their smartphones. Apple's Siri functions as an individual digital assistant, assisting users with a variety of responsibilities, like sending messages and scheduling reminders. According to Covington et al. (2016), chatbots on social media platforms such as Facebook help businesses manage consumer enquiries and provide quick responses, which in turn increases user engagement.

1.6.4 Challenges and future implications

Challenges keep arising despite the benefits AI provides. The desire to achieve an improved level of understanding and contextual awareness in AI responses constitutes a goal that continues to be pursued. There is an urgent requirement for thoughtful discussion on privacy problems that are associated with the sharing of data and user interactions. As AI continues to progress, the trajectory indicates a preference for interactions that are more sophisticated and human-like, with multimodal interfaces that include speech, text, and graphics.

1.6.5 Transforming user-centric engagement

The integration of virtual assistants, chatbots, and communication networks enhances the dynamics of human-machine interaction. The process of integration serves to augment the accessibility, efficiency, and personalisation of services. As advancements in AI technologies continue to unfold, AI-driven entities are poised to achieve heightened levels of efficiency. This transformative potential holds the promise of revolutionising user-centric engagement across various domains.

1.7 SMART CITIES AND IOT CONNECTIVITY

The simultaneous use of AI and communication networks is ushering in an entirely new era of smart cities, which has been defined as the reconfiguration of urban environments into intelligent ecosystems that are linked. Not only does this integration improve the quality of life in metropolitan areas, but it also contributes to the advancement of sustainability, efficiency, and

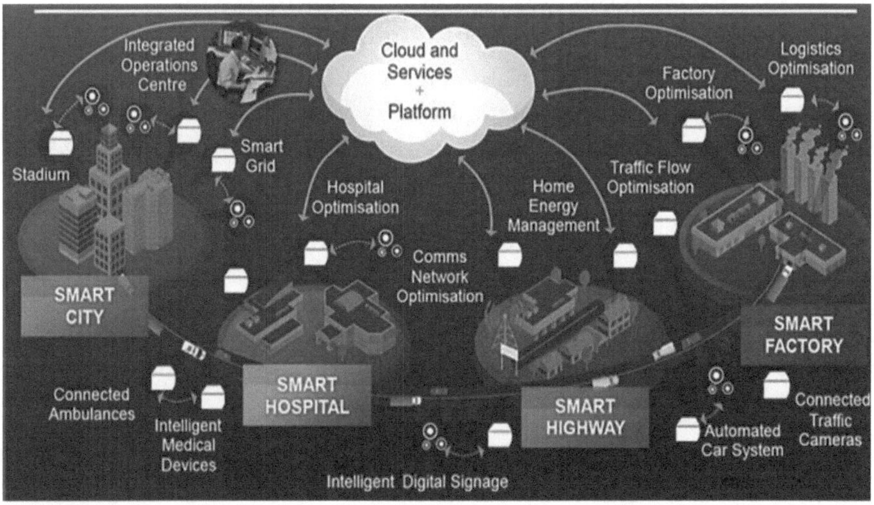

Figure 1.6 Smart cities and IoT connectivity.

innovation. The incorporation of IoT connections into smart cities is seen in Figure 1.6.

1.7.1 Role of AI and communication networks in smart cities

Smart cities depend on AI and communication networks to enable decision-making based on data and efficient use of resources. City administrators can monitor infrastructure, traffic, energy use, and more using real-time data from sensors, electronic devices, and people in general. Using this data, AI systems optimise the distribution of resources, improve services, and boost quality of life (Albino et al., 2015).

1.7.2 IoT devices and their impact on urban living

The integration of IoT devices plays a crucial role in the materialisation of the overarching concept of smart cities. The network of interconnected devices, encompassing a diverse range of smart metres and environmental sensors, engenders a comprehensive web of data sources that furnish real-time insights into the intricate dynamics of the urban landscape. IoT devices have emerged as a pivotal technology, facilitating the optimisation of waste management processes, the implementation of intelligent transportation systems, and the establishment of highly responsive energy grids. As an exemplification, intelligent traffic signals adeptly adapt their timings according to the prevailing traffic flow, thereby mitigating congestion and curbing emissions (Atzori et al., 2010).

1.7.3 Data-driven decision-making for improved city services and sustainability

The confluence of AI, communication networks, and IoT devices synergistically enables the facilitation of data-driven decision-making processes in the realm of urban planning and service delivery. AI algorithms are employed to meticulously scrutinise the copious data streams emanating from IoT devices. Through this process, these algorithms adeptly discern intricate patterns and discernible trends, thereby providing invaluable insights that inform and shape urban strategies. This methodology effectively augments urban services through the utilisation of predictive analytics, energy optimisation techniques, and real-time responsiveness mechanisms for emergency management systems (Caragliu et al., 2011).

1.7.4 Challenges and future outlook

Smart cities pose distinct and intricate challenges, encompassing apprehensions regarding the privacy of data, vulnerabilities in cybersecurity, and the imperative for resilient infrastructure. Given the ongoing advancements in AI technologies, it is imperative to prioritise the resolution of these challenges. The forthcoming era of smart cities is poised to witness the advent of cutting-edge technologies such as AI-powered autonomous vehicles, energy-efficient architectural structures, and interconnected healthcare systems. These advancements hold the potential to significantly augment citizen well-being while simultaneously fostering sustainable urban development.

1.7.5 Pioneering urban transformation

The convergence of AI, communication networks, and IoT technologies is playing a pivotal role in the transformation of urban landscapes into smart cities, which are emerging as epicentres of ground-breaking advancements, enhanced operational effectiveness, and environmentally conscious practices. As urban centres adopt and integrate these cutting-edge technologies, they undergo a transformative process, morphing into intricate and adaptive ecosystems that efficiently allocate resources, enhance service delivery, and foster a heightened sense of connectivity, thereby elevating the overall quality of life experienced by their inhabitants.

1.8 FUTURE DIRECTIONS AND CHALLENGES

Considering the convergence between AI and communication networks, we stand at a critical juncture where the trajectory of human social interactions teeters on the precipice of a captivating transformation. In this era brimming with potential, as we witness the emergence of novel trends, a myriad

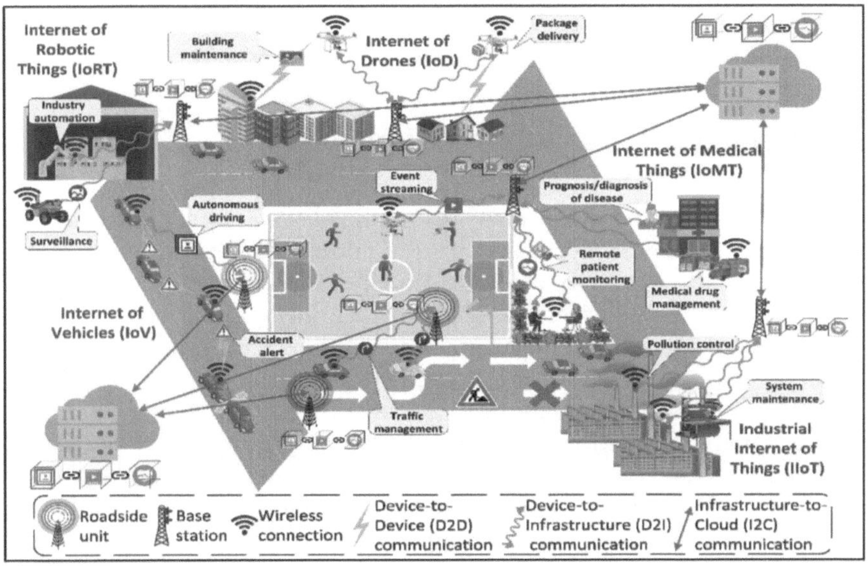

Figure 1.7 Future directions of AI and communication networks.

of obstacles presents itself, demanding our careful consideration. Our pursuit of innovation must transcend mere novelty as we strive to achieve ground-breaking advancements that are not only transformative but also ethically sound and equitable in nature. Figure 1.7 presents a comprehensive depiction of the potential trajectories of advancement in the domains of AI and communication networks. This visual representation allows us to envision a future that harmoniously integrates technological progress with the critical ethical considerations that underpin our society.

1.8.1 Emerging trends in AI and communication networks for social activities

In the forthcoming era, we are poised to witness a multitude of paradigm-shifting trends that hold the potential to revolutionise various aspects of our existence. The ongoing advancement of AI-powered personalisation is poised to further enhance the efficacy of content delivery, seamlessly tailoring it to the unique preferences of individuals in a dynamic and instantaneous manner. The burgeoning proliferation of connected devices through the IoT is poised to exert unprecedented pressure on communication networks, necessitating their evolution to effectively cater to the escalating demands for data transmission and reception. The convergence of AI and virtual reality holds the potential to revolutionise the realm of social interactions, transcending the limitations imposed by physical boundaries. This integration has the capacity to create immersive experiences that engender a profound sense of presence, thereby redefining the way individuals engage with one another in virtual environments (Zhao et al., 2021).

1.8.2 Challenges such as algorithm bias, network security, and digital divide

The rapid progress in the field of AI, coupled with the development of sophisticated communication networks, undoubtedly introduces a set of formidable challenges. The phenomenon of algorithmic bias, wherein AI systems inadvertently perpetuate unjust biases during the decision-making process, necessitates the implementation of diligent mitigation strategies. The imperative of safeguarding network security in light of escalating cyber threats is of utmost significance, considering the expanding dependence on interconnected technologies. Addressing the issue of the digital divide, which results in the exclusion of specific demographics from reaping the benefits of AI owing to restricted access and inadequate skill sets, necessitates the implementation of comprehensive and inclusive strategies (DiMaggio et al., 2001).

1.8.3 Speculation on the evolving landscape of human interactions in the AI era

The advent of AI is poised to fundamentally transform the intricate tapestry of human interactions. The potential for chatbots and virtual assistants to undergo a transformative evolution towards becoming heightened empathetic and emotionally attuned entities is a subject of considerable interest within the realm of AI research. Such advancements have the capacity to significantly augment their existing capabilities, thereby expanding their role in fostering companionship and promoting mental well-being. The potential for social media to harness the power of AI to foster substantial and constructive conversations, while simultaneously mitigating the detrimental effects of echo chambers and fostering a more inclusive range of viewpoints, is a topic of great interest and significance. The intricate interplay between human-generated and AI-generated content will inevitably engender profound enquiries regarding the notions of authenticity and originality, thereby exerting a discernible impact on society's cognitive framework for perceiving and ascribing value to information (Suler, 2016).

1.9 CONCLUSION

The confluence of AI and communication networks has created a conduit between the existing situation and a future era full of revolutionary possibilities. Our extensive study of this symbiotic relationship has revealed a captivating terrain of ground-breaking ingenuity, seamless interconnectedness, and profound ethical deliberations, where AI-powered communication networks represent a transformative epoch in human social engagements. After looking at AI, we explored machine learning, NLP, and computer vision. We examined the cognitive foundations that allow AI to decipher, augment, and possibly even revolutionise human communication. Learning and assimilating this information is the foundation for building sophisticated AI-driven communication networks. As we explored social media, virtual assistants, and chatbots, we saw that AI has given us many benefits in data analysis and pattern recognition. These include customised content, online interaction redesign, and digital community formation. This connectivity extends to smart cities and telemedicine solutions. Data-driven insights improve urban life and healthcare access. However, this major transition has complex ethical and privacy issues. Responsible AI integration requires balancing personalisation and privacy, reducing algorithmic biases, and maintaining openness and responsibility. Telehealth technologies powered by AI have changed the healthcare industry by allowing medical consultations to transcend geographical borders. These revolutionary technologies have revolutionised healthcare access and delivery by using AI to put services at people's fingertips. Remote social connections will benefit from virtual and augmented reality technology in the future. These innovative technologies create immersive experiences that blur the lines between the actual and digital worlds, enhancing our social connections. As we look ahead at the great possibilities, we face formidable hurdles that require our unflinching devotion. These problems include algorithmic bias, network security, and the digital divide. To overcome these obstacles, we must create and adopt breakthrough AI solutions that will define the future. The magnitude of these difficulties should motivate us to advance technology and strengthen inclusion. Ingenuity and responsibility unite at the crossroads of AI and communication networks to revolutionise human interaction. Based on our shared knowledge and understanding, this expedition leads us to a future where AI-powered communication networks empower individuals, strengthen communities, and elevate human connection in ways that were previously unthinkable.

REFERENCES

Albino, V., Berardi, U., & Dangelico, R. M., Smart Cities: Definitions, Dimensions, Performance, and Initiatives. *Journal of Urban Technology, 22*(1), 3–21, 2015.

Andrews, J. G., Buzzi, S., Choi, W., Hanly, S. V., Lozano, A., Soong, A. C. K., & Zhang, J. C., What Will 5G Be? *IEEE Journal on Selected Areas in Communications, 32*(6), 1065–1082, 2014.

Atzori, L., Iera, A., & Morabito, G., The Internet of Things: A Survey. *Computer Networks, 54*(15), 2787–2805, 2010.

Caragliu, A., Del Bo, C., & Nijkamp, P., Smart Cities in Europe. *Journal of Urban Technology, 18*(2), 65–82, 2011.

Covington, P., Adams, J., & Sargin, E., Deep Neural Networks for YouTube Recommendations. *Proceedings of the 10th ACM Conference on Recommender Systems*, 191–198, 2016.

DiMaggio, P., Hargittai, E., Celeste, C., & Shafer, S., Social Implications of the Internet. *Annual Review of Sociology, 27*, 307–336, 2001.

Ekstrand, M. D., Riedl, J., & Konstan, J., Collaborative Filtering Recommender Systems. *Foundations and Trends® in Human–Computer Interaction, 4*(2), 81–173, 2011.

Gawlikowski, J., Tassi, C.R.N., Ali, M., Lee, J., Humt, M., Feng, J., Kruspe, A., Triebel, R., Jung, P., Roscher, R. and Shahzad, M., A survey of uncertainty in deep neural networks. *Artificial Intelligence Review, 56*(Suppl 1), 1513–1589, 2023.

Haykin, S., *Neural Networks and Learning Machines*. Pearson, 2019.

Li, M., Zhang, S., Ge, Y., Gao, F. and Fan, P., Joint channel estimation and data detection for hybrid RIS aided millimeter wave OTFS systems. *IEEE Transactions on Communications, 70*(10), 6832–6848, 2022.

Ma, C., Ma, Y., Pan, L., Li, X., Yin, C., Zong, R. and Zhang, Z., Automatic diagnosis of multi-task in essential tremor: Dynamic handwriting analysis using multi-modal fusion neural network. *Future Generation Computer Systems, 145*, 429–441, 2023.

Marston, S., Li, Z., Bandyopadhyay, S., Zhang, J., & Ghalsasi, A., Cloud Computing - The Business Perspective. *Decision Support Systems, 51*(1), 176–189, 2011.

McCarthy, J., A Proposal for the Dartmouth Summer Research Project on Artificial Intelligence, 1955. Retrieved from http://www-formal.stanford.edu/jmc/history/dartmouth/dartmouth.html

Mikolov, T., Chen, K., Corrado, G., & Dean, J., Efficient Estimation of Word Representations in Vector Space. *arXiv preprint arXiv:1301.3781*, 2013.

Narottama, B., Mohamed, Z. and Aïssa, S., Quantum Machine Learning for Next-G Wireless Communications: Fundamentals and the Path Ahead. *IEEE Open Journal of the Communications Society*, 2023.

Newell, A., Shaw, J. C., & Simon, H. A., Elements of a Theory of Human Problem Solving. *Psychological Review, 67*(3), 153–182, 1960.

Pang, B., & Lee, L., Opinion Mining and Sentiment Analysis. *Foundations and Trends® in Information Retrieval, 2*(1–2), 1–135, 2008.

Russell, S. J., & Norvig, P., *Artificial Intelligence: A Modern Approach*. Pearson, 2016.

Russell, Stuart J., and Peter Norvig. Artificial intelligence: a modern approach. Pearson, 2016.

Shi, W., Cao, J., Zhang, Q., Li, Y., & Xu, L., Edge Computing: Vision and Challenges. *IEEE Internet of Things Journal, 3*(5), 637–646, 2016.

Stallings, W., *Data and Computer Communications*. Pearson, 2016.

Suler, J., The Online Disinhibition Effect. *Cyberpsychology, Behavior, and Social Networking, 7*(3), 321–326, 2016.

Tanenbaum, A. S., & Wetherall, D., *Computer Networks*. Pearson, 2018.

Thompson J, Ge X, Wu HC, Irmer R, Jiang H, Fettweis G, Alamouti S., 5G wireless communication systems: Prospects and challenges [Guest Editorial]. *IEEE Communications Magazine*, *52*(2), 62–64, 2014.

Turing, A. M., Computing Machinery and Intelligence. *Mind, 59*(236), 433–460, 1950.

Wasserman, S., & Faust, K., *Social Network Analysis: Methods and Applications*. Cambridge University Press, 1994.

Chapter 2

AI-based intelligent mobility in vehicular ad hoc networks

Bhagwati Sharan, Megha Chhabra, Anil Kumar Sagar, Parma Nand, Rani Astya, Vivek Kumar Singh, and Subrata Sahana

2.1 INTRODUCTION

Vehicular ad hoc networks (VANETs) are subcategories of mobile ad hoc networks, or MANETs [1, 2], that facilitate communications for messages from nearby vehicle-to-vehicle (V2V), and vehicles to roadside infrastructures (V2I). Wireless network technology contains these kinds of networks. Ad hoc networks allow nodes to communicate information without relying on base stations. VANET differs from conventional networks due to its features, and its topology is constantly changing [3]. These networks are mobile. While traveling, each node – a vehicle – on the road is regarded as a host, and its function is to transfer packets to the subsequent node. To get better and more accurate results, VANETs are used with an optimized MANET protocol [4]. Decentralized networks are produced by VANETs. VANETs are widely used and feature a number of difficult characteristics, such as very high mobility. The nodes' (vehicles') very rapid mobility causes their location and speed to be dynamic. A dynamic network topology is the result. Nodes often link and disengage from one another. A significant concern in putting the VANET environment into practice is the methods for creating routing protocols that will provide safe, effective, and economical packet transfer between nodes [5]. The first car will use location management protocols to broadcast its current position to another vehicle that is nearby at a certain place. It may also respond to requests for its current position. Certain methods that depend on the number of cars on a certain path assume that the end users' tolerated delay from the vehicle's intolerable delay would be met by an efficient route modal. The function of roadside units (RSU) in facilitating vehicle communication is similar to that of routers. A data insertion via a node may result in data packets that are sent to end-user cars with little latency and minimal data loss. These packets are then transported to RSUs. Data distribution is possible using both single-hop and multi-hop methods. To provide data concerning congestion and barrier information, a lot of work has been done, and data distribution methods have been developed [6]. There are mainly two kinds of communication that are based on latency in VANETs. One of these is described

 DOI: 10.1201/9781003541363-2

as "which could tolerate the delay" in [7], while the other is described as "which could not tolerate the delay." Advertisements, a vehicle's estimated arrival time, the amount of goods remaining in a business, etc. are some instances of information that can wait a little while. Examples of "where the delay could not be tolerated" include notification of toxic attacks, air traffic control, and railway crossings. This article describes routing protocols for VANET, generates trace files, uses NSG2.1 to analyze them, and adds a case study for the possible applications of artificial intelligence to enhance network performance.

This chapter provides:

- A description of VANETs and an exploration of routing protocols
- The generation of vehicular nodes
- An analysis of the performance of the network with the trace file
- A case study of Artificial Intelligence to improve network performance

2.2 LITERATURE SURVEY

W. Liang et al. [8] discussed the topic in their review article titled "VANETs: designs, examined issues, procedures, tests, and trends." They examined three issues, reviewed all-purpose research techniques, and ended with the investigation of issues and upcoming trends in VANETs. This paper also made comparisons with other articles, including the outline of layered architecture for VANETs. A well-organized survey of vehicle ad hoc networks was completed. The authors additionally provided a detailed investigation of VANET issues/obstacles and upcoming trends, which will be useful for future systematic VANET research.

I. Achour et al. [9], in their research article, proposed a simple and effective protocol for information dissemination in VANET. SEAD is a reliable data distribution mechanism that may be used for safety applications as well as other types of applications. The goal of this protocol is to address the complex situation of the broadcast storm in a scalable vehicular network. A density-awareness system was created without the use of beacons for this purpose. Every automobile is indirectly aware of the density of the adjacent cars when given an updated redundancy ratio. The SEAD protocol is intended to be adaptable to various kinds of applications and to function in adaptive mode by just tweaking a single constraint. In comparison to S1PD, SEAD gained a significant advantage by offering huge packet delivery ratio (PDR) with a short end-to-end delay (EED) latency while optimizing bandwidth use.

N. S. Divya et al. [10], in their exploration article, recommended an RSUDP that exploits the utilization of RSUs in VANETs. In RSUDP, the RSUs are connected for information trade. In the RSUDP convention, the model was upgraded, which brought about better results, such as throughput. The significance of the proposed directing convention is represented by reproduction and correlation discoveries that increase QoS. The performance of three gateway-finding strategies is examined using NS2. To provide an internet connection with vehicular nodes, the source code of ad hoc on-demand distance vector (AODV) in NS2 is first extended. It has a variety of applications, including traffic, such as two-way traffic, analyzing real-time GPS maps, and successfully navigating road intersections.

Y. Jiang et al [11] proposed an enhanced message dissemination technique relying on a locking chain in VSNs. This study presents an efficient data-sharing strategy in VSNs based on blockchain. The authors originally presented a random hidden validation process on the basis of establishing trust-level relations to allow earlier information transfer between elements in VSNs by merging an ID-based signature technique with an assumed name-generating technique. Then, a blockchain-dependent data-sharing scheme was developed, where the sign method and the consensus protocol provide data safety and traceability. The results of the study and simulation trial showed that VAB could outperform existing methods. The outcomes of the proposed model have been compared with the performance of KPSD, LIAP, and IMAEP.

C. Wang et al. [12] compiled a complete study of blockchain applications in the IoV. IoVs and the blockchain were first introduced. Then the grouping of blockchain technology with IoVs was examined from seven perspectives to illustrate how blockchain has been applied to IoVs. Finally, upcoming research directions in the field of integration were planned to be discussed and published in Wireless Com Network in 2021. Blockchain technology is used to provide a data storage environment that is decentralized, transparent, immutable, and safe. As a result, various proposed scenarios from recent works were grouped and described, together with blockchain-dependent IoV security, faith management, certificate administration, and privacy-protecting in IoV.

H. Shahwani et al. [13] provided a thorough literature survey of existing data and systems in VANET. The authors' article also focused on delivering a basic summary of data dissemination in VANET, highlighting potential possibilities and obstacles, and looking at current research in reputable resources that are interrelated to information dissemination in VANET.

2.3 INTRODUCTION TO VANETS

The idea of wireless vehicle communication gained popularity around 1980 with the introduction of the VANETs, an ad hoc network infrastructure. These days, VANETs are used in several applications, including improving traffic flow, lowering traffic congestion, and enhancing driver comfort while operating a vehicle [14].

Figures 2.1, 2.2, and 2.3 show the different components of VANETs: roadside unit (RSU), on-board unit (OBU), and trusted authorities (TA).

A Roadside unit (RSU): The main use of an RSU in the VANET's architecture is computation. It is located mainly along a roadside, and there may be specific fixed locations, such as parking lots and crossings. It has many network devices that are specially used for dedicated short-range communications (DSRC), whose functionality is centered on the IEEE 802.11p radio technology, as shown in Figure 2.1 [15, 16].

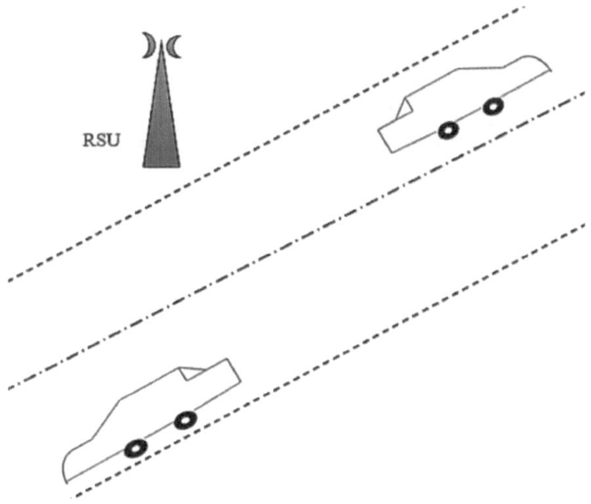

Figure 2.1 Roadside unit (RSU).

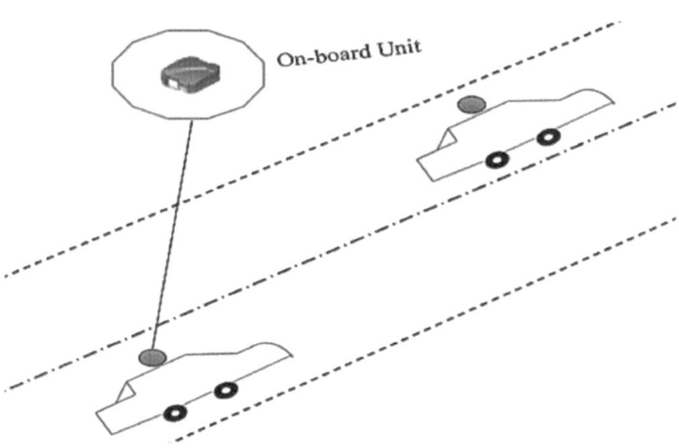

Figure 2.2 On-board unit (OBU).

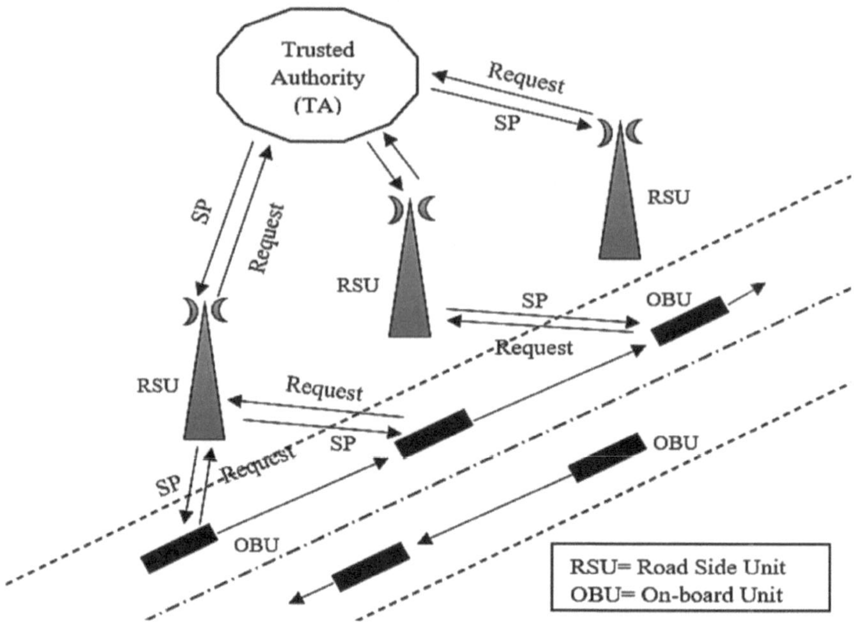

Figure 2.3 Trusted authority (TA).

B On-board unit (OBU): This device is based on GPS tracking, and its main aim is to share vehicle information with the RSUs and the other on-board units, as shown in Figure 2.2. It is created with many electronic elements like a resources command processor (RCP) and sensors, and it has a user interface and read/write elements to process the stored data [17]. The purpose of using an on-board unit is to wirelessly link roadside devices or several on-board units together. The OBU does this by sending signals back and forth. The automobile battery provides the necessary electricity to run the on-board device [18].

C Trusted authority (TA): This is the brain of the VANETs, whose main task is to handle the complete system. It looks after the roadside units, on-board units, and register records of the automobile. It also manages the security mechanisms that are present inside VANETs. As shown in Figure 2.3 [19], it monitors the cars, user ID, and on-board unit ID to ensure that nobody damages the vehicle. Compared to other VANET components, it uses more energy to function. When necessary, the data from RSUs and OBUs will assist us in identifying suspicious or malicious activity, which may also aid in identifying attackers. However, storing this data requires a significant amount of storage [20].

2.4 NAVIGATING VANETS: A STUDY OF ROUTING PROTOCOLS

The vehicles in VANETs have very high mobility, leading to the main challenge of selecting the exact routing protocol. Packets are transferred in VANETs between vehicles that are moving at great speeds. To increase or reduce the density of vehicles on the road, the difficult task is choosing the most suitable routing protocol [21, 22]. Routing protocols are classified in the sections below.

2.4.1 Broadcasting-based protocols in VANETs

Many broadcast protocols have been reported in the literature. These are divided into two major groups as per the scattering of information packets within the VANETs [23, 24]. Both are described as follows.

- *Sing-hop broadcasting protocols*: In these broadcasting protocols, vehicles transmit a huge amount of information to the opposite vehicle in the VANET. Printed intervals should not be too short or too long, as shown in Figure 2.4.
- *Multi-hop broadcasting protocols*: In these protocols, flooding is used for packet broadcasts between all networks in VANET. However, due to high packet collisions and the absence of quantified collisions, pure flooding is inefficient. As networks are thick and information is wasted across channels, the network's quantifiable reduction will occur, leading to a rise in redundancy. Packet collision is another significant drawback, which may occur when many cars in nearby areas send packets at the same time. A broadcasting storm is the term used to describe this. Figure 2.4 [25] illustrates the three groups into which multi-hop broadcasts are separated.

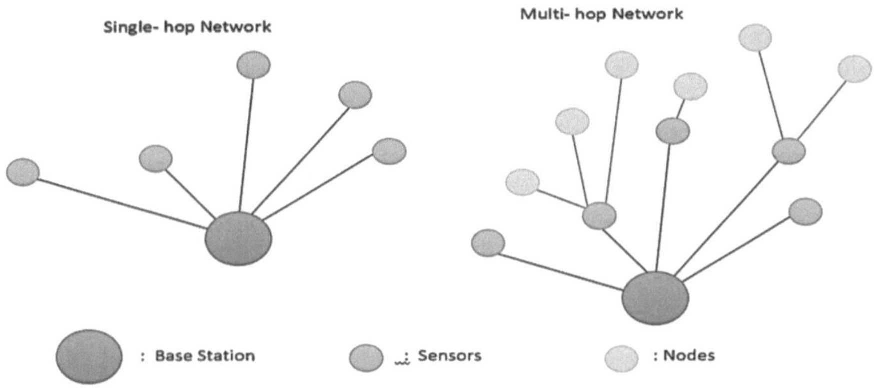

Figure 2.4 Single-hop and multi-hop network.

2.4.2 Geo-based protocols

Here, a source vehicle will transmit to the other vehicles in the network by using its geographical location. The load balancing routing protocol (LBRP) estimates and configures the path as per data available about the node positions. Therefore, designing the routing tables is not necessary [23]. There are three main components available for this protocol: beaconing, position, and sending services. The weakness of this protocol finding the vehicle position, as it requires GPS. In addition, sometimes the satellite waves are very weak when the vehicle passes in an area like a tunnel or high-density area, or if moving at a high speed on a highway. The most common examples are greedy perimeter stateless routing (GPSR) and the distance routing effect algorithm for mobility (DREAM) [26].

2.4.3 Geo-cast routing protocols

Geo-cast routing protocols have two major zones: zone of relevance (ZOR) and zone of forwarding (ZOF) [27]. ZOR is devoted to the part of the node available in the region. These protocols aim to establish communication between the vehicles available in ZOR. Additionally, when the source vehicle requests to communicate with another vehicle that is not in the ZOR of that vehicle, then the vehicle will be part of ZOF. The responsibility of those vehicles present in ZOF is to then forward the information to the ZOR. As the zone varies frequently, link connection and disconnection take place repeatedly.

2.4.4 Cluster-based routing protocols

In these protocols, automobiles having similar kinds of features, such as speed and direction, are combined to create a single cluster. Additionally, when an automobile node requests to transfer with the other node in that same cluster, then a direct path will be followed to transfer the data for local communication. But when the vehicle node requests to communicate from the cluster outsider node, then it demands the value of the cluster head for the destination. For network designers, it is a good option for considering scalability. But its problem is a delay in traffic. The best example of this is the clustering for open inter-vehicular communication network (COIN) [28].

2.4.5 Topology-based routing protocols

The two major types of topology-based routing protocols are proactive and reactive [29]. There is also a third type, known as hybrid. We will explain each one in the sections that follow. Which routing protocol is used depends on the topology.

2.4.5.1 Proactive protocols

These protocols compute the path frequently, and so the routing table will be updated regularly. This protocol uses the Bellman-Ford algorithm, in which each node has the information associated with the next, upcoming node. One advantage of these protocols is that it is very easy to track the route of the delivered packets. A few examples are optimized link state routing (OLSR) and destination-sequenced distance vector (DSDV) protocols [30].

OLSR is fundamentally an advanced form of link state protocol technique. It broadcasts any alteration in topology to all the nodes in that network chain that leads to an increment in the network overhead. OLSR generally takes care of two types of messages, "Hello," and a message to govern the topology in VANETs. "Hello" messages are utilized to calculate the data about link connection status (connected or disconnected). On the other hand, the topology-based control message is used to telecast their neighbor's info by using a multipoint relay (MPR) selected list. Overload can be reduced by using this MPR concept for pure link protocol. The DSDV protocol is an altered version of the Bellman-Ford algorithm.

2.4.5.2 Reactive protocols

In this kind of protocol, node information is unknown. It contains information on the nodes that are found along the journey. These protocols include dynamic source control routing (DSR), dynamic MANET on demand (DYMO), and ad hoc on-demand distance vector (AODV) [31]. In the VANET network, the AODV protocol is often used. It contains the information about the end stop node's next hop. When a route's demand declines, it also expires and is replaced with a new route based on demand. A source node will consult its routing table and begin the search for that route when it seeks to send data to the last specified node, according to AODV. The packet will be transmitted to the intended destination as soon as the route data are discovered there. If not, the originating node will broadcast the request for route discovery to its neighbors. One kind of mobile ad hoc protocol suitable for high mobility traffic is AODV. The process of finding a route will happen as needed. DSR is an inventive routing technology that was primarily developed for multi-hop VANETs. There is no need for administration since it is a self-organizing and configuring protocol. The protocol mentioned has two purposes: route maintenance and discovery. Both proactive and reactive applications of the DYMO routing system are possible. However, the route discovery approach is available anytime it is needed.

2.4.5.3 Hybrid protocols

A combination of proactive protocol and reactive routing protocol is used to minimize the overhead and delay occurrence by sharing information about topology periodically. Using the hybrid protocol technique, the efficiency and scalability of the network are enhanced. However, one disadvantage of the hybrid approach is the high latency for directing novel routes. The zone routing protocol (ZRP) is the best example for these protocols [32].

2.5 INTRODUCTION TO NSG2.1

A secure message dissemination in VANETs using the blockchain method was suggested. In this study, the proposed system's primary components are RSU, the source node (vehicle), the trusted authority (TA), the blockchain algorithm, and VANET.

Ubuntu is a Debian Linux-based operating system. Debian Linux was founded in 1993 by Ian Murdock, a computer specialist. Debian is famous among developers and administrators. It was not easy for regular people to utilize Debian Linux daily. Ubuntu has made several improvements to Debian Linux to make it more useful to the public. According to statistics, Ubuntu is the most widely used Linux operating system [33].

Ubuntu comes with the GNOME desktop. The purple-shade user interface of the Ubuntu GNOME desktop is very popular among Linux users. Ubuntu's minimum hardware specifications are a dual-core processor running at 2 GHz, with 4 GB of RAM and 25 GB of storage. Ubuntu's official website is https://ubuntu.com.

Network simulation is the most effective approach for network parameters evaluation. This is primarily helpful in improving novel communication frameworks and construction protocol layouts for networks. NS2 is the most popular open discrete-event simulator and the most practical, adaptable, and effective design handling network applications.

The network simulation tool is a solution for the creation of complex network scenarios. Scripts written in TCL (Tool Command Language) are commonly utilized in the NS2 network simulator. The network simulation utilizes TCL scripts to configure wired and wireless scenarios and then runs such TCL scripts with NS2 software to generate simulation results for that network [34].

2.5.1 Reading the trace file

Table 2.1 Format of a trace file

Event	Time	From node	To node	Pkt type	Pkt size	Flags	fid	Source address	Destination address	Seq. number	Pkt ID

r: Received (at the node)
d: Drop (at queue)
+ : Enqueue (at queue)

- : Dequeue (at queue)
Source address: node.port (0.0)
Desti. address: node.port (49.0)

Each following line begins with an occasion (+, -, d, r) descriptor followed by the reproduction time (in a flash) of that occasion, and from and to the hub, which distinguishes the connection on which the occasion happened. Have a look at Figure 2.7 in the "Organization components" area to see where in a connection each sort of occasion is followed. The following data in the line before banners (shown as "- - - - - -" since no banner is set) is parcel type and size (in Bytes), shown in Table 2.1.

2.5.2 NAM in NS2 highlights

- Provides a visual representation of a huge organization.
- Directly executes from the TCL script.
- Provides options to play, pause, stop, display a speed limiter, package the screen office, and more.
- The NAM chart provides a graphical representation of the parcel flow, indicating when the packages are dropped off or picked up.
- The TCL script's workstream is included in NAM papers.
- To execute a NAM, utilize the accompanying order: *Execnam$filename*

2.5.3 Network specification

A network of 120 nodes was proposed with the following specific parameters, as listed in Table 2.2.

2.5.3.1 Performance evaluation metrics

The different parts of execution assessment measurements in VANETs are given below [2, 35–38].

Table 2.2 Network specification

Specification	*Details*
Network type	Grid
No. of nodes	8x15
Horizontal and vertical distance	200x200
Agent	TCP and TCPSink
Traffic generator	FTP
Start time	1 s
Stop time	30 s
Channel	Wireless
Packet size	

- Packet delivery ratio (PDR)

  PDR is the level of effectively conveyed bundles (Pd) to the absolute number of sent parcels (Ps) in that system. Numerically, it can be well represented as seen in (2.1).

$$PDR = (Pd / PS) * 100 \tag{2.1}$$

- End-to-end delay (EED)

  EED refers to the time it takes for packages to travel from their point of origin to their intended destination inside the company. While TA discusses the parcel's appearance season, TS deals with the source time. In terms of numbers, it can function as seen in (2.2).

$$EED = \Sigma TA - TS \tag{2.2}$$

- Packet loss ratio (PLR)

  PLR can be determined by taking the proportion of dropped parcels (Pd) to the complete sent bundles (PS). Numerically, it tends to be represented by (2.3).

$$PLR = (Pd / PS) * 100 \tag{2.3}$$

- Network reputation factor (NRF)

  NRF indicates the past performance of the organization and helps in giving a high degree of certainty to the client.
- Overhead transmission (OH)

  This considers the level of immersion of the network that is utilized. It is the proportion of the extra steering parcels to the bundles received at the collector. It is conversely relative to the organization's execution. Numerically, it is displayed in (2.4).

$$OH = (\text{Additional routing Packets}) / \text{Packets received at the receiver} \tag{2.4}$$

- Throughput

  Throughput is utilized to quantify the successful conveyance of pieces per second at the beneficiary's end. Numerically, it can be illustrated by (2.5).

$$\text{Throughput} = (8 \times \text{Number of received packets}) / \text{Tmac} + \text{Trout} + \text{Tagg} \tag{2.5}$$

where,

$$T_{mac} = \text{number of retrials} * \text{backoff delay} \tag{2.6}$$

$$T_{agg} = (T_{enc} + T_{dec}) * \text{data rate} \tag{2.7}$$

$$\text{Back off Delay} = \text{Tmac}(\text{MAC back off delay time}),$$
$$\text{Trout}(\text{Time required for route discovery}),$$
$$\text{Tenc}(\text{Data bits encoding time}), \text{and}$$
$$\text{Tdec}(\text{Data bits decoding time})$$

2.6 ARTIFICIAL INTELLIGENCE IN VANETS

Artificial Intelligence (AI) is now a promising tool that may be used to increase network efficiency, safety, and performance when integrated into VANETs. Some examples of how AI may be used in VANETs are detailed in the sections that follow.

2.6.1 *Traffic management*

J. Grover et al. [39] presented the issue of malevolent nodes that broadcast deceptive or false signals in order to negatively impact VANET applications and suggested the use of machine learning to detect and categorize various VANET misbehaviors according to the behavioral and physical characteristics of every node that broadcasts safety packets. Machine learning can also be used in VANET for many purposes, including data control, traffic monitoring, routing, misbehavior detection, driver assistance, and collision alerts. A security structure to identify rogue from trustworthy nodes has been developed. WEKA has been used to evaluate the suggested framework and demonstrates that J-48 and random forest classifiers perform better than other classifiers.

H. El-Sayed et al. [40] suggested an entity-centric trust architecture that combines decision tree (DT) classification and artificial neural networks (ANN) for ensuring secure data transmission and identifying malicious vehicles in real-time scenarios, which are important for road safety and traffic control. The DT classification model creates a standard rule to calculate trust, but if the trust value is not satisfactory, the ANN adjusts the vehicular nodes. The model uses role- and distance-based metrics such as Euclidean distance to evaluate trust. The model also applies a new direct and recommended trust assessment approach to calculate trust. The proposed model is simple, reliable, and effective compared to other entity-centric trust models.

Machine learning (ML) techniques have been used to get optimal network performance [41] and have also been examined from many angles, including traffic, safety, and communication. Applications of the vehicular network that may be used to study machine learning include traffic monitoring, routing, driver assistance, collision alert, misbehavior recognition, and data control from various angles. It turns out that traffic management software makes extensive use of techniques linked to neural networks.

S. S. Sepasgozar and S. Pierre [42] presented an architecture to detect traffic of the network in virtual automotive networks, or VANETs. Based on both road and network traffic, the model predicts the network traffic flow with ML and DL approaches. To forecast the network traffic flow, the model uses a gated recurrent unit to choose pertinent characteristics from the data using random forest. The model performs better than other methods in prediction error and execution time.

A vehicle-consensus routing management scheme (VCRMS) for VANET has been proposed. This is a network of intelligent automobiles that interact with roadside equipment and one another to provide a range of traffic and vehicle management applications [43]. DL techniques have been applied to match the data from the vehicles and the roadside devices. To remove irrelevant data and gather pertinent traffic statistics, the strategy makes use of both past and present choices. The plan reduces data congestion while improving the traffic and vehicle control application's efficiency. The study shows that for varying vehicle speeds, the suggested strategy produces superior results in terms of information latency, application response, and traffic.

2.6.2 *Intelligent routing*

An artificial intelligence-based centralized routing strategy study has been presented for mobility prediction in the vehicular network to improve route planning and increase the dependability and availability of the vehicular network. The artificial neural network technology allows the SDN controller to accurately predict vehicle movement. Using the network data, the SDN controller determines the best routing paths for switches (i.e., BS and RSU). In situations when the source and destination vehicles are located in the same switch area, the routing is determined separately by the RSUs or the BS to minimize the overall vehicle service time. Through vehicle-to-vehicle or vehicle-to-infrastructure communication, vehicle requests are sent from the source vehicle to the destination vehicle and are managed by the BS or RSUs [44].

The study in [45] addresses the challenge of effectively transferring data between vehicles due to higher mobility, which led to network topology changing continuously in urban scenarios through the development of efficient clustering routing (ECRDP). The research also proposes a novel clustering approach that selects the cluster heads – the entities in charge of overseeing data transfer both inside and across clusters – by using density peak clustering (DPC) and particle swarm optimization (PSO). Additionally, the authors propose a maintenance phase in which vehicles are moved to other clusters, and cluster heads are replaced depending on the reliability of the links connecting the vehicles. The proposed plan is tested in the study using a MATLAB simulation in an actual metropolitan setting. The study demonstrates that, in comparison to other designs, the recommended plan improves network speed, stability, and efficiency.

The study in [46] addresses the challenge of increasing the number of vehicles on the road with an ML approach. This study also demonstrates the contributions of ML approaches in developing autonomous vehicles in the automobile sector. Multiple sensors and cameras may be used in conjunction with ML algorithms as per the need and safety of the driver and the passengers.

The research in [47] investigates the use of improved deep reinforcement learning (IDRL) for routing in VANETs. The study addresses the challenging issues of routing and control overhead in VANETs, which arise from the network's dynamic and diversified structure. To lower control overhead and improve the routing route, the study suggests a novel routing method that makes use of IDRL.

The study in [48] investigates the use of an intelligent cluster-based routing protocol for VANETs. The issue of data interchange between cars in VANETs, which might be hampered by malevolent or fast-moving vehicles, is the focus of this research. The study suggests a method that groups the cars into clusters with cluster heads (CH) by using improved clustering techniques and an ANN model to identify hostile vehicles.

2.6.3 *Resource management*

To improve the efficiency of next-generation automotive networks, the study in [49] provides an integrated framework that dynamically orchestrates networking, caching, and computing resources. Within the paradigm, the proposed approach of resource allocation is a joint optimization issue, taking into account not just networking advantages but also caching and processing benefits. The way these three elements interact adds a significant layer of complexity to the system. An innovative deep reinforcement learning strategy to deal with this problem, this study illustrates the efficacy of the strategy using simulations by changing system parameters.

An ML-based approach [50] was presented to enhance VANETs to facilitate the intelligent mobility of vehicles. This article also addresses the issue of malevolent vehicles breaking into networks and causing disruptions. The proposed method for smart vehicles can forecast and make use of trustworthy data. The solution is tested by a simulation in the study and compared to other models. The study demonstrates how the solution may enhance network security, efficiency, and performance.

High vehicle mobility causes frequent changes to communication linkages and network architecture. Improving system throughput and resolving latency concerns are important. A machine learning contributed path selection (MLCPS) has been proposed to solve this issue. The proposed model determines routing pathways with reasonably high transmission capacity for packet forwarding, forecasting vehicle movements, and retaining road information inside RSUs. Furthermore, using the estimated destination location and anticipated transmission delays in both directions, MLCPS helps determine the transmission route between two RSUs and finds significant improvements in metrics like packet delivery ratio and end-to-end latency [51].

2.6.4 *Security and privacy*

Traditional trajectory protection approaches tend to focus solely on one aspect – either geographical or semantic protection. To address this, an optimized privacy scheme based on differential privacy and reinforcement learning within VANETs was developed. To maintain the balance between geolocation obfuscation and semantic security, the method dynamically assigns the privacy budget for each point along the vehicle itinerary. We achieve a fair trade-off between privacy and usefulness by doing this, thereby reducing the danger of both geographical and semantic location leaking [52].

The possibility of security breaches makes it difficult to ensure safe data exchange in wireless communication. A unique approach to handling security problems has been proposed using federated learning while maintaining data privacy. Before exchanging data with RSUs, cars in the first layer employ local differential privacy measures for their data. Model updates and RSU training are part of the second layer. Performance analysis demonstrates that this framework outperforms conventional methods in terms of efficiency and security against intrusions. This method improves data-sharing in VANETs by fusing FL with local differential privacy [53].

A decentralized privacy-preserving deep learning (DPDL) model has been developed to maintain data security and privacy preservation in VANETs and to diminish network overhead and congestion delay. The decentralized VANETs (DVANETs) architecture was developed, in which computing workloads are divided from centralized cloud services to edge computing (EC) nodes. Blockchain technology was implemented to create a reliable and safe data transmission system between cars, roadside equipment, and EC nodes [54].

Implementing an intrusion detection system (IDS) is a viable solution for preventing such attacks in VANETs. However, coping with the substantial volume of vehicular data, which continues to grow in urban environments, poses a critical challenge for IDSs. Therefore, a novel machine learning model that enhances IDS performance was developed. This approach combines an RF model with a posterior recognition mechanism with improvements in detection accuracy and efficiency [55].

2.7 RESULTS AND DISCUSSIONS

The network scenario generator (NSG) was used to create a scenario with JAVA. NSG is built on the JAVA programming language and may be executed on any platform. NSG can generate TCL scripts for NS2 that are both wired and wireless.

A few key functions of NSG2 are:

1. Allows wired and wireless nodes
2. Allows connections between nodes
3. Allows links to be made
4. Allows agents to be made
5. Allows applications to be made

A grid-type network of 120 nodes was created in our project, shown in Figure 2.5. Node 0 was labeled as the source node, node 119 as the destination node, and the rest of the nodes as intermediate nodes. TCP agent was applied at node 0, and TCPSink agent at node 119. The horizontal and vertical distances were considered as 200x200. FTP had been placed on the source node 0. The simulation lasted from 1 second to 30 seconds. The selected channel was wireless, having a maximum queue size of ten packets, and the drop-tail technique was applied to the buffer system. Drop-tail is

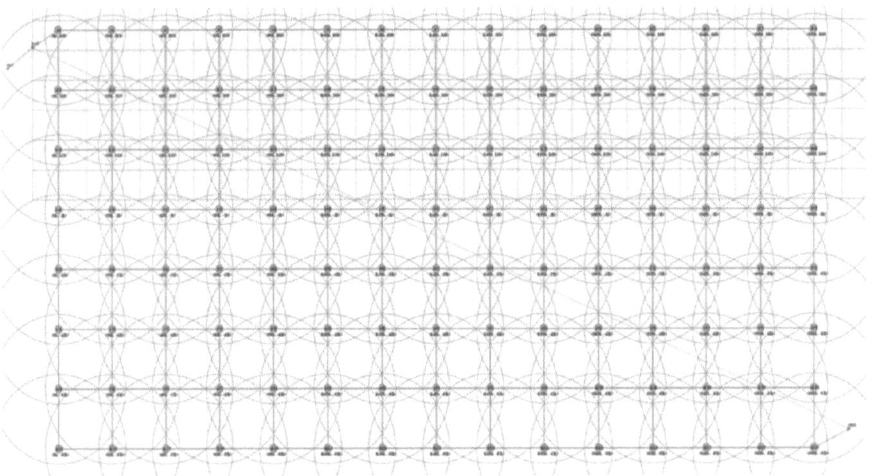

Figure 2.5 Node creation and necessary connections.

a technique in which it drops the incoming packets when they exceed the number of packets permitted throughout the buffer. When we simulated the .tcl file, two additional files were created there. One of them was a trace file (.tr extension), and the other one was a nam file (.nam file).

The name for the trace file and nam file will be generated after the simulation of .tcl in NS2. The names given for the trace file and nam file are 120.tr and 120.nam respectively. We considered a simulation time of 30 seconds, as shown in Figure 2.6.

Here, we selected the channel type as the wireless channel, the propagation model as TwoRayGround, the queue type as drop-tail, and the maximum number of packets in the queue as ten packets (Figure 2.7). When the number of packets increases to more than ten, it simply discards them because we are using the drop-tail technique.

Figure 2.6 shows the dissemination of the packets from source node 0 to destination node 119 with several intermediate nodes. This visualization is possible only with the help of the .nam file created during the simulation of the .tcl file. Running these TCL scripts creates a NAM trace file which could be utilized as an input to NAM, as well as a trace file named "out.tr" for our simulation analysis. The NAM file is important in NS2 projects because, without it, the output cannot be seen in NS2. This is an NS2 GUI tool, thus it's an important part of the game and a key priority for us. The NAM file is a distinct software that comes with the NS2 simulator and is used to read an input file and visually depict network events (Figure 2.8).

The trace file has 12 sections, each of them holding a piece of specific information. To extract the useful information from the trace file, we need

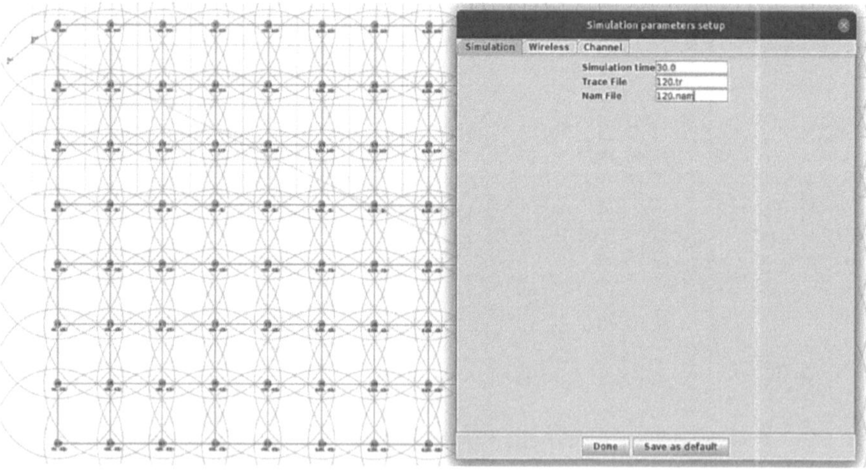

Figure 2.6 Fixing the name for trace and nam file.

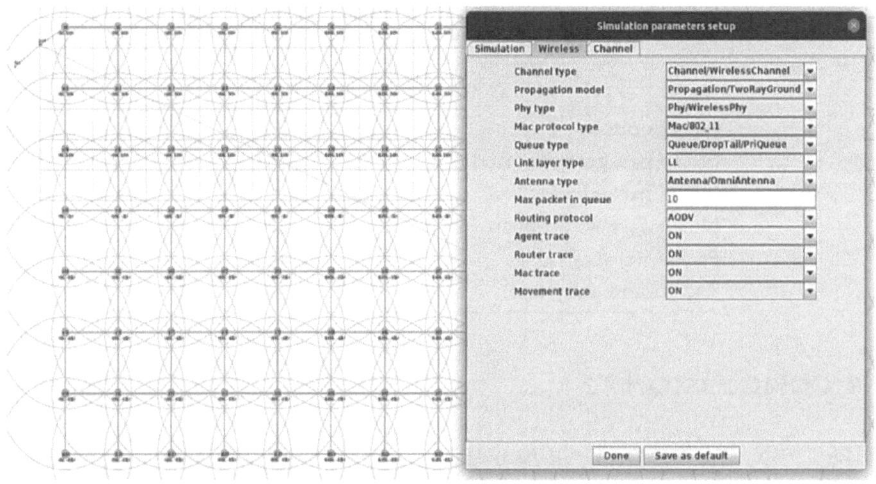

Figure 2.7 Buffer/queue setup for simulation.

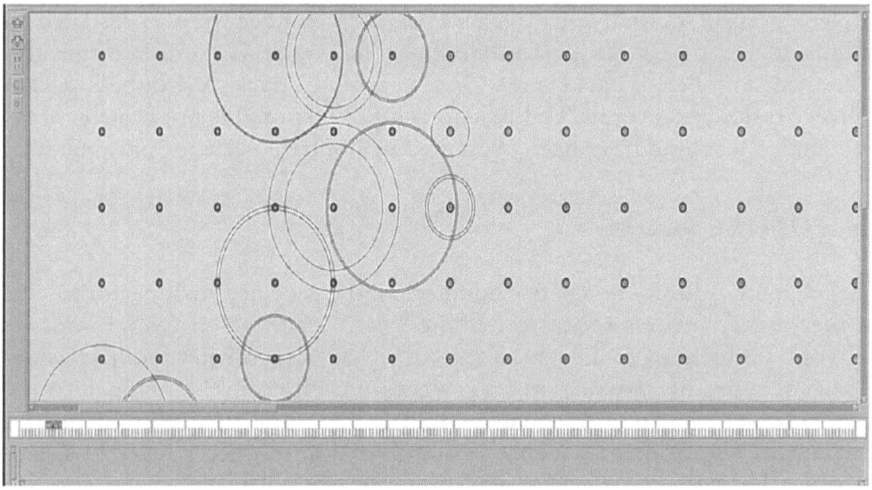

Figure 2.8 NAM file simulation from TCL script.

to create the .awk file for statistical analysis. The awk file should call in between the finish procedure, with the command written as: exec awk -f 120pdr.awk 120.tr & exec awk -f 120.awk 120.tr.

In the simulation, we obtained the number of generated packets as 241, out of which the number of received packets was 112. Thus, the obtained PDR was 46.473. The total packet drop in that network is 27, and the obtained end-to-end delay is 206.81 ms, as shown in Table 2.3.

Table 2.3 Generated packets, received packets, PDR, dropped packets, and end-to-end delay

Outputs generated from network	
No. of nodes	120
No. of packets generated	241
No. of packets received	112
No. of packets dropped	27
Packet delivery ratio	46.47
End-to-end delay (ms)	206.81

2.8 CONCLUSION

A review of the existing technique for VANETs was proposed. This chapter has been divided into sections to make it easy to understand the concepts involved in VANET and NSG2.1 used to generate the network scenario. A TCL file was generated. Upon stimulation, NAM files and trace files were obtained. AWK file was created and linked with the trace file. All steps of implementation for analyzing the network performance were represented by obtaining many network performance metrics, such as number of packets generated, number of packets received, number of packets dropped, packet delivery ratio, and average end-to-end delay. The possible application areas of machine learning have been presented to improve network parameters.

2.9 FUTURE SCOPE

This chapter is limited to performing an analysis of network performance. In the future, various scenarios should be studied, which would improve network performance. The first scenario that researchers could consider is several types of network attacks, where output could be mapped to the original case. The second scenario could be the implementation of various blockchain algorithms to secure communication, reduce packet loss, and enhance the authentication system in the network. In this way, a detailed comparative study would be possible. This chapter also shows the possible inclusion of machine learning for improving network parameters. We, the authors, will address these above-mentioned limitations in our next study, but other researchers are also welcome to do the same.

REFERENCES

1. M. Rath, B. Pati, and B. K. Pattanayak, "An Overview on Social Networking: Design, Issues, Emerging Trends, and Security," *Social Network Analytics*. Elsevier, pp. 21–47, 2019. doi: 10.1016/b978-0-12-815458-8.00002-5.

2. B. Sharan, M. Chhabra, and A. K. Sagar, "State-of-the-Art: Data Dissemination Techniques in Vehicular Ad-hoc Networks," *2022 9th International Conference on Computing for Sustainable Global Development (INDIACom)*, New Delhi, India, 2022, pp. 126–131, doi: 10.23919/INDIACom54597.2022.9763249.
3. Sandeep N. Kugali, "Vehicular ADHOC Network (VANET):-A Brief Knowledge," *International Journal of Engineering Research*, vol. V9, no. 6. ESRSA Publications Pvt. Ltd., Jun. 29, 2020. doi: 10.17577/ijertv9is060784.
4. Huaqun Guo, F. B. A. Thani, Wai Choong Wong, and Yongdong Wu, "An Optimized Routing Protocol for Vehicular ad Hoc Networks," *TENCON 2010 - 2010 IEEE Region 10 Conference. IEEE*, Nov. 2010. doi: 10.1109/tencon.2010.5685996.
5. S. Malik and P. K. Sahu, "A Comparative Study on Routing Protocols for VANETs," *Heliyon*, vol. 5, no. 8. Elsevier BV, p. e02340, Aug. 2019. doi: 10.1016/j.heliyon.2019.e02340.
6. B. Dwivedy, A. K. Bhola, and S. Yadav, "Cluster Based Multi-Hop Data Dissemination Protocol in V2V Networks using Whale Optimization Technique," *2019 International Conference on Automation, Computational and Technology Management (ICACTM)*. IEEE, Apr. 2019. doi: 10.1109/icactm.2019.8776703.
7. J. Tian and Q. Han, "The Priority Assignment of Messages Effects on Delay Performance in VANET," *Journal of Physics: Conference Series*, vol. 1187, no. 4. IOP Publishing, p. 042106, Apr. 2019. doi: 10.1088/1742-6596/1187/4/042106.
8. W. Liang, Z. Li, H. Zhang, S. Wang, and R. Bie, "Vehicular Ad Hoc Networks: Architectures, Research Issues, Methodologies, Challenges, and Trends," *International Journal of Distributed Sensor Networks*, vol. 2015, Hindawi Publishing Corporation, 2015, Article ID 745303, 11 pages, doi: 10.1155/2015/745303.
9. I. Achour, et al. SEAD: *A Simple and Efficient Adaptive Data Dissemination Protocol in Vehicular ad-Hoc Networks*, Wireless Networks, 2015, doi: 10.1007/s11276-015-1050-9.
10. N. Sree Divya, Dr. Veeramallu Boddu. "An Efficient Data Dissemination Method for VANET with the Help of RSUs," *International Journal of Advanced Science and Technology*, vol. 29, no. 6, 2010–2021, 2020. Retrieved from http://sersc.org/journals/index.php/IJAST/article/view/13468
11. Y. Jiang, X. Shen, S. Zheng, "An Effective Data Sharing Scheme Based on Blockchain in Vehicular Social Networks," *Electronics*, vol. 10, p. 114, 2021. doi: 10.3390/electronics10020114.
12. C. Wang, et al. "A Survey: Applications of Blockchain in the Internet of Vehicles," *Journal on Wireless Communications and Networking*, vol. 77, 2021. doi: 10.1186/s13638-021-01958-8.
13. H. Shahwani, et al. "A Comprehensive Survey on Data Dissemination in Vehicular Ad Hoc Networks," *Vehicular Communications*, 2021, issn = 2214-2096, doi: 10.1016/j.vehcom.2021.100420.
14. N.-W. Lo and H.-C. Tsai, "A Reputation System for Traffic Safety Event on Vehicular Ad Hoc Networks," *EURASIP Journal on Wireless Communications and Networking*, vol. 2009, no. 1. Springer Science and Business Media LLC, Dec. 2009. doi: 10.1155/2009/125348.

15. A. Guerna, S. Bitam, and C. T. Calafate, "Roadside Unit Deployment in Internet of Vehicles Systems: A Survey," *Sensors*, vol. 22, no. 9. MDPI AG, p. 3190, Apr. 21, 2022. doi: 10.3390/s22093190.
16. M. Mao, P. Yi, Z. Zhang, L. Wang, and J. Pei, "Roadside Unit Deployment Mechanism Based on Node Popularity," *Mobile Information Systems*, vol. 2021. Hindawi Limited, pp. 1–11, Jun. 24, 2021. doi: 10.1155/2021/9980093.
17. Sheikh, Liang, and Wang, "A Survey of Security Services, Attacks, and Applications for Vehicular Ad Hoc Networks (VANETs)," *Sensors*, vol. 19, no. 16. MDPI AG, p. 3589, Aug. 17, 2019. doi: 10.3390/s19163589.
18. Q. Yang, L. Wang, W. Xia, Y. Wu, and L. Shen, "Development of On-Board Unit in Vehicular Ad-Hoc Network for Highways," *2014 International Conference on Connected Vehicles and Expo (ICCVE)*, 2014, pp. 457–462, doi: 10.1109/ICCVE.2014.7297589.
19. M. S. Sheikh and J. Liang, "A Comprehensive Survey on VANET Security Services in Traffic Management System," *Wireless Communications and Mobile Computing*, vol. 2019. Hindawi Limited, pp. 1–23, Sep. 15, 2019. doi: 10.1155/2019/2423915.
20. Z. El-Rewini, K. Sadatsharan, D. F. Selvaraj, S. J. Plathottam, and P. Ranganathan, "Cybersecurity challenges in vehicular communications," *Vehicular Communications*, vol. 23. Elsevier BV, p. 100214, Jun. 2020. doi: 10.1016/j.vehcom.2019.100214.
21. M. Rizwan Ghori, A. Safa Sadiq, and A. Ghani, "VANET Routing Protocols: Review, Implementation and Analysis," *Journal of Physics: Conference Series*, vol. 1049. IOP Publishing, p. 012064, Jul. 2018. doi: 10.1088/1742-6596/1049/1/012064.
22. S. Malik and P. K. Sahu, "A Comparative Study on Routing Protocols for VANETs," *Heliyon*, vol. 5, no. 8. Elsevier BV, p. e02340, Aug. 2019. doi: 10.1016/j.heliyon.2019.e02340.
23. S. Boussoufa-Lahlah, F. Semchedine, and L. Bouallouche-Medjkoune, "Geographic Routing Protocols for Vehicular Ad Hoc NETworks (VANETs): A Survey," *Vehicular Communications*, vol. 11. Elsevier BV, pp. 20–31, Jan. 2018. doi: 10.1016/j.vehcom.2018.01.006.
24. K. Abboud and W. Zhuang, "Stochastic Modeling of Single-Hop Cluster Stability in Vehicular Ad Hoc Networks," *IEEE Transactions on Vehicular Technology*, vol. 65, no. 1, pp. 226–240, Jan. 2016, doi: 10.1109/TVT.2015.2396298.
25. S. Latif, S. Mahfooz, B. Jan, N. Ahmad, Y. Cao, and M. Asif, "A Comparative Study of Scenario-Driven Multi-Hop Broadcast Protocols for VANETs," *Vehicular Communications*, vol. 12. Elsevier BV, pp. 88–109, Apr. 2018. doi: 10.1016/j.vehcom.2018.01.009.
26. M. Garcia de la Fuente and H. Labiod, "Performance Analysis of Position-Based Routing Approaches in VANETS," *2007 9th IFIP International Conference on Mobile Wireless Communications Networks*, 2007, pp. 16–20, doi: 10.1109/ICMWCN.2007.4668172.
27. Y.-S. Chen, Y.-W. Lin, and S.-L. Lee, "A Mobicast Routing Protocol in Vehicular Ad-Hoc Networks," *Mobile Networks and Applications*, vol. 15, no. 1. Springer Science and Business Media LLC, pp. 20–35, May 26, 2009. doi: 10.1007/s11036-009-0176-3.

28. S. Vodopivec, J. Bešter, and A. Kos, "A Survey on Clustering Algorithms for Vehicular Ad-Hoc Networks," *2012 35th International Conference on Telecommunications and Signal Processing (TSP)*, 2012, pp. 52–56, doi: 10.1109/TSP.2012.6256251.
29. O. Salman, R. Morcel, O. Al Zoubi, I. Elhajj, A. Kayssi, and A. Chehab, "Analysis of Topology Based Routing Protocols for VANETs in Different Environments," *2016 IEEE International Multidisciplinary Conference on Engineering Technology (IMCET)*, 2016, pp. 27–31, doi: 10.1109/IMCET.2016.7777421.
30. E. E. Akkari Sallum, G. dos Santos, M. Alves, and M. M. Santos, "Performance Analysis and Comparison of the DSDV, AODV and OLSR Routing Protocols under VANETs," *2018 16th International Conference on Intelligent Transportation Systems Telecommunications (ITST)*, 2018, pp. 1–7, doi: 10.1109/ITST.2018.8566825.
31. B. Hamid and E.-N. E. Mokhtar, "Performance analysis of the Vehicular Ad hoc Networks (VANET) Routing Protocols AODV, DSDV and OLSR," *2015 5th International Conference on Information & Communication Technology and Accessibility (ICTA)*, 2015, pp. 1–6, doi: 10.1109/ICTA.2015.7426885.
32. A. Setiabudi, A. A. Pratiwi, Ardiansyah, D. Perdana, and R. F. Sari, "Performance Comparison of GPSR and ZRP Routing Protocols in VANET Environment," *2016 IEEE Region 10 Symposium (TENSYMP)*, 2016, pp. 42–47, doi: 10.1109/TENCONSpring.2016.7519375.
33. V., Vimal, "Introduction to Ubuntu Linux", TY - BOOK, SN- 978-93-5406-177-6.
34. F. S. Ghodichor and B. Madan, "Network Analyzer and Report Generation Tool for NS-2 using TCL Script," *International Research Journal of Engineering and Technology (IRJET)*, vol. 4, no. 06, June 2017, e-ISSN: 2395 -0056, p-ISSN: 2395-0072.
35. W. Z. Khan, N. M. Saad, and M. Y. Aalsalem, "An Overview of Evaluation Metrics for Routing Protocols in Wireless Sensor Networks," *2012 4th International Conference on Intelligent and Advanced Systems (ICIAS2012)*, 2012, pp. 588–593, doi: 10.1109/ICIAS.2012.6306083.
36. M. F. Khan, E. A. Felemban, S. Qaisar, and S. Ali, "Performance Analysis on Packet Delivery Ratio and End-to-End Delay of Different Network Topologies in Wireless Sensor Networks (WSNs)," *2013 IEEE 9th International Conference on Mobile Ad-hoc and Sensor Networks*. IEEE, Dec. 2013. doi: 10.1109/msn.2013.74.
37. W. Mansouri, K. B. Ali, F. Zarai, and M. S. Obaidat, "Radio Resource Management for Heterogeneous Wireless Networks," *Modeling and Simulation of Computer Networks and Systems*. Elsevier, pp. 767–792, 2015. doi: 10.1016/b978-0-12-800887-4.00027-4.
38. X. Zhang, G. Neglia, and J. Kurose, "Network Coding in Disruption Tolerant Networks," *Network Coding*. Elsevier, pp. 267–308, 2012. doi: 10.1016/b978-0-12-380918-6.00010-x.
39. J. Grover, N. K. Prajapati, V. Laxmi, and M. S. Gaur, "Machine Learning Approach for Multiple Misbehavior Detection in VANET," *Advances in Computing and Communications*. Springer Berlin Heidelberg, pp. 644–653, 2011. doi: 10.1007/978-3-642-22720-2_68.

40. H. El-Sayed, H. A. Ignatious, P. Kulkarni, and S. Bouktif, “Machine Learning Based Trust Management Framework for Vehicular Networks,” *Vehicular Communications*, vol. 25. Elsevier BV, p. 100256, Oct. 2020. doi: 10.1016/j.vehcom.2020.100256.
41. S. Khatri, et al., “Machine Learning Models and Techniques for VANET Based Traffic Management: Implementation Issues and Challenges,” *Peer-to-Peer Networking and Applications*, vol. 14, no. 3. Springer Science and Business Media LLC, pp. 1778–1805, Sep. 17, 2020. doi: 10.1007/s12083-020-00993-4.
42. S. S. Sepasgozar and S. Pierre, “Network Traffic Prediction Model Considering Road Traffic Parameters Using Artificial Intelligence Methods in VANET,” *IEEE Access*, vol. 10. Institute of Electrical and Electronics Engineers (IEEE), pp. 8227–8242, 2022. doi: 10.1109/access.2022.3144112.
43. J. Gao, G. Manogaran, T. N. Nguyen, S. Kadry, C.-H. Hsu, and P. M. Kumar, “A Vehicle-Consensus Information Exchange Scheme for Traffic Management in Vehicular Ad-Hoc Networks,” *IEEE Transactions on Intelligent Transportation Systems*, vol. 23, no. 10. Institute of Electrical and Electronics Engineers (IEEE), pp. 19602–19612, Oct. 2022. doi: 10.1109/tits.2021.3130087.
44. Y. Tang, N. Cheng, W. Wu, M. Wang, Y. Dai, and X. Shen, “Delay-Minimization Routing for Heterogeneous VANETs With Machine Learning Based Mobility Prediction,” *IEEE Transactions on Vehicular Technology*, vol. 68, no. 4. Institute of Electrical and Electronics Engineers (IEEE), pp. 3967–3979, Apr. 2019. doi: 10.1109/tvt.2019.2899627.
45. K. Kandali, L. Bennis, O. E. Bannay, and H. Bennis, “An Intelligent Machine Learning Based Routing Scheme for VANET,” *IEEE Access*, vol. 10. Institute of Electrical and Electronics Engineers (IEEE), pp. 74318–74333, 2022. doi: 10.1109/access.2022.3190964.
46. I. Seth, K. Guleria, and S. N. Panda, “Introducing Intelligence in Vehicular Ad Hoc Networks Using Machine Learning Algorithms,” *ECS Transactions*, vol. 107, no. 1. The Electrochemical Society, pp. 8395–8406, Apr. 24, 2022. doi: 10.1149/10701.8395ecst.
47. P. Upadhyay, et al., “An Improved Deep Reinforcement Learning Routing Technique for Collision-Free VANET,” *Scientific Reports*, vol. 13, no. 1. Springer Science and Business Media LLC, Dec. 8, 2023. doi: 10.1038/s41598-023-48956-y.
48. M. ul Hassan et al., “ANN-Based Intelligent Secure Routing Protocol in Vehicular Ad Hoc Networks (VANETs) Using Enhanced AODV,” Sensors, vol. 24, no. 3. MDPI AG, p. 818, Jan. 26, 2024. doi: 10.3390/s24030818.
49. Y. He, F. R. Yu, N. Zhao, H. Yin, and A. Boukerche, “Deep Reinforcement Learning (DRL)-based Resource Management in Software-Defined and Virtualized Vehicular Ad Hoc Networks,” *Proceedings of the 6th ACM Symposium on Development and Analysis of Intelligent Vehicular Networks and Applications*. ACM, Nov. 21, 2017. doi: 10.1145/3132340.3132355.
50. G. P. K. Marwah, et al., “An Improved Machine Learning Model with Hybrid Technique in VANET for Robust Communication,” *Mathematics*, vol. 10, no. 21. MDPI AG, p. 4030, Oct. 30, 2022. doi: 10.3390/math10214030.
51. M. V. Gudur, S. M, P. P, N. L, M. B. Neelagar, and S. B, “Machine Learning based Routing approach and Resource Management in Vehicular Adhoc

Networks," *2023 International Conference on Recent Trends in Electronics and Communication (ICRTEC)*. IEEE, Feb. 10, 2023. doi: 10.1109/icrtec56977.2023.10111882.

52. X. Chen, T. Zhang, S. Shen, T. Zhu, and P. Xiong, "An Optimized Differential Privacy Scheme with Reinforcement Learning in VANET," *Computers & Security*, vol. 110. Elsevier BV, p. 102446, Nov. 2021. doi: 10.1016/j.cose.2021.102446.
53. M. Min, W. Wang, L. Xiao, Y. Xiao, and Z. Han, "Reinforcement Learning-Based Sensitive Semantic Location Privacy Protection for VANETs," *China Communications*, vol. 18, no. 6. Institute of Electrical and Electronics Engineers (IEEE), pp. 244–260, Jun. 2021. doi: 10.23919/jcc.2021.06.019.
54. J. Chen, K. Li, and P. S. Yu, "Privacy-Preserving Deep Learning Model for Decentralized VANETs Using Fully Homomorphic Encryption and Blockchain," *IEEE Transactions on Intelligent Transportation Systems*, vol. 23, no. 8. Institute of Electrical and Electronics Engineers (IEEE), pp. 11633–11642, Aug. 2022. doi: 10.1109/tits.2021.3105682.
55. H. Bangui, M. Ge, and B. Buhnova, "A Hybrid Machine Learning Model for Intrusion Detection in VANET," *Computing*, vol. 104, no. 3. Springer Science and Business Media LLC, pp. 503–531, Aug. 23, 2021. doi: 10.1007/s00607-021-01001-0.

Chapter 3

Intelligent human-centric network (IHCN) applications

K. Deepa and D. Devasena

3.1 INTRODUCTION

The evolution of networks has played a significant and vital role in communication over the years starting from postal services to telecommunication networks followed by the internet revolution. The internet revolution changed the way of long-distance communication. It also paved the way for real-time conversations and fostered global connectivity. It also led to social networks that connected individuals worldwide where mobile networks led the way for constant communication. With technological progression, IoT allowed devices to communicate, thereby influencing daily activities. Today, human interactions are empowered by the rise of intelligent networks and redefine the communication model. The complexity of these networks and services demands intelligence, and conventional communication networks struggle to handle the complexity that affects the user experience and may lead to security issues. Thus, it is essential to integrate intelligence into the communication networks to address the challenges. Integrating artificial intelligence (AI) enhances the user experience and ensures reliable, secure, and fast connections. To engage the user in the developing interconnected digital world, intelligence integration alone isn't sufficient. Human-centric methods are also essential.

AI and machine learning (ML) play a vital role in the design of any intelligent system where the system behavior is based on data. In today's scenario, contact with AI and ML systems is rapidly increasing, from Alexa and Google Assistant to self-driving cars and many more. Considering the impact that intelligent systems can have on people's lives, it is important to develop these systems with care and consideration [1]. As the systems interact with humans, relying on algorithmic advancement in AI and ML is inadequate. Therefore, there is a need for a new technique that assists in analyzing data from the human perspective. Data-driven AI has shown remarkable success in almost all fields, yet there are some limitations in terms of deliberative intelligence. This leads to the need for intelligent human-centric network (IHCN) applications. Nowak states that "Human-centric AI focuses on collaborating with humans, enhancing human capabilities, and

 DOI: 10.1201/9781003541363-3

empowering humans to better achieve their goals" [2]. IHCN is one such concept that combines cutting-edge technologies, data analysis, and human-centric design principles to construct a network infrastructure that handles the preferences and demands of the user. Therefore, there is a need for a new technique that assists in analyzing data from the human perspective. IHCN is one such concept that combines cutting-edge technologies, data analysis, human-centric design principles, and AI algorithms to construct a network infrastructure that handles the preferences and demands of the user. IHCN is a branch of AI and ML that guides the development of intelligent systems. Being human-centric does not mean humans are "centered." It refers to how people view various situations. The word "people" has different meanings in the context it is being used. In engineering, it is considered the influential factor or human factor; in management, people are perceived as resources or capital [3]. In IHCN, people are viewed as people. In IHCN, the strengths of both AI and humans are combined responsibly for the long-term benefit of the user. To achieve a truly intelligent human-centric network with AI, the following factors need to be addressed.

1. IHCN systems should be designed to address biases in both data and training algorithms to ensure unbiased outcomes for the end users.
2. IHCN systems should be designed to address data access issues.
3. IHCN systems should be guided by ethical principles and must be transparent.
4. IHCN systems should incorporate mechanisms that can switch between being automatic and human-centric, especially in critical decision-making processes.

IHCN systems must focus on transparency, collaboration, and responsible technology usage to create and contribute to a more equitable and sustainable future. Although IHCN supports good decision-making, there are a few challenges in its development and deployment.

1. AI technology is continually evolving, which impacts its reliability. The rapid pace of development may introduce challenges about its stability and reliability.
2. The dependency between machines and humans may lead to challenges in terms of trust and transparency.
3. Deployment of IHCN systems beside humans in random environments may lead to inconsistent decisions and new challenges.
4. Creating and maintaining ethical AI systems may be challenging, as the ethical constraints are complex and dynamic.
5. Incorporating various sensors, platforms, and protocols can lead to compatibility issues.
6. Designing energy-efficient IHCNs and optimizing the algorithms while uploading performance will be demanding.

7. Maintaining better infrastructure and identifying significant business models for IHCNs may lead to financial complexities.

3.2 KEY TERMS AND CONCEPTS RELATED TO IHCN

In the field of IHCN, it is important to understand a few key terms and concepts necessary to build a robust and efficient system. This section aims to provide a comprehensive explanation of fundamental key terms and concepts that form the backbone of IHCN.

1. **Intelligent systems/networks:** Intelligent systems/networks are capable of making autonomous decisions by analyzing, adapting, and learning from data. They incorporate AI and ML algorithms for predictions and decision-making. They play a vital role in IHCN for personalizing and automation.
2. **Human-machine interaction:** This refers to the interaction between a human and a machine through a user interface. The communication mode may include voice control, text-based control, and virtual and augmented reality. It holds significant importance in everyday life where more and more devices are being connected and tasks are automated.
3. **Huma-centered AI:** This is also referred to as human-centric AI or human-centered artificial intelligence. This is an AI approach that focuses on deploying human abilities in systems rather than eliminating them, and emphasizing human well-being, values, and needs.
4. **Use-centered design:** This is a systematic process that focuses on user-friendly objectives, user preferences, and operational timelines at each phase of the design. It relies on user demand, priorities, experiences, and user satisfaction.
5. **Context-aware computing:** The system can sense and adapt itself to the physical environment. It may adapt and respond to factors like user location, user needs, identity, and activities. This concept plays an important role in mobile and global computing environments. An example is augmented reality-based applications.
6. **Mixed-initiative interaction:** This approach emphasizes creating collaborative and effective interactions between agents (humans and an AI system) aiming to create mutually beneficial outcomes.
7. **System thinking design:** It is a comprehensive approach recognizing the mutual dependence of components within the AI system, focusing on feedback, relationships, and emerging characteristics. Conversely, system design is the practical approach to system integration to achieve specific goals. In combination, they form an interdependent relationship where system design implements designated structures

and feedback methods to create an adaptive system. System thinking guides the development of an adaptive system. This coordinated approach manages system complexity, fostering iterating development and assisting ethical approaches to yield solutions that are aligned with user needs within a broader social framework.

8. **Human-centric IoT:** This focuses on designing and integrating IoT solutions to meet human needs and enhance user experience. This is possible by developing human-centric IoT-based intelligent networks, which can lead to proactive services.

3.3 STAGES OF THE IHCN DEVELOPMENT PROCESS

The development process involves four stages. Figure 3.1 depicts the four stages of network development.

1. **Understanding user requirements** [4]: Understanding user requirements is the fundamental step in developing any system, service, or protocol that is user-centric. Research must be conducted in a broader context for a better understanding of the requirements, behaviors, and preferences of the user.
2. **Defining the user context:** Instead of creating networks based on assumptions, the research helps to create a better user-centric design. This requires collaboration between designers, developers, and users. Defining the network requirement factors like network traffic patterns, security concerns, privacy, legal works, cultural diversity, etc. must be considered.

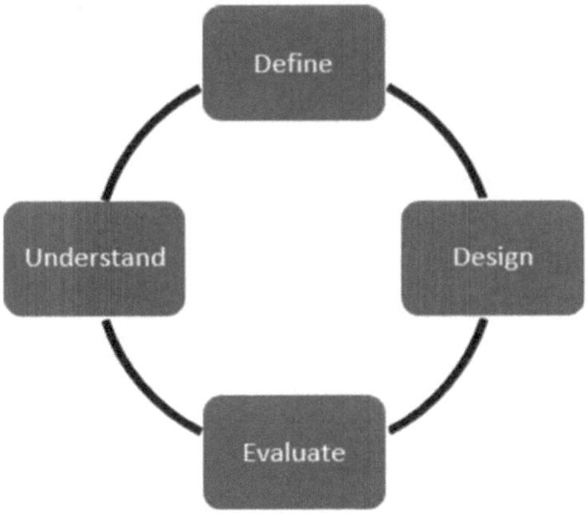

Figure 3.1 Stages of the IHCN development process [4].

3. **Designing the user requirements:** The next phase involves the design of the network prioritizing user requirements. Prototypes, drawings, etc. must be developed to fulfill the user requirements.
4. **Evaluate the design:** Once a design is done, the next step is to evaluate the design. The network performance can be validated by testing the network performance under different conditions including device types, network traffic, user load, etc. The prototypes may be presented to the users for validation and feedback.

3.4 UNDERSTANDING THE IHCN DESIGN: A CASE STUDY ON SMART CITY NETWORKS

To design any IHCN, requirements, preferences, and feedback from users must be the primary consideration. To understand the design, let's consider the case study of a smart city network. To design an IHCN-based smart city network, the preferences and needs of the people living in the city are prioritized. The first step in designing an IHCN network is to identify the key stakeholders, which include residents of that city, local businesses, city authorities, and technology providers. Users within the smart city context should also be grouped into, e.g., residents, tourists, and service providers. Also, it is very important to understand the demographics of the city [5]. The network is designed to address the needs of the people in the smart city, so the socio-demographic environment must also be considered. This can be done by conducting extensive research to understand user needs, behaviors, etc. This may include conducting surveys and interviews focusing on socio-economic factors, environmental conditions, and involvement of government. A "user" should cover all categories of people. Once the research is done, the next step is to define clear goals. For a smart city, enhancing public safety, efficient transportation, better ICT structure, energy, waste management, etc. may be considered as goals. The goals can be set focused on people, framework, collaboration, and innovation. In any IHCN network, the primary focus must be on people, which enables to development of policies and initiatives to meet their needs. The next step is to design the user requirements [6]. For example, if the network is designed to handle traffic updates, then a city route plan can be made considering parameters like route scheduling, predictive analysis reports, emergency services, and traffic laws. When designing the network, factors like scalability, reliability, sustainability, data security, and privacy should be considered. The last step is to evaluate the designed network. Apart from performance evaluation, parameters like accessibility, scalability, reliability, data security, cost, and regulatory compliance must be considered.

3.5 IHCN TO MANAGE TRAFFIC IN A SMART CITY

Smart city traffic networks focus on collecting real-time data to avoid congestion by syncing their traffic signals, controlling the speed limits guiding the drivers in current traffic, and alerting them of faster routes [7]. This can be achieved through a network of sensors, cameras, and GPS data to monitor real-time traffic. These systems collect data on traffic volume in a particular area, queue length, traffic density, lane usage, weather conditions, incident/accident reports, public and emergency transportation, air quality and pollution, etc. To manage traffic congestion, a diverse approach in combination with data-driven strategies and advanced technology is required. Smart cities use the real-time traffic data generated from sensors to identify the areas of congestion and reroute the traffic to less congested areas. IHCNs play a significant role in traffic management systems by integrating advanced technologies with human-centric approaches to improve the overall efficiency of the system. The following are a few differences in how IHCN can contribute to managing traffic when compared to traditional networks. Apart from traffic, the major hindrance faced by the drivers is highway and road construction. IHCN can use unmanned air vehicles to provide real-time alerts for the drivers and emergency crews [8].

1. **Real-time traffic monitoring:** Live traffic conditions are continually monitored through a network of cameras and sensors to collect data on variables like speed, congestion, traffic volume, and road conditions. As IHCN integrates human intelligence in the network control process, it can interfere when necessary to coordinate the multi-modal traffic flow. Moreover, IHCN can provide real-time visualization of the traffic route and simulation with a geographic view for the user and traffic controllers. Apart from traffic, the major hindrance faced by the drivers is highway and road construction. IHCN can use unmanned air vehicles to provide real-time alerts for the drivers and emergency crews [9]. IHCN can use the sonification technique to translate the traffic data into audible sound patterns to create situational awareness [10]. The network traffic data collected from sensors are mapped to particular sonic tributes like volume or tempo. For example, heavy traffic can be mapped to high-pitched sounds. As the traffic events are in real-time, the user can listen to the network rather than relying on visual representations for better situational awareness. This varies from the fully automated networks that solely depend on predefined algorithms.
2. **Predictive data analysis (PDA)**: PDA uses real-time present and past data to forecast traffic conditions and events in a particular area to optimize traffic flow. IHCN can enhance the PDA by incorporating

human expertise and advanced technologies in road traffic management. IHCN ensures that all relevant real-time data is aggregated from the various sensors, data management systems, and traffic cameras. It also refines the predictive analysis system that is already being used. Content delivery time in highly congested areas can be improved through IHCN. It can also be used to model the shortest path to optimize the traffic flow. It can be used to forecast congestion through ML algorithms [11]. For example, IHCN can predict the heavy traffic congestion that is likely to occur on a highway due to various combinations of factors like road work, heavy rain, or accidents. The IHCN will send a warning to the traffic management system about its predictions and provide suggestions to drivers. Traffic managers can validate and implement alternative measures to divert traffic. IHCN guides the traffic control system to take proactive measures.

3. **Incident management:** IHCN can be a useful tool in incident management in traffic management systems, as it emphasizes human-centric decision-making for complex situation analysis. It integrates and aggregates data from its deployed sensor network to provide a comprehensive overview of the traffic situation. It incorporates advanced analytics and AI techniques to analyze the situation, helping the traffic managers to make decisions quickly. Based on the data analysis, it allows for automatic lane configurations, adjustment of traffic signals, and diversions. IHCN can adopt intelligent zone-based content pre-caching, which can be used to optimize the data flow in different zones within a city. Smart algorithms in the system segregate the information based on priority within a specific zone of the city. The city is divided into sub-regions with unique traffic characteristics considering the neighboring boundaries and highways. Content pre-catching is processing the information in advance rather than on-demand [12]. The pre-cached data includes camera images, sensor data, traffic signal timings, parking details, road closure alerts, and weather conditions. As well as incident management, implementing this system has various advantages, such as reducing latency, improving smoother traffic flow, allocating resources efficiently, and enhancing safety.
4. **Smart parking solutions**: IHCNs can be utilized for smart parking solutions to address parking inefficiencies in crowded areas. They can use the long-range wide-area network (LoRaWAN) communication protocol for this application. LoRaWAN sensors can be used to detect the availability of parking spaces, and the data must be transmitted to the LoRaWAN gateways. These gateways must be deployed throughout the city strategically. The gateway acts as a bridge between the centralized server and the sensor network. The server aggregates the data from the gateways and processes to determine the availability of

parking spaces. A smartphone app can be developed to make the system user-friendly to check real-time parking space availability, reserve parking lots, and make payments [13, 14]. The data can be used for predictive analysis to understand the pattern of parking usage to improve the system's efficiency. IHCN-based smart parking solutions can be integrated with payment apps for convenient payments. IHCN can also use information-centric networking techniques for smart parking solutions. This approach focuses on retrieving content or data rather than connecting with end devices. In smart parking solutions, the parking spaces can be considered objects. Each parking lot is assigned a unique content identifier. Users may request the parking lot information based on their current location, and the network will retrieve information about neighboring parking lots. The parking lots are modeled as nodes or edges in this approach, and the central server provides responses based on the query raised by the nodes [15]. This can help drivers to find available parking lots quickly, thereby reducing traffic congestion.

5. **Integration of public transportation:** IHCNs can be used to integrate public transport to make it more accessible for passengers. These networks connect various modes of public transport, thereby prioritizing passenger needs. They can use IHCN-based vehicular communications for this purpose. IHCN focuses on content retrieval, not on host-based communication. In vehicle management systems, information like transit updates and traffic conditions may be treated as objects. Each public transport vehicle is assigned a unique identifier based on location to fetch details about objects like updates on delays, route changes, multi-modal information, and service updates [16]. Incorporating IHCN-based vehicular communications in IHCNs can lead to more efficient and connected transit systems to promote sustainable mobility solutions.

3.6 CHALLENGES IN DEPLOYING IHCN TO MANAGE TRAFFIC IN A SMART CITY

There are several challenges in deploying IHCN to manage traffic in a smart city. These challenges can be categorized as technical, social, and security facets.

3.6.1 Technical challenges [17]

1. Deploying various IoT devices, sensors, cameras, and communication networks and developing the necessary infrastructure to collect real-time data is a challenging task. As the traffic management system relies on a network of interconnected devices, ensuring seamless

integration can be a complex task. It is also essential to develop an efficient data integration system. This requires substantial investment, cooperation, and coordination from various stakeholders.
2. Developing AI and ML algorithms for data analysis and predictions on real-time traffic data is a tedious task, as factors like pedestrian movement, congestion, vehicular networks, and weather reports need to be considered. Ensuring consistent connectivity and reliable communication in densely populated areas may be challenging.
3. In the traffic management system, massive data is generated in real-time. Managing this large volume of sensitive data leads to significant privacy and security concerns. Efficiently managing this huge amount of data is a challenging task. Therefore, it is necessary to implement a robust data protection system and follow strict rules and regulations to gain public trust.
4. Smart cities are dynamic, and new technologies/devices are often incorporated,. Designing an IHCN that scales to accommodate such a system is crucial. Maintaining a high level of reliability in these systems is also a challenging task.
5. Finding ways to optimize energy in the large network and incorporating renewable energy sources is a challenging task.

3.6.2 Social challenges

1. Creating awareness among citizens about the IHCN traffic management system will be a challenging task. As the network deploys extensive sensor networks for data collection, this can raise privacy concerns. Therefore, citizen data literacy must be improved to understand the system [18].
2. The IHCN-based traffic management system must benefit all citizens, including communities on low incomes. It must be ensured that the technology does not create any social inequalities that could lead to other challenges. There is also a risk of creating a digital divide due to a lack of data or technological literacy [19]. When deploying these systems, it is important to understand and respect the cultural norms of people.
3. Decisions made based on traffic data must be transparent and made with fairness considering ethics in mind. Also, in case of emergencies like natural disasters or any accident, the system must be adaptable to minimize disruptions and ensure public safety.

3.6.3 Security challenges

1. As the network collects huge amounts of data from various sources, ensuring the security of the data both in rest and motion is critical to prevent data theft, unauthorized access, and data loss. The data

management system in IHCN is susceptible to cyber-attacks, which can disrupt the entire system and lead to safety risks. Therefore, implementing robust measures to protect the data from cyber-attacks is essential [20].

2. Designing these systems to be resilient and able to recover quickly from such attacks is essential for maintaining the city's operations. Security should be an essential part of planning, managing, and designing the IHCN for a smart traffic management system.

Thus, the deployment of IHCN to manage traffic in a smart city is a transformative step toward creating a more efficient and sustainable urban environment. IHCN employs advanced technologies like AI, ML, IoT, and image processing algorithms to optimize the traffic flow within the city. Further, the integration of predictive analysis helps in proactive traffic management, which can reduce traffic-related accidents and improve the transportation infrastructure. Incorporating sonification techniques, smart traffic lights, smart vehicular networks, etc. may facilitate the needs of all types of people including senior citizens and persons with disabilities. IHCN also provides an open ground for startups and the development of new strategies in terms of technology, services, and solutions. This in turn stimulates job opportunities for technological advancement. The implementation of IHCN in a smart city to manage traffic enhances quality of life and improves the safety culture, paving the way toward a brighter and connected future for cities and their residents.

3.7 A REAL-TIME EXAMPLE

IHCN for traffic management can be seen in many places around the world.

3.7.1 Las Vegas, USA

IHCN networks are implemented in Las Vegas for traffic management to enhance the user experience. This is a business model led by Cisco with an approximate investment of four million dollars from its project Black Jack. This network is a unique example of next-gen smart mobility for any city. The state-of-the-art network is a combination of advanced traffic signal controls and high bandwidth communications. The system utilizes real-time data analysis to manage traffic flow and provide support for connected and autonomous cars, leading to futuristic multi-modal traffic management. The system design also includes remote sensing technology, which uses light detection and ranging (LIDAR) to assess pedestrian behavior like crowd management, path analysis, and crowd density. This is one of the key features enabled in the IHCNs of Las Vegas toward equity. The city is separated into intersections. DSRC transceivers and LIDAR sensors are installed in each intersection, enabling vehicle-to-vehicle communication. The data analysis enables traffic engineers to deliver a variety of services coordinated across intersections for customer satisfaction [21].

3.8 UNDERSTANDING THE IHCN DESIGN: A CASE STUDY ON SUPPLY-CHAIN MANAGEMENT

Business operations are no exception for automation, which is revolutionizing operations. To improve customer satisfaction, it is important to incorporate new techniques and strategies in business to enhance supply-chain management. Incorporating IHCNs in supply-chain management will optimize the interactions between stakeholders, including suppliers and customers [22]. IHCN relies on real-time data analysis, which can provide transparency in the aspect of the supply chain from purchase to delivery. The AI and ML techniques allow predictive analysis, which helps in making better decisions related to inventory, demand forecasting, transportation, etc. It may also aid in risk management, inventory optimization, scalability, and adaptability to changing market conditions. IHCNs promote better collaboration and communication between supply-chain partners.

3.9 IHCN FOR THE SUPPLY CHAIN IN AN E-COMMERCE WEBSITE

IHCs can be incorporated into the supply-chain management system for an e-commerce website to create a more efficient customer-based supply chain. These networks prioritize the role of human expertise to support and enhance the performance of the work. The Japanese initiative "super smart society," a concept of society 5.0, focuses on IHCN to achieve customer satisfaction [23]. IHCNs in supply management prioritize the well-being of workers and customers. IHCNs also align the technological advancements promoting sustainability and social responsibility in the manufacturing industry [24]. Creating IHCNs for supply chains in an e-commerce website like those of Amazon, Flipkart, and Blinkit involves advanced technologies, logistics, and customer-centric strategies. Let's imagine a case study on revolutionizing supply chains toward IHCN. In any e-commerce business, like Amazon, the supply-chain strategy focuses on three main factors: customer passion, operational excellence, and innovative practices. Incorporating IHCN in supply-chain management aids in the following.

1. **Demand forecasting:** IHCN can be implemented for demand forecasting to serve the customers in a better way. Demand forecasting is

one of the critical aspects of business planning. The future demand for a product is predicted, and the production process is streamlined accordingly. Data is collected from various sources like customer feedback, market trends, and buying patterns. AI and ML techniques along with human expertise can be used for data analytics to predict future demand. IHCN can adopt a federated learning technique for demand prediction. This is a collaborative approach that permits multiple parties to build a predictive model without sharing their data. This learning approach helps in improving prediction accuracy while maintaining data privacy and security [25]. Long short-term memory is a neural network-based approach for demand forecasting in an e-commerce business model that can predict and analyze consumer behavior toward a product [25].

2. **Handling transportation and logistics:** IHCN can improve the efficiency of supply-chain management in an e-commerce business handling transportation and logistics. The network can provide real-time visibility and tracking, as the network is equipped with GPS and IoT devices to track the shipment. It can also be used for route optimization. IHCN has all the data related to weather, road status, highway construction, etc. which helps in planning efficient delivery routes using software. It can also be utilized for supply-chain optimization [26]. IHCN aids in the creation of digital supply networks. These networks are AI-enabled, capture real-time data, and are capable of analyzing and making automatic decisions that can optimize costs and service quality [27].
3. **Order processing:** IHCN in a supply management system can enhance order processing efficiency. IHCN employs real-time data analysis techniques for data integration and visibility. IHCN forecasting algorithms can predict inventory levels, reducing stock-outs or overstocking. IHCN in supply-chain management can improve the speed and accuracy of order processing through intelligent automation tools that provide recommendations to the operators. AI-powered chatbots or virtual assistants can assist customers in placing orders, tracking status, complaint handling, etc.

Incorporating IHCN into supply management for an e-commerce business can optimize operations, enhance customer satisfaction, and achieve sustainable growth. IHCN allows companies to make decisions based on real-time data analysis improving their supply-chain visibility. Apart from inventory management and demand forecasting, this approach enables businesses to respond to dynamic market trends and customer preferences. Implementation of IHCN paves the path for continuous improvement in the organization. It also fosters the culture of adapting and learning new AI-based business models.

3.10 CONCLUSION

In conclusion, IHCN represents an advanced model in terms of technology and connectivity. IHCN can pave the way for pioneering developments from smart homes to health care systems, where these networks can redefine the user experience. The concept of IHCN emphasizes the need for the implementation of these technologies and prioritizing their adoption by human users in their daily routines. These applications are driven by advancements in AI, ML, and data analytics, leading to an AI being able to meet the needs and preferences of users. In education, the IHCN network can enhance the learning environment, while in business it can optimize processing and yield, and maximize productivity. When implemented, IHCN networks can revolutionize the way tasks are executed. As the networks are driven by AI and ML algorithms focusing on user-centric design, they can streamline tasks, automate processes, and enhance productivity. The human-centered approach involves the use of human feedback to enhance the performance of the network in practical use. Adopting this technology requires a clear framework and the development of a sustainable and transparent environment for humans who are interacting with the technology. The human-centric intelligent approach will reform almost all sectors, from healthcare to smart homes, and transportation to entertainment. They are capable of delivering proactive solutions to improve the quality of life of people. These networks redefine the way we interact with technology and adopt them daily. However, it is important to address ethical concerns and ensure that the applications benefit society. Currently, IHCN has diverged to multiple interpretations that require a coherent knowledge of the developing concept and its practical values. In the future, the implementation and development of IHCN networks will likely lead to a unified human-machine interface and obtaining better efficiency across professional domains. IHCN unlocks a new way to live in the digital era.

REFERENCES

1. Ben Shneiderman. “Human-Centered Artificial Intelligence: Reliable, Safe & Trust-Worthy”. In: *International Journal of Human–Computer Interaction* 36.6 (2020), pp. 495–504.
2. Andrzej Nowak, Paul Lukowicz, and Pawel Horodecki. “Assessing Artificial Intelligence for Humanity: Will ai be the Our Biggest Ever Advance? Or the Biggest Threat [Opinion]”. In: *IEEE Technology and Society Magazine* 37.4 (2018), pp. 26–34.
3. Jan Auernhammer. “Human-centered AI: The Role of Human-centered Design Research in the development of AI”. In: *Synergy - DRS International Conference 2020*, 11–14 August (2020).

4. ionos. *www.ionos.com/digitalguide/websites/web-development/human-centered-design*. 2020. url: www.ionos.com/digitalguide/websites/web-development/human-centered-design/.
5. Maroula N Alverti et al. “A Human Centric Approach on the Analysis of the Smart City Concept: The case study of the Limassol city in Cyprus”. In: *Advances in Geosciences* 45 (2018), pp. 305–320.
6. www.interaction-design. *What is Human-Centered Design*. 2021. url: https://www.interaction-design.org/literature/topics/human-centered-design.
7. Jodi Joseph Asiag - Head of Content Communications. *5 Ways Smart Cities Use Traffic Data for Traffic Management*. 2021.url: https://otonomo.io/blog/smart-cities-use-traffic-data-for-traffic-management/.
8. Charles Malveaux PhD et al. “Real-Time Work Zone Traffic Management via Unmanned Air Vehicles”. In: (2020).
9. Leizhen Wang et al. “Human-Centric Multimodal Deep (HMD) Traffic Signal Control”. In: *IET Intelligent Transport Systems* 17.4 (2023), pp. 744–753.
10. Mohamed Debashi and Paul Vickers. “Sonification of Network Traffic Flow for Monitoring and Situational Awareness”. In: *PLoS One* 13.4 (2018), e0195948.
11. Akashdeep Bhardwaj et al. “IIoT: Traffic Data Flow Analysis and Modeling Experiment for Smart IoT Devices”. In: *Sustainability* 14.21 (2022), p. 14645.
12. Hui Guo, Lan-lan Rui, and Zhi-peng Gao. “A Zone-Based Content Pre-Caching Strategy in Vehicular Edge Networks”. In: *Future Generation Computer Systems* 106 (2020), pp. 22–33.
13. The Things Industries. *Smart parking on LoRaWAN*. 2021. url: https://www.thethingsindustries.com/news/smart-parking-on-lorawan/.
14. Juan Ram´on Santana et al. “LoRaWAN-based Smart Parking Service: Deployment and Performance Evaluation”. In: *Proceedings of the 19th ACM International Symposium on Performance Evaluation of Wireless Ad Hoc, Sensor, & Ubiquitous Networks* 2022, pp. 107–114.
15. Boubakr Nour et al. “A Survey of Internet of Things Communication Using ICN: A Use Case Perspective”. In: *Computer Communications* 142 (2019), pp. 95–123.
16. Ikram Ud Din et al. “Information-Centric Network-Based Vehicular Communications: Overview and Research Opportunities”. In: *Sensors* 18.11 (2018), p. 3957.
17. Auwal Alhassan Musa et al. “Sustainable Traffic Management for Smart Cities Using Internet-of-Things-Oriented Intelligent Transportation Systems (ITS): Challenges and Recommendations”. In: *Sustainability* 15.13 (2023), p. 9859.
18. HMKKMB Herath and Mamta Mittal. “Adoption of Artificial Intelligence in Smart Cities: A Comprehensive Review”. In: *International Journal of Information Management Data Insights* 2.1 (2022), p. 100076.
19. Auriol Degbelo et al. “Opening Up Smart Cities: citizen-Centric Challenges and Op- Opportunities from GIScience”. In: *ISPRS International Journal of Geo-Information* 5.2 (2016), p. 16.
20. Chen Ma. “Smart City and Cyber-Security; Technologies Used, Leading Challenges and Future Recommendations”. In: *Energy Reports* 7 (2021), pp. 7999–8012.

21. Traffic Mangement. *Three More Smart Cities in Traffic Manage- Ment: Las Vegas, Dubai and Shenzhen.* 2021. url: https://www.intertraffic.com/news/traffic-management/three-more-smart-cities-in-traffic-management-las-vegas-dubai-and-shenzhen.
22. Meraqui. *Why a Human-Centric Supply Chain Management Is Key to an Effective Business?* 2023. url: https://www.linkedin.com/pulse/why-human-centric- supply-chain-management-key-effective-business/.
23. Dimitrios Serpanos. "The Cyber-Physical Systems Revolution". In: *Computer* 51.3 (2018), pp. 70–73.
24. Praveen Kumar Reddy Maddikunta et al. "Industry 5.0: A Survey on Enabling Technologies and Potential Applications". In: *Journal of Industrial Information Integration* 26 (2022), p. 100257.
25. Juntao Li et al. "Demand forecasting of e-commerce enterprises based on hori- zontal federated learning from the perspective of sustainable development". In: *Sustainability* 13.23 (2021), p. 13050.
26. Amine Belhadi et al. "Artificial Intelligence-Driven Innovation for Enhancing Sup- Ply Chain Resilience and Performance Under the Effect of Supply Chain Dynamism: An Empirical Investigation". In: *Annals of Operations Research* 333 (2021), pp. 1–26.
27. CETDIGIT. *Smart Selling in ECommerce and Online Retail.* url: https://www.cetdigit.com/smart-selling-in-ecommerce-and-online-retail.

Chapter 4

Mathematical modeling for vehicle density and traffic flow with intelligent vehicular communication

Raju Singh Gaur

4.1 INTRODUCTION

There is no need to analyze the historic traffic patterns of a particular area because the fundamental requirements of travelers are simply that the traffic should always be safe [1], rapid [2], and environmentally friendly.

These requirements are addressed by the concept of intelligent transportation systems (ITSs) [3], which reduce traffic-related complexity in terms of the frequency of accidents, the effects of vehicle congestion, and traffic emissions/pollution. ITS is regarded as an independent discipline, as it covers wide research interests. The wide scope of ITS in traffic management has led to it being a well-established research subject for researchers and scholars. Many researchers and scientists are involved in resolving the complexity of transportation. To develop ITSs to be more advanced in the future means that the internet, networks, devices, and equipment must be more capable and efficient in acquiring data concerning the vehicle [4, 11] to assist the driver in making better decisions [17]. This information will also help in the development of driverless vehicles [15].

The enormous rise in the number of vehicles is a direct result of population growth, and this has led to traffic congestion. Traffic congestion increases not only capital expenses (CapEx) but also operational expenses (OpEx) and increases other traffic-related complexities during traveling for both drivers and passengers. This decreases the quality and pleasure of a journey and increases time and costs. Slow traffic flow can have a negative impact on economic and social development. Air pollution, noise pollution, accidents, and other traffic-related issues caused by traffic congestion have an adverse effect on many social elements in addition to lowering environmental quality. In modern industrialized cities, traffic congestion is a major problem. To resolve this, researchers and scientists are studying the area of vehicular networks. Analysis of the performance of networks by researchers will lead to networks becoming more efficient and accurate, which will increase the mobility of vehicles and reduce traffic congestion [14, 16]. This chapter covers the following topics: roadside units (RSU), autonomous vehicles, vehicle-to-vehicle (V2V) transmission [3], vehicle-to-infrastructure

DOI: 10.1201/9781003541363-4

(V2I) transmission [4], vehicle-to-everything (V2X) transmission [5, 6], and vehicle-to-pedestrian (V2P) transmission [7].

4.2 PURPOSE OF DEVELOPING SOFTWARE-DEFINED VEHICLE NETWORKS

An increasing density of vehicles causes traffic congestion. This congestion causes collisions to occur, which then causes even more traffic congestion. To avoid this situation, a network that is advanced enough to make better decisions when forwarding packets is required, and that can also improve service quality and resource utilization. The principles of the software-defined vehicle network (SDVN) are able to provide better solutions for improving data traffic flow. SDVNs not only work in simple conditions like plane areas but also in more complex conditions like highways, bridges, and tunnels. The use of traditional methods can cause errors in complex conditions, but SDVNs undoubtedly offer better advantages and results in these conditions. Software-defined networks (SDNs) have the features of providing programmability and survivability to traditional networks by separating the control plane from the data plane. Data flow management between source and destination nodes [2] is more effectively managed with the flexible and dependable SDN. For the fast development of ITSs, the internet of vehicles (IoVs) plays an important role. SDN will furnish reliability, survivability, and flexibility to the mobility networks and also enhance the network performance in vehicular ad hoc networks (VANETs). To develop mobility networks [9], intelligent software-defined internet of vehicles (iSDIoVs) has been introduced [10]. Sometimes the network connection of moving vehicles may fail. In that case, different transportation networks, such as SDVNs, iSDIoVs [12], and ITSs, can manage network complexity [8] and make the impassable road passable [13].

4.3 OBJECTIVES OF VANET

ITS focuses on both individual intelligence and systematic intelligence equally. The ITS offers individual intelligence for intelligent vehicles and vehicles that are driven by drivers, while systematic intelligence is implemented with vehicular networks. Intelligent transportation can be achieved only by replacing the vehicles on the road with intelligent vehicles. The ITS has the ability to build communication between all users of the road and vehicles. Communication bridges enable information to be exchanged between them. The packets of information can be analyzed at very fast speeds in order to avoid vehicle collisions and traffic jams.

The network that is spontaneously established during information exchange between the vehicles is known as a VANET [18]. Vehicles are viewed as mobile nodes within the VANET. The VANET is a crucial component of ITS and is the best network for establishing communication between vehicles. The main objective of VANET is to improve road safety by minimizing traffic accidents. The type of application known as VANET provides a variety of services to drivers and passengers, such as connection packages, safety features, and convenience in the form of data and information. VANET extends safety on the road and provides infotainment to travelers by establishing communication between vehicles, and between RSUs and vehicles. Cloud-based software, traffic control, information dissemination, lane exchange and parking facilitation, and continuous internet access are all gradually added to it. Complications from these add-on features include erratic traffic patterns, poor resource allocation, network maintenance, incorrect information dissemination, and quality of amenity (QoA).

To resolve these issues, SDNs have been integrated into VANETs. Alternatively expressed, VANET comprises SDN and becomes the structure of SDVN. This application of the advanced network allows vehicle-to-everything services (V2X).

4.4 VANET INFRASTRUCTURE

VANET requires various segments to perform efficiently. The first segment is on-board units (OBUs), which are within the vehicles to send or receive the information required to perform communication. The second segment is RSUs, which develop infrastructure and make the network more powerful and reliable for communication. The third segment is dedicated short-range communication (DSRC), which VANET uses for communication in vehicles. In fact, VANET itself has no infrastructure.

VANET is now popular as an advanced technology and develops vehicle-to-vehicle, vehicle-to-RSU, and vehicle-to-pedestrian communication, as shown in Figure 4.1 as V2V, V2R, and V2P respectively. Availability and reliability of the performance of VANET architecture are totally based on mobile and wireless LAN networks. The internet of things (IoT) and IoV are branches of VANET architecture. These services are helpful in establishing various kinds of interactions and conducting intelligent communication between vehicles, which improves driving support.

VANET architecture offers various services:

1. Vehicle transmission to vehicle
2. Infrastructure transmission to vehicle
3. Roadside unit transmission to vehicle
4. Pedestrian transmission to vehicle
5. Sensor transmission to vehicle
6. Personal device transmission to vehicle

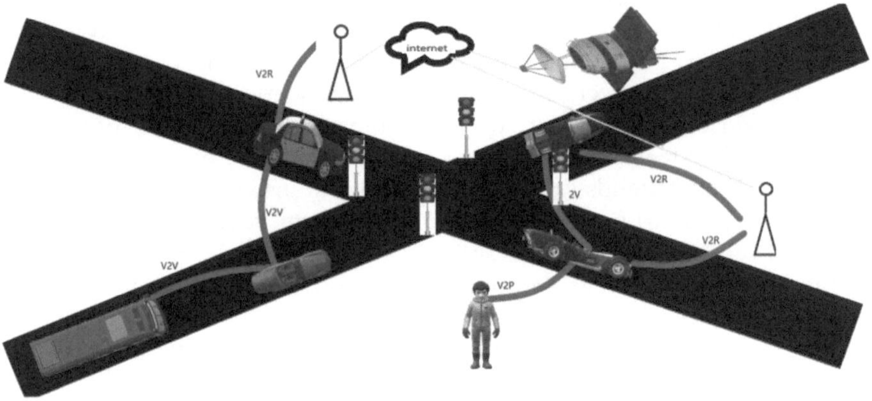

Figure 4.1 Types of communication in VANET.

4.5 SECURITY CHALLENGES RELATED TO VANET

Day by day, traveling is becoming more dangerous because of traffic congestion and the possibility of collisions. VANET security is extremely important so that hackers cannot alter or delay critical information transmission.

A very important question asked by Jeremy Blum and Azim Eskandarian is "A wireless network of intelligent vehicles can make a highway travel safer and faster but can hackers use the system to cause the accidents?" in their article "Threat of Intelligent Collisions." This question from researchers and scientists has required car manufacturers to provide protection to travelers by offering VANET security. VANET uses electromagnetic waves for communication. These waves have a negative impact on the environment. Hence, the deployment of VANETs in vehicles should be considered by governments and manufacturers.

4.6 ATTACKS ON VANET

Data transmission between different vehicles can be distributed by malicious nodes. Trust management is a challenging task between vehicles. There are security threats such as integrity, availability, and confidentiality of the network. Malicious actors jeopardize security in different ways:

1. Attackers collect information like email, address, etc. from the vehicular network.
2. When a vehicle connected to the network transmits data to another vehicle, attackers can access and intercept the data without authorization and use it for their own purposes.

3. If the attack is ongoing, the perpetrator may dispute that they sent or received any messages that caused the traffic to move more slowly.
4. Attackers may intercept and modify the data; thus vehicles receive manipulated information.
5. An attacker may damage the network and obtain information that the attackers can misuse for profit, thus affecting the legal users.
6. The attacker may shut down the whole network or degrade the network's performance by launching attacks on RSUs.

4.7 MATHEMATICAL MODELING

A mathematical model can detail V2V, V2I, V2X, and V2P transmission technologies in a better way and have the capacity to ensure the reliability of the network connections. Mathematical modeling offers accurate calculations to assist ITSs to perform more efficiently to execute autonomous driving and traffic management, and reduce traffic collisions.

In this section of the chapter, a simulation algorithm of mathematical modeling for improved traffic flow is introduced. There are theories discussed below that can make it easier to understand traffic complexity, and the simulation algorithm is suggested to resolve these complexities.

(a) **Kinetic theory**

In this section, the LWR model of kinetic theory is discussed, which was presented by Lighthill and Whitham, and by Richards. Traffic flow Q is analyzed in a small local area where the concentration of vehicles is denoted by k. The vehicles move at speed v, at time t, and space x. Then,

$$Q = kv \tag{4.1}$$

Further, according to the LWR model, a relationship is obtained in terms of traffic flow, concentration, time, and space, which is represented by

$$\frac{\partial k}{\partial t} + \frac{\partial Q(k)}{\partial k}\frac{\partial k}{\partial x} = 0 \tag{4.2}$$

The microanalysis through kinetic theory analyzes the data second by second.

(b) **Queuing theory**

Queuing theory helps to describe traffic flow. In a queue, the speed of the vehicle is dependent upon the surrounding elements i.e., traffic flow and control parameters. If n number of vehicles exit the queue in time t, then the total number of vehicles exiting from the queue in that particular span of time t can be represented

$$\Delta = nt \quad (4.3)$$

According to the law of motion, when a body moves in one direction

$$v = u + at \quad (4.4)$$

If the vehicles move with a uniform speed in a queue i.e., acceleration a = 0, then the speed of the vehicle u at the point of entry in the queue is equal to the speed of the vehicle v at the point of exit from the queue. This is represented in equation form as

$$v = u \quad (4.5)$$

From Equation (4.4), if acceleration, a, increases, then the speed v at the point of exit from the queue will increase. In the case of deceleration, the speed v at the point of exit from the queue will decrease for a given time t.

The relationship between the concentration of vehicles ϕ is inversely proportional to the speed v of the vehicles at the point of exit from the queue. This can be depicted in equation form as,

$$\phi = \frac{1}{v} \quad (4.6)$$

Equation (4.6) states that if the speed v of the vehicles increases, the density of vehicles in the queue will decrease, and if the speed v of the vehicles decreases, the density of the vehicles in the queue will increase.

(c) **Vehicle mobility model**

According to the law of motion, acceleration can be defined as,

Acceleration = final speed – initial speed / time

$$a = \frac{v - u}{t} \quad (4.7)$$

In this section, we look at two events in which time t is the same in both events. In the first event, the final speed v (speed at the point of exit from the congestion) of the vehicle is greater than the initial speed u (speed at the point of entry into the congestion) of the vehicle. In the second event, the final speed v (speed at the point of exit from the congestion) of the vehicle is less than the initial speed u (speed at the point of entry in the congestion) of the vehicle. Both the above events can be represented in equation form as,

According to the first event, where $v > u$, Equation (4.4),

$$a = \frac{v-u}{t}$$

will give a positive value, which shows the increased speed of vehicles because of acceleration. This shows that vehicle congestion reduces gradually.

According to the second event, where $v < u$, Equation (4.4),

$$a = \frac{v-u}{t}$$

will give a negative value, which shows the decreased speed of vehicles because of deceleration. This shows that vehicle congestion increases gradually.

Acceleration can be calculated by the following equation,

$$a(f) = a[1-(v_i / v_i(d))^4] \tag{4.8}$$

where a(f) denotes free acceleration, v_i(d) denotes the desired speed, and v_i denotes the speed of the ith vehicle.

$$a(o) = -a\left(\frac{\Upsilon}{\Delta S_i(t)}\right)^2 \tag{4.9}$$

Where a(o) denotes the on-spot acceleration, $\triangle$ denotes the gap between the two vehicles, and S_i denotes the location of the vehicle at any time t.

(d) **Arrival-to-departure theory**

In the event that a local area has traffic congestion, the time interval between a vehicle's arrival and departure from the congestion is referred to as arrival-to-departure (a2d) timing. Traffic congestion increases a2d timing. Therefore, the time taken to remove arrival-to-departure congestion is the total duration lapsed (L_t) in the congestion, which includes waiting time (W_t) and transmission time (T_t). This can be represented in equation form as,

$$\Sigma(L_t) = \Sigma(W_t) + \Sigma(T_t) \tag{4.10}$$

Several heterogeneous challenges like vehicle density, RSUs, and lighting management affect the a2d lapsed time. These challenges work as a trigger to increase or decrease a2d lapsed time.

The pipe rule is applied to describe the time, speed, and position of vehicles on the road.

$$\Delta S^{pfh}\left(v_i\right) = \mathrm{D} + \mathrm{T}v_i + \Delta v_i^2 \tag{4.11}$$

S^{pfh} denotes the protection from harm gap between two vehicles, D denotes the length of the vehicle, T denotes the safe time, v_i is the velocity of the vehicle, and $\triangle$ denotes the modifying parameter for deceleration.

(e) **Time-speed theory**

This theory comprises a model in which the location of the vehicle at the starting point of the congestion is l_0, and the location of the vehicle at the end point of the congestion is l_z.

$(l_z - l_0)$ can be assumed to be a finite local area, and u is the speed of the vehicle. Then time t can be calculated by the following equation:

$$t = \frac{l_z - l_0}{u} \tag{4.12}$$

In the above equation, $(l_z - l_0)$ is finite, and time t consumed in congestion is inversely proportional to the speed of the vehicle.

According to Equation (4.10), the required time t to cross the congestion area is higher when the speed of the vehicle is lower, and the required time t to cross the congestion area is lower when the speed of the vehicle is higher.

4.8 CONCLUSION

In this chapter, traffic blockages responsible for traffic jams were analyzed. To resolve these issues, several mathematical simulation models were suggested. Lack of right-of-way priority creates congested locations and consumes a lot of time. There are many triggers that can cause congestion. Under such conditions, mathematical modeling – which is based on the actual vehicle network framework – becomes extremely important. Triggers that cause congestion are real and based on vehicle density and the information received from V2V, V2I, V2X, and V2P transmission, RSU commands, and road light control. Areas that are impassable due to traffic congestion can be streamlined by using mathematical simulations. In this chapter, applying a mathematical approach has proven to be more reliable, useful, and necessary for the free-flow transportation of vehicles and to provide a complete solution for the critical concern of road safety in real situations for both drivers and passengers.

In this chapter, the location of congestion was not limited. Therefore, the mathematical models can be used in any location for traffic congestion-related problems.

REFERENCES

1. Geng, R., Wang, X., and Liu, J. (2018). A software defined networking-oriented security scheme for vehicle networks. *IEEE Access*, 6, 58195–58203.
2. Djahel, S., Doolan, R., Muntean, G. M. et al (2015). A communications-oriented perspective on traffic management systems for smart cities: challenges and innovative approaches. *IEEE Commun. Surv. Tut*, 17, 125–151.
3. Zadobrischi, E., Cosovanu, L., Avatamanitei, S., Cailean, A. (2019). Complementary radiofrequency and visible light systems for indoor and vehicular communications. In *Proceedings of the 2019 23rd International Conference on System Theory, Control and Computing (ICSTCC)*, Sinaia, Romania, 419–423.
4. Chaqfeh, M., Lakes, A., Jawhar, I. (2014). A survey on data dissemination in vehicular ad hoc networks. *Veh. Commun.* 1, 214–225.
5. Singh, R., Kathuria, K., Sagar, A. K. (2018). *Secure routing protocols for wireless sensor networks.* Proceedings of the 4th international conference on computing, communication and automation (ICCCA), 1–5.
6. Cailean, A. M., Dimian, M., Popa, V. (2020). Noise-adaptive visible light communications receiver for automotive applications: a step toward self-awareness. *Sensors*, 20, 3764.
7. Wantoro, J., Mustika, I. W. (2014). *M-aodv+: an extension of aodv+ routing protocol for supporting vehicle-to-vehicle communication in vehicular ad hoc networks.* In Proceedings of the 2014 IEEE International Conference on Communication, Networks and Satellite (COMNETSAT), pp. 39–44, Jakarta, Indonesia.
8. Kenney, J. B. (2011). Dedicated short-range communications (DSRC) standards in the United States. *Proc. IEEE*, 99, 1162–1182.
9. Wu, X., Subramanian, S., Guha, R., White, R. G., Li, J., Lu, K. W., Bucceri, A., Zhang, T. (2013). Vehicular communications using DRCS: challenges, enhancements, and evolutions. *IEEE J. Sel. Areas Commun.*, 31, 399–408.
10. Ligo, A. K., Peha, J. M., Ferreria, P., Barros, J. (2018). Throughput and economics of DRCS-based internet of vehicles. *IEEE Access*, 6, 7276–7290.
11. Li, R., Z., Zhou, X., Ding, G., Chen, Y., Wang, Z., Zhang, H. (2017). Intelligent 5G: when cellular networks meet artificial intelligence. *IEEE Wirel. Commun.*, 24, 175–183.
12. Samaras, N.S. (2016). Using basic manet routing algorithms for data dissemination in vehicular ad hoc networks (VANETs). In *Proceedings of the 2016 24th Telecommunication Forum (TELFOR)*, pp. 1–4, IEEE, Belgrade, Serbia.
13. Jiang, W., Han, B., Habibi, M. A., Schotten, H. D. (2021). The road towards 6G: a comprehensive survey. *IEEE Open J. Commun. Soc.*, 2, 334–366.
14. Abadi, M., Barham, P., Chen, J., Chen, Z., Davis, A., Dean, J., Devin, M., Ghemawat, S., Irving, G., Isard, M. et al. (2016). TensorFlow: a system for large-scale machine learning. *In OSDI*, 16, 265–283.

15. Koziol, M. (2019). *MWC Barcelona 2019; On the road to self-driving cars, 5G will make us better drivers.* IEEE Spectrum Press, New York, NY, USA.
16. Letaief, K. B., Chen, W., Shi, Y., Zhang, J., Zhang, Y. J. A. (2019). The roadmap to 6G: AI empowered wireless networks. *IEEE Commun. Mag.*, 57, 84–90.
17. Cailean, A., Dimian, M. (2016). Toward environmental-adaptive visible light communications receivers for automotive applications. *A Review. IEEE Sens. J.*, 16, 2803–2811.
18. Dixit, M., Kumar, R., Sagar A. K. (2016). *VANET: architectures, research issues, routing protocols and its applications.* Proceedings of the international conference on computing, communication and automation (ICCCA), 555–561.

Chapter 5

The impact of Artificial Intelligence on energy optimization for cellular network base stations

Mamta Punjabi and Ratnalata Gupta

5.1 INTRODUCTION

A significant number of efforts have been made in recent years to promote a sustainable, energy-saving, and comfortable life in this digital age. This research in [1] examines the correlation between user communication energy usage, overall system energy consumption, communication user categories, the number of communication users, the activation of base station dormancy technology, and the resultant energy efficiency resulting from energy-saving approaches. The primary focus remains on ensuring optimal communication quality for users [2]. The emerging power system is environmentally friendly, low-carbon, secure, manageable, adaptable, efficient, intelligent, user-friendly, open, and interactive. This aligns with the essence of the energy internet, indicating the trajectory for power grid transformation and advancement [3]. This chapter focuses on sleep issues in urban heterogeneous cellular networks, analyzing the sleep probability and power consumption model of distance-aware micro base stations in sleep mode. To minimize system power consumption, it proposes a collaborative sleep scheme involving macro and micro base stations [4]. This chapter [5] offers a comprehensive review and tutorial on network traffic prediction, covering autoregressive moving average models (ARMA, ARIMA, SARIMA) and artificial neural network approaches (RNN, LSTM, GRU, CNN). The mathematical foundations of each technique are presented in a self-contained manner, providing readers with a complete understanding of their operations. Thus, the details of the data analysis steps are explained. Moreover, we assess the performance of the models using a genuine 4G LTE cellular network dataset sourced from Kaggle [6]. Additionally, we showcase the efficiency of the proposed model in conserving energy for cellular base stations. Finally, the proposed data model's performance and the support vector regression (SVR)-based model have been compared. Finally, based on experimental observations, the conclusion of the work is reported.

DOI: 10.1201/9781003541363-5

5.2 LITERATURE REVIEW

The section analyzes the recent works carried out in the domain of traffic load forecasting. Thus, the different data sources, applications, and techniques have been investigated.

To provide reliable, sustainable, and secure smart city services, I. A. Ridhawi et al. [7] suggest enabling collaboration with IoT devices. J. Li et al. [8] propose a proactive load-adjusting approach which is alluded to as E2LB dependent on the anticipated street traffic. A CNN is applied to get the relationship and then determine the adjusting issue utilizing NLP and CNN. E2LB ensures continuous scheduling. B. Mahdy et al. [9] evaluate the ML and statistical techniques that are suitable for predicting traffic load. The public dataset is used, which contains traffic loads of base stations. Based on a clustering algorithm, the base stations are grouped. Time series data clustering will mitigate the dissimilar behavior problem. The RNN provides better accurate results.

S. Narejo et al. [10] introduce a technique of internet traffic prediction. The model is prepared using ANN, which contains four hidden layers. The pre-training network is also used. The model is focused on the topology of the network and provides higher accuracy. The study demonstrates fewer RMSE0.028.

Discussion: The summary of the techniques is highlighted in Table 5.1. The table provides the key insights of the research articles. Based on the obtained methods and techniques, we also propose an ML model for proactive scheduling techniques to preserve energy in smart city cellular network base stations. This is discussed in the next section.

5.3 PROPOSED WORK

The smart city introduces convenient services and sustainability but consumes significant energy through its heterogeneous networks and computational infrastructures. Energy preservation has become a major concern. This chapter aims to develop a proactive resource scheduling technique to conserve energy in cellular networks within the smart city infrastructure. Therefore, the proposed work contributes in the following manner:

1. Prepares the technique for predicting the upcoming workload on each cell
2. Prepares the technique to predict the consolidated load on an entire network
3. Designs an algorithm for scheduling cellular towers so that energy is conserved

Table 5.1 Key highlights of collected literature

Year	*Area of applications*	*Algorithms*	*Dataset*	*Results*	*Highlight*
2019[11]	Secure and reliable communication in smart cities	SVM, DBN, extreme gradient boost	Simulation-based data	NA	Resource capabilities are defined using ontologies, and willingness using nature, attitude, and awareness. Reduces unnecessary traffic.
2018[12]	System for managing vehicle traffic	CNN, NLP	Private taxi and bus services in Beijing	NA	I2V communication boosts vehicular services. Enables efficient cooperation. Guarantees real-time scheduling, with a small number of operations.
2019[13]	Traffic congestion in mobile networks	Time series and spatial grouping	The C2TM City Mobile Traffic Map	NA	Predicting load on BS for network optimization and reducing energy consumption. Explores ML methods on a traffic load dataset. Different behaviors of BS affect forecasting. Clustering can mitigate this problem.
2017[14]	Internet traffic prediction	DBN	Time Series Data Library	RMSE0.028	Provides accurate predictions and validates choice for hidden layer size selection.
2018[15]	Prediction of network traffic	Classic time series, ANN, and wavelet transform-based	Real network traces	Reasonable tradeoff between performance and cost overheads	Analyzes predictions that are highly accurate, low in complexity, and high in power consumption. Compares accuracy, cost of computation, and power.
2019[16]	Network traffic congestion in the data center	The concept of convolutional neural networks.	ONTS dataset and ISP data	MAE6.72, MSE15.97	Predict short-term traffic changes across a data center network. The non-linear regularities of CNN can be exploited. With multi-resolution inputs, CNNs can be designed to increase accuracy.

(Continued)

Table 5.1 (*Continued*) Key highlights of collected literature

Year	*Area of applications*	*Algorithms*	*Dataset*	*Results*	*Highlight*
2020[17]	IoT traffic prediction	ANN	IoT traffic generator	MSE1.1145e- 05, SSE 1.0030e-04, MAE0.0015, MAPE6.18%	IoT traffic prediction using NARX feedback neural networks.
2018[18]	Prediction of traffic in data centers	CNN, TNTF, And GRU are all media outlets.	The data collected by Yahoo!	RMSE:GRU (14.3%) and CNN (13.0%)	Predicting traffic would be helpful for load balancing and resource allocation. Only temporal features without communications weaken QoS.
2019[19]	Cellular traffic in metropolis	GNN	Cellular usage dataset covering 1.5 million users and 5,929 cell towers	Consistently outperforms the state-of-the-art time-series-based approaches	It is challenging due to temporal and spatial dynamics. Revealed dependency between distant cell towers.
2019[20]	The urban traffic forecasting system	Deep-meta, learning platform (ST- Meta Net)	The METR-LA dataset and TDrive dataset are two separate datasets.	TDriver (MAE 16.9$\pm$0.13, RMSE 34.0$\pm$ 0.25), METR-LA (MAE 3.05 $\pm$0.02, RMSE 6.25$\pm$0.02)	Public safety is challenging due to complex correlations. Diversity of correlation based on location. They model diverse correlations and temporal correlations.
2020[21]	City road collapse under vibrating load	NA	NA	NA	Several factors cause city roads to collapse, Including underground pipeline seepage and traffic overload.
2020[22]	Intelligent transportation system	V-CLIQUE, ANN	Taxi location and density	NA	Nonlinearity, time variations, and uncertainty make prediction difficult. Analyzes the GPS data. In real-time, ANN path planning is capable of accurately predicting traffic flow.

(*Continued*)

Table 5.1 (*Continued*) Key highlights of collected literature

Year	*Area of applications*	*Algorithms*	*Dataset*	*Results*	*Highlight*
2017[23]	Electric load forecasting and traffic management	Bayesian spatiotemporal Gaussian Process	FDOT, MDMS	rMSEP 31.8%	Traditional systems do not provide observability due to limited coverage and high costs. Employs spatiotemporal interdependency among different networks.
2018[24]	Road traffic flow	Seasonal ARIMA	ED occupancy	RMSE (4.23), MAE (3.25)	Analyzes the flow of traffic on regional roads to determine public activity. There is a relationship between human activity levels and the circadian pattern of medical emergencies.
2019[25]	Intelligent transportation systems	G-VSPA	Simulation-based data	Optimal performance while having a lower complexity	Determines how to place V2X services. A greedy-based algorithm named G-VSPA was developed. Maintains QoS requirements and guarantees them.
2017[26]	Intelligent transportation systems	Review	NA	NA	The feasibility and challenges, potential, and synergies of implementing are discussed.
2021[27]	Intelligent transportation systems	Secure key management, blockchain	Simulation-based data	Allows SMs to flexibly fit various traffic levels	Secure key management schemes are considered for heterogeneous networks. Blockchain is used as a key management framework.
2018[28]	Intelligent transportation systems	NA	NA	NA	With big data, ITS is more efficient, safer, and profitable. The ITS sector should focus on big data analytics as well as data sources, collection methods and analytics techniques, platforms, and

In proactive scheduling techniques, the scheduling process is divided into three phases:

1. *Data analysis:* Application-centric data is analyzed in this phase. It is possible to identify the valuable features and dependencies using analysis.
2. *Prediction:* As part of the ML algorithms, features and dependencies are utilized to train the algorithms and perform predictions.
3. *Scheduling:* For different application-centric purposes, a plan is prepared based on the predicted future trends.

This study focuses on conserving energy in the base stations of cellular networks. Next, we discuss and describe all three phases of the proposed proactive energy-saving technique in detail.

5.4 DATA ANALYSIS

The impact of data analysis on applications is substantial, and to achieve improved and acceptable prediction accuracy, it is essential to enhance data analysis. Furthermore, obtaining an appropriate dataset is crucial for optimal performance. Therefore, we have acquired a suitable dataset for predicting cellular network traffic, obtained from Kaggle [29]. Base station traffic data identified as 4G cell traffic is identified as cellular network traffic. This data can be interpreted as follows. One cell serves several mobile devices at one time. During an hour, the total amount of data that a cell serves to its users is called its traffic load. As an example, consider Cell 001803, which serves 50 subscribers. Every individual utilizes around 10 MB of data per month. The rate of cell phone traffic per hour (x) is calculated as 50 * 10 = 500Mb. The dataset comprises traffic statistics from 57 cells collected over a year, 24 hours a day. The statistics reflect variations in workload, based on time, specific occasions, and special circumstances. From the analyzed dataset of 57 cells, there are 497,545 instances of traffic data. Notably, 9,520 instances are allocated for testing purposes. The dataset covers network activity from October 23, 2017, to October 22, 2018, and includes information such as traffic volume, cell name, date, and hour. Table 5.2 displays a raw dataset sample used for testing and training.

Table 5.2 Sample dataset

	Date	*Hour*	*Cell name*	*Traffic*
	10/23/2017	7	Cell_001803	15.13867
1	10/23/2017	2	Cell_002303	2.05516
2	10/23/2017	7	Cell_004083	71.55308
3	10/23/2017	0	Cell_003781	557.98491
4	10/23/2017	3	Cell_000112	0.98166

Next, data preprocessing is performed to improve the quality of the dataset. The aim is to transform the data in such a way that the ML algorithm can suitably take training.

Therefore, attributes have been identified based on their data types. A cell-name attribute belongs to an object type; thus for suitable representation and processing, we have transformed the cell name into a unique identifier. Next, we proceeded to modify the date property to the time stamp as an index. Finally, the min-max method will be employed to normalize the hour and traffic data, utilizing the following equation:

$$norm = \frac{val - min}{max - min}$$

In Table 5.3, we can see the pre-processed data. The entire dataset contains the combined data for all the cell towers. Therefore, the behavior of individual cell towers is complex to identify. In this context, it is necessary to group the data based on the cell tower identity. Figure 5.1 demonstrates the dataset cell traffic patterns for both scenarios. Figure 5.1 (A) demonstrates the traffic pattern for the entire network. Figure 5.1 (B) shows the grouped data of the individual cell towers. To group cell tower traffic, we used a simple Python grouping method based on cell ID and sorted the data before use. Visualizing the traffic data revealed that traffic behavior varies across cell towers, with some consistently experiencing high workloads and others low workloads. Training a machine learning model with the entire dataset may impact performance, so we propose grouping cells with similar behavior before training. Our behavior-based traffic clustering algorithm uses a customized *k*-means clustering iteration to categorize cell traffic based on their load behavior, adapting the traditional *k*-means approach, which uses Euclidean distance for calculations.

Input: Traffic Dataset D with five clusters, number of clusters K
Output: Clustered Data[*centroids,index*]

Process:
1. D_N=*ReadDataset(D)*
2. D_n=*SortBy('Date')*
3. *for(i=1;i<n;i++)*
    a. *if(count<24)*
        i. *TDj,count=Di*
        ii. *count++*
    b. *else*
        i. *Count=0*
        ii. *J=j+1*
  c. Endif
4. *endfor*
5. [*centroids*,]=*KMeans.Cluster(TD,k)*
6. Return[*centroids,index*]

Table 5.3 Pre-processed data for training

	Date	*Hour*	*Cell name*	*Traffic*
0	10/23/2017	0.419697	20	0.907664
1	10/23/2017	0.697424	32	0.716659
2	10/23/2017	0.097365	44	0.995249
3	10/23/2017	0.000000	33	1.000000
4	10/23/2017	0.950412	1	0.310994

To achieve this, we introduced five distinct groups for cell traffic patterns: (1) underloaded, (2) normally loaded, (3) moderately loaded, (4) heavily loaded, and (5) overloaded. As a result, our algorithm utilizes k=5, as specified in the behavior-based traffic clustering algorithm. Additionally, the classical *k*-means has worked with the 2d structured attributes However, since we have time series data, our approach is to convert daily data into 24-hour intervals before feeding it to the k-mean core method. We must therefore classify 365 sequences for every cell tower. Following data modification, the traditional *k*-means clustering technique was used for behavior-based traffic grouping. Finally, the grouped data is again transformed into the time series data by a reverse transformation process as seen in the behavior-based traffic clustering algorithm.

5.5 PREDICTION

To conduct predictions using the pre-processed data from the preceding section, we developed a unified prediction model, as illustrated in Figure 5.2. The model operates on three distinct sets of training datasets. First is an exhaustive dataset with standardized characteristics. Second is grouped data from individual cell towers, providing a breakdown of the workload for each specific cell. Last is grouped data based on the behavior of cell towers. For training machine learning algorithms for cellular traffic prediction, we have employed three unique training scenarios.

Scenario 1: Using the complete dataset for training

Our goal in this situation is to forecast the total network's combined workload. As a result, we trained machine learning models using the complete dataset. We employed the KNN method for this work. We also used the SVR algorithm to compare the results. Both classification and regression issues can be solved using KNN. KNN takes into account the *K* closest samples in the feature space in both scenarios. Furthermore, the result is contingent

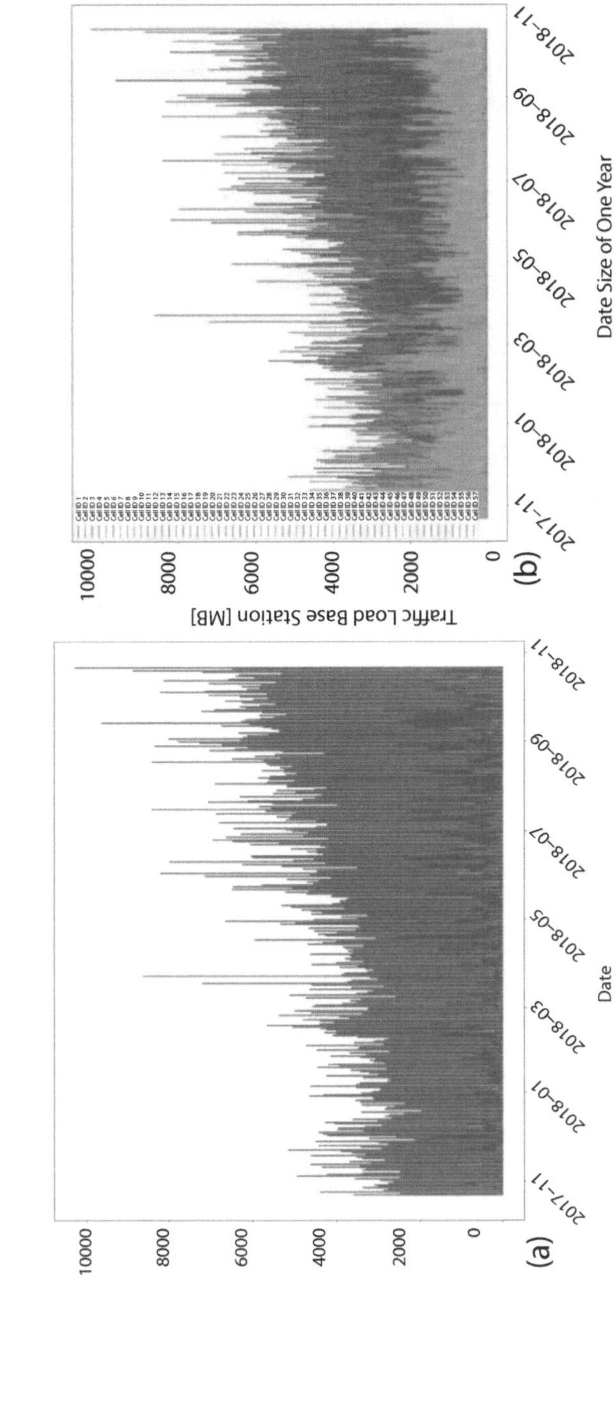

Figure 5.1 Demonstrate the traffic sample patterns in terms of (A) Mixed Data for entire network (B) grouped data for individual cell tower.

upon the needs of the application, specifically the classification prediction. Its ability to generate predictions without training is a benefit. KNN only requires three easy steps. The first step determines the distance between a fresh data point and each training point. In the second step, the K-nearest data points are chosen, and *K* can be any integer. The third step gives most of the *K* data points a value or allocates the data point to the class. In this case, the regression method we employ is KNN. As a result, we train and validate the KNN regression using the data split. The suggested algorithm for training the consolidated traffic prediction algorithm shows how the prediction approach operates. We use the complete dataset as input into our method. The pre-processed data is then arranged chronologically. Currently, the algorithm determines the mean value of the consolidated workload for every date. Subsequently, we divide the training and testing samples to train the SVR and KNN algorithms.

Input: Traffic-Samples D
Output: Predicted trends of the P network

1. D_n=*readDataset(D)*
2. G_m=*GroupBy(Dn,Date)*
3. *for(i=1;i<m;i++)*
    a. GDo = Gm[i]
    b. for(*j*=1;*j*<length(GDo);*j*++)
        i. $S=GD_j+S$
    c. *endfor*
    d. *M* = *S*/*O*
    e. *TD*.(*M*)
4. *endfor*
5. $[TD_{train'}]=TD.(70\%, 30\%)$
6. $T_{model}=KNN.(TD_{train})$
7. $P=T_{model'}(TD_{test})$
8. *returnP*

Scenario 2: Grouped data based on cell towers

Predicting the traffic pattern for individual cell towers is the goal of this experiment. In this instance, we group the data based on cell names first, using the pre-processed data. The date and hour have been used to index the grouped data. Since there are 57 cells in the dataset, the algorithm must be trained. A cycle is executed for every group thereafter. Here, training and testing are done in a single group at a time. We used 30% of the sample for testing and 70% for training in our experiments. The model was trained on the training samples, and its performance was verified by the model's ability to predict the test samples using the test dataset. Predicting and validating the workload of individual cells details the model training and testing procedure for every cell.

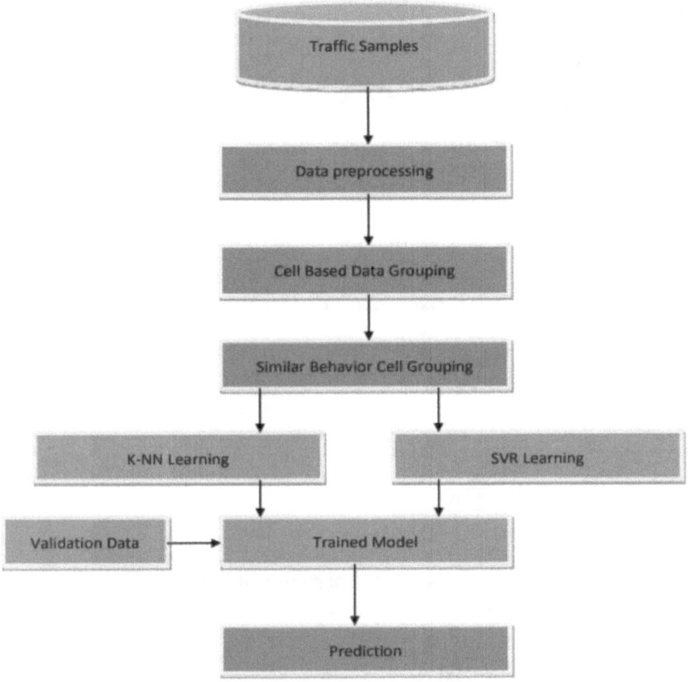

Figure 5.2 Proposed cellular traffic prediction model.

Input: A data set D
Output: T is expected to be a predetermined workload trend

Process: 1. *R=ReadDataset(D)*
2. *P=preprocessData(R)*
3. G_n=*GroupData(P,CellName)*
4. *T* = []
5. *for(i=1;i<n;i++)*
a. [*Train*,]=G_i.(70%,30%)
b. *TrainedModel=KNN.(Train)*
c. T_i=*TrainedModel.(Test)*
d. *T*.append (*Ti*)
6. *endfor*
7. Return *T*

Scenario 3: Grouping data based on behavior

During this stage, we train the prediction algorithms using the prepared data based on the suggested technique. We implemented a modified version of the *k*-means data representation algorithm when training it. This process has the benefit of shortening the system's overall training time due to the smaller training sample sizes. The trained model can also forecast the traffic for a specific cell tower. The behavior-based traffic clustering algorithm was applied to the prepared dataset. We assessed the model's performance based on the scenarios that have been described. The results section describes the implemented models' comparative performance.

5.6 SCHEDULING

The prepared predictive data modeling aims to maintain base station energy consumption by leveraging machine learning prediction power. In this context, Figure 5.3 proposes and illustrates an energy-saving model.

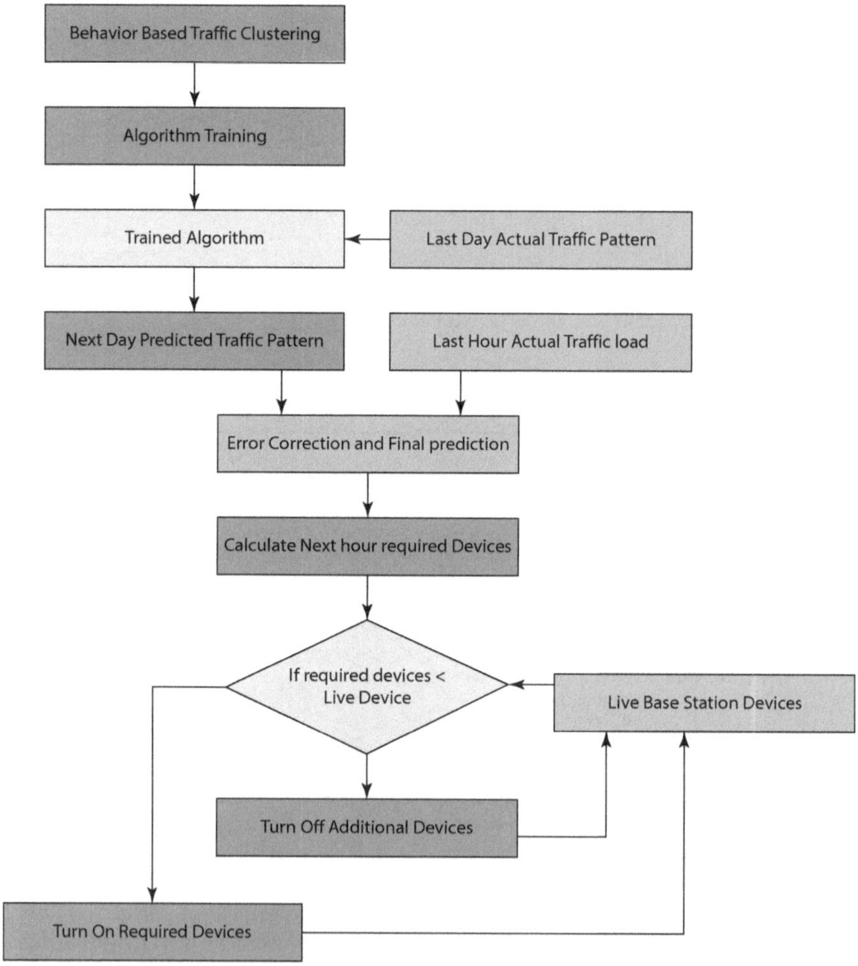

Figure 5.3 Proposed energy-saving model.

The overall model of energy saving in mobile communication base stations is shown in the diagram. A machine learning algorithm is trained using aggregated behavioral data in this model. Following algorithm training, we input the traffic pattern from the previous day to obtain a prediction for the traffic of that specific base station for the following day. Next, the traffic patterns based on hours are taken into account, and the predicted pattern's obtained error is adjusted accordingly. To understand the energy-saving mechanism of the model, the anticipated traffic is represented by the symbols $P_{24} = \{p1, p2, p3, \ldots ,24\}$. And if the actual value for an hour is Ai, we compute the error by using: $E = Pi–Ai$

All of the projected remaining hours are adjusted using the obtained error value. The next part of the model is supplied by the modified predicted trend. In accordance with the final projected value, that component computes the necessary devices for serving the mobile users. Thus, consider a target base station equipped with N devices capable of effectively servicing the traffic load L_{max}. As a result, the base station's D devices can each be utilized to manage workload:

$$D_{Load} = \frac{L_{max}}{N}$$

Therefore, the number of devices required for the projected workload P is:

$$ND = \frac{L_{\max}}{N} * P$$

However, we have added a margin of 10% to account for differences in workload. Therefore, the final number of devices is required:

$$ND = \frac{L_{\max}}{N} * P + P * 0.1$$

After that, we use a decision function to conserve energy as defined in the given function:

$$f = \int \begin{matrix} if\ ND > Live\ then\ turn\ off\ devices \\ if\ ND < Live\ then\ turn\ on\ devices \end{matrix}$$

As per the specified function, if the number of required devices is below the threshold ND, we deactivate additional devices; otherwise, we will activate the required number of devices. This section explains the prediction model and its role in helping mobile network base stations reduce network power consumption. Next, we examine the outcomes and analysis, with a focus on the energy savings and performance achieved by the predictive methods employed in the creation of predictive base station equipment management technology.

5.4 RESULTS ANALYSIS

The section provides an overview of the evaluation of energy efficiency in the mobile base station model. The discussion focuses on two distinct experimental cases: performance evaluation of the machine learning algorithm, and evaluation of the energy-saving capacity of the model. We begin by analyzing the performance and efficiency of machine learning methods, and then we demonstrate how the proposed model conserves energy.

5.4.1 Performance of ML algorithms

Appraising predictive algorithms involves evaluating numerous performance measures. However, in this case, our main focus is on accuracy and training time. Two machine learning algorithms that are employed, KNN and SVR, have the highest prediction accuracy, as shown in Figure 5.4.

This performance evaluation led to the final selection of the machine learning algorithm to be incorporated into the energy-saving model. ML algorithms are represented in Figure 5.4 (A) by their percentage accuracy, while Figure 5.4 (B) shows their error. An evaluation of performance is conducted based on three prepared traffic samples with distinct example scenarios for training. As a result, the cells exhibit behavior clustering, where an algorithm-based approach provides comparatively better accuracy and lower error rates. A total of three different numbers of neighbors (k = 5, 7, and 21) were used to measure error and accuracy. In comparison to the other two k values, k = 5 offers the most accurate predictions.

In order to achieve the proposed model's high level of accuracy, up to 99.99% of the network's load is predicted accurately, which is higher than the accuracy achieved for individual cells. Moving to time complexity, the duration required for model training is measured for three different scenarios and is shown in Figure 5.5, with time measured in minutes. The results

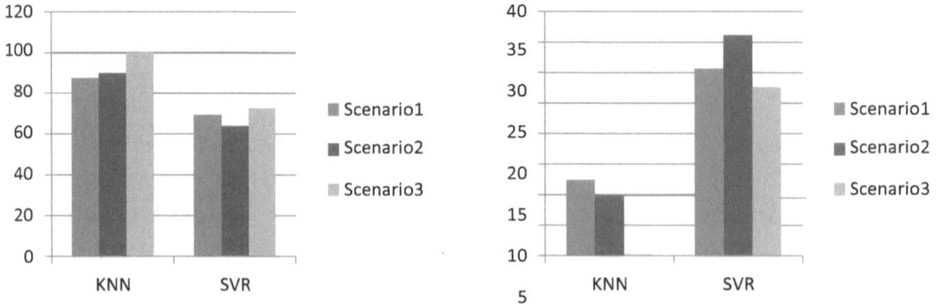

Figure 5.4 Proposed data model for the prediction of cellular traffic. It performs well, showing the results of connection (A) accuracy (%) and (B) error (%).

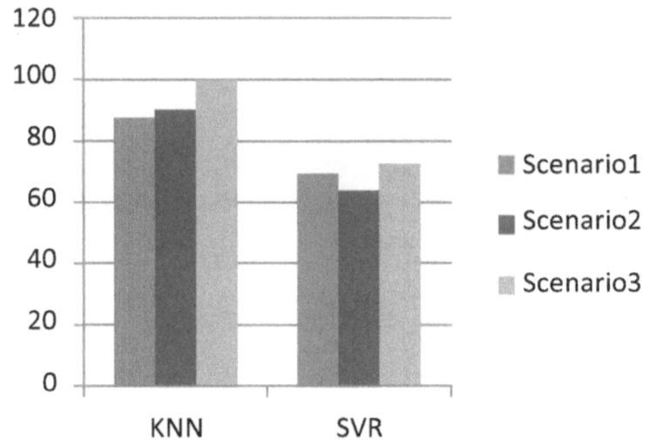

Figure 5.5 Training time for both models in three different scenarios.

indicate that KNN techniques exhibit a shorter training time compared to SVR-based models. Furthermore, training and validating the model using grouped data based on cell identity proves to be computationally more demanding than the alternative scenario. Ultimately, SVR is identified as a more resource-intensive algorithm compared to the KNN techniques.

5.4.2 Power consumption

The first step in measuring power consumption is to compute the cell tower's peak workload. The assumption is that, traditionally, *NDp* devices are used continuously to handle this workload. This context dictates that there is a power consumption rate of 10 units every hour for each device. Thus, the conventional power usage is calculated in the following way:

$$PowerT = NDp * 10$$

In addition, the proposed approach uses *NDpro* devices, which leads to a power consumption:

$$Powerpro = NDpro * 10$$

Thus, the percentage amount of energy saving is measured in the form:

$$S = \frac{power_T - power_{pro}}{power_T} * 100$$

Figure 5.6 illustrates the power saving achieved by three different cell towers using the proposed approach. By implementing the proposed technique, energy savings of 20–28% can be achieved while maintaining normal cell tower operations, it still has an impact.

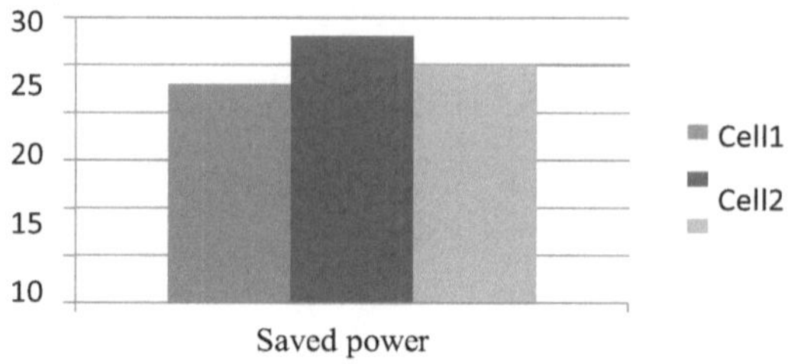

Figure 5.6 Saved amount of power.

5.5 CONCLUSION

Utilizing a proactive method, cell tower power consumption will be minimized in a smart city by overcoming the load on the towers. This chapter inspected the predictive capabilities of lazy learning approaches and their application in how they help decrease power usage in cellular network base stations. First, the workload of cell towers was analyzed and predicted. Then, a power optimization method was applied to preserve energy in the towers. After experimental analysis, we found the following crucial findings.

First, using a behavior-based clustering process yields superior results and enhances the learning process with entire datasets. Second, KNN regression provides more accurate workloads compared to SVR models. Finally, the proposed model can conserve energy by 28% more than the traditional method of cell tower operation. Minimizing the power consumption of cell towers in a smart city is critical for cost savings, sustainability, and reducing environmental impact. The influence of AI on energy optimization for cellular network base stations is favorable and holds great promise for environmental sustainability and the telecommunications industry.

REFERENCES

1. Y. Hou, H. Gong, X. Wang, W. Gao, D. Liu, X. Xiao, and K. Xu, "Energy Saving of Base Station System for Power Private Wireless Network Based on D2D Communication," in *Proc. of the 2023 6th International Conference on Artificial Intelligence and Big Data (ICAIBD)*, pp. 597–602, May 2023.
2. Bo Li, Minyou Chen, Haiwang Zhong, Ziming Ma, Dongran Liu, and Gang He, "Review of long-term planning of new power system with high proportion of renewable energy", *Proceedings of the CSEE*, vol. 43, no. 2, pp. 555–581, 2023.
3. Zhang Zhigang and Chongqing Kang, "Challenges and prospects of building a new power system under the goal of carbon neutrality", *Proceedings of the CSEE*, vol. 42, no. 8, pp. 2806–2819, 2022.

4. L. Li and W. Meng, "Collaborative base station sleeping solution design in heterogeneous cellular network," *2022 27th Asia Pacific Conference on Communications (APCC), Jeju Island, Korea, Republic of*, 2022, pp. 231–235, doi: 10.1109/APCC55198.2022.9943603.
5. G. O. Ferreira, C. Ravazzi, F. Dabbene, G. C. Calafiore and M. Fiore, "Forecasting network traffic: a survey and tutorial with open-source comparative evaluation," *in IEEE Access*, vol. 11, pp. 6018–6044, 2023, doi: 10.1109/ACCESS.2023.3236261.
6. https://www.kaggle.com/naebolo/predict-traffic-of-lte-network
7. I. Al Ridhawi, S. Otoum, M. Aloqaily, Y. Jararweh, and T. Baker, "Providing secure and reliablecommunication for next generation networks in smart cities," *Sustainable Cities and Society*, vol. 56, Article 102080, 2020.
8. J. Li, G. Luo, N. Cheng, Q. Yuan, Z. Wu, S. Gao, and Z. Liu, "An end-to-end load balancer based on deep learning for vehicular network traffic control," *IEEE Internet of Things Journal*, vol. 6, no. 1, pp. 953–966, 2018.
9. B. Mahdy, H. Abbas, H. S. Hassanein, A. Noureldin, and H. Abou-zeid, "A clustering-driven approach to predict the traffic load of mobile networks for the analysis of base stations deployment," *Journal of Sensor and Actuator Networks*, vol. 9, no. 4, Article 53, 2020.
10. S. Narejo and E. Pasero, "An application of internet traffic prediction with deep neural network," *Multidisciplinary Approaches to Neural Computing*, pp. 139–149, 2018, doi: 10.1007/978-3-319-56904-8_14
11. M. F. Iqbal, M. Zahid, D. Habib, and L. K. John, "Efficient prediction of network traffic for real-time applications," *Journal of Computer Networks and Communications*, 2019, doi: 10.1155/2019/4067135
12. A. Mozo, B. Ordozgoiti, and S. Gomez-Canaval, "Forecasting short-term data center network traffic load with convolutional neural networks," *PLoS One*, vol. 13, no. 2, Article e0191939, 2018.
13. A. R. Abdellah, O. A. K. Mahmood, A. Paramonov, and A. Koucheryavy, "IoT traffic prediction using multi-step ahead prediction with neural network," *in Proc. 2019 11th International Congress on Ultra-Modern Telecommunications and Control Systems and Workshops (ICUMT)*, 2019, pp. 1–4.
14. X. Cao, Y. Zhong, Y. Zhou, J. Wang, C. Zhu, and W. Zhang, "Interactive temporal recurrent convolution network for traffic prediction in data centers," *IEEE Access*, vol. 6, pp. 5276–5289, 2017.
15. X. Wang, Z. Zhou, F. Xiao, K. Xing, Z. Yang, Y. Liu, and C. Peng, "Spatio-temporal analysis and prediction of cellular traffic in metropolis," *IEEE Transactions on Mobile Computing*, vol. 18, no. 9, pp. 2190–2202, 2018.
16. Z. Pan, Y. Liang, W. Wang, Y. Yu, Y. Zheng, and J. Zhang, "Urban traffic prediction from spatio-temporal data using deep meta learning," *in Proc. of the 25th ACM SIGKDD International Conference on Knowledge Discovery & Data Mining*, July 2019, pp. 1720–1730.
17. Y. Wang, G. Shi, X. Tian, C. Li, and H. Cheng, "Experimental study on city road collapse under vibrating load," *Shock and Vibration*, vol. 2020, pp. 1–13, 2020.
18. D. Zhu, H. Du, Y. Sun, and N. Cao, "Research on path planning model based on short-term traffic flow prediction in intelligent transportation system," *Sensors*, vol. 18, no. 12, Article 4275, 2018.

19. M. Gilanifar, H. Wang, E. E. Ozguven, Y. Zhou, and R. Arghandeh, "Bayesian spatiotemporal Gaussian process for short-term load forecasting using combined transportation and electricity data," *ACM Transactions on Cyber-Physical Systems*, vol. 4, no. 1, pp. 1–25, 2019.
20. J. Rauch, U. Hübner, M. Denter, and B. Babitsch, "Improving the prediction of emergency department crowding: a time series analysis including road traffic flow," *in dHealth*, January 2019, pp. 57–64.
21. A. Moubayed, A. Shami, P. Heidari, A. Larabi, and R. Brunner, "Edge-enabled V2X Service Placement for Intelligent Transportation Systems," *arXiv preprint* arXiv:2001.06288v1 [eess.SP], January 13, 2020.
22. A. Richter, M. O. Löwner, R. Ebendt, and M. Scholz, "Towards an integrated urban development considering novel intelligent transportation systems: urban development considering novel transport," *Technological Forecasting and Social Change*, vol. 155, Article 119970, 2020.
23. A. Lei, H. Cruickshank, Y. Cao, P. Asuquo, C. P. A. Ogah, and Z. Sun, "Blockchain-based dynamic key management for heterogeneous intelligent transportation systems," *IEEE Internet of Things Journal*, vol. 4, no. 6, pp. 1832–1843, 2017.
24. L. Zhu, F. R. Yu, Y. Wang, B. Ning, and T. Tang, "Big data analytics in intelligent transportation systems: a survey," *IEEE Transactions on Intelligent Transportation Systems*, vol. 20, no. 1, pp. 383–398, 2018.
25. X. Pan, W. Zhou, Y. Lu, and N. Sun, "Prediction of network traffic of smart cities based on DE-BP neural network," *IEEE Access*, vol. 7, pp. 55807–55816, 2019.
26. A. Al-Dweik, R. Muresan, M. Mayhew, and M. Lieberman, "IoT-based multifunctional scalable real-time enhanced road side unit for intelligent transportation systems," in *Proc. of the 2017 IEEE 30th Canadian Conference on Electrical and Computer Engineering (CCECE)*, April 2017, pp. 1–6.
27. A. Lamssaggad, N. Benamar, A. S. Hafid, and M. Msahli, "A survey on the current security landscape of intelligent transportation systems," *IEEE Access*, vol. 9, pp. 9180–9208, 2021.
28. A. Dairi, F. Harrou, Y. Sun, and M. Senouci, "Obstacle detection for intelligent transportation systems using deep stacked autoencoder and k-nearest neighbor scheme," *IEEE Sensors Journal*, vol. 18, no. 12, pp. 5122–5132, 2018.
29. K. Sultan, H. Ali, and Z. Zhang, "Call detail records driven anomaly detection and traffic prediction in mobile cellular networks," *IEEE Access*, vol. 6, pp. 41728–41737, 2018.

Chapter 6

Semantic scrutiny of wireless sensors, Internet of Medical Things, and blockchain in mobile healthcare applications

Unveiling intelligent networks for patient privacy, security, and resilience

Bhupinder Singh and Christian Kaunert

6.1 INTRODUCTION

Wireless sensors included in medical equipment or wearables are critical for continuous and real-time monitoring of patient vitals. The IoMT has heralded a new era of networked medical equipment, while blockchain technology provides unparalleled data transaction security. The semantic scrutiny of these technologies, particularly in the context of patient-centric applications, remains a critical yet underexplored domain. This chapter aims to unravel the semantic intricacies of these technological forces and their collective contribution to the establishment of intelligent networks [1]. Through a meticulous examination of wireless sensor functionalities, IoMT implications, and blockchain applications, we seek to delineate the nuanced impact on patient privacy, fortify the security infrastructure, and bolster the overall resilience of mobile healthcare applications. This unveiling of intelligent networks not only addresses current challenges but also sets the stage for a more sophisticated, secure, and patient-centric future in the realm of digital healthcare [2].

The relevance of this integration goes beyond technological advancement; it fundamentally addresses the most pressing issues of patient privacy. By employing semantic scrutiny and advanced security protocols, these intelligent networks allow for a nuanced understanding of health data. Patients can have confidence that their sensitive information is not only secure but is also subject to sophisticated privacy measures, ensuring responsible and ethical use of their healthcare data. The predictive and preventive capabilities of this intelligent network are transformative [3].

DOI: 10.1201/9781003541363-6

6.2 SEMANTIC SCRUTINY IN HEALTHCARE: WIRELESS SENSORS, IOMT, AND BLOCKCHAIN IN MOBILE HEALTHCARE APPLICATIONS

The scrutiny of healthcare processes, particularly in the context of mobile applications, is undergoing a transformative phase with the integration of wireless sensors, the Internet of Medical Things (IoMT), and blockchain technologies. Wireless sensors, strategically deployed in various healthcare settings, enable the continuous monitoring of patient vitals and the collection of real-time data. This scrutiny at the sensor level not only enhances the precision of medical diagnostics but also contributes to a proactive healthcare approach by allowing early detection of anomalies and timely interventions [4].

The advent of the IoMT further amplifies the scrutiny capabilities in healthcare. Blockchain technology adds an extra layer of scrutiny by addressing critical issues related to data security, integrity, and privacy in mobile healthcare applications. The scrutiny brought about by wireless sensors, IoMT, and blockchain in mobile healthcare applications represents a paradigm shift toward a more precise, interconnected, and secure healthcare ecosystem. This scrutiny is not merely about surveillance but rather about leveraging advanced technologies to scrutinize healthcare data intelligently, leading to improved patient outcomes, enhanced privacy, and resilient healthcare networks [5].

6.2.1 Semantic interoperability in healthcare data exchange

Interoperability in healthcare data exchange refers to the ability of different healthcare information systems, devices, and applications to seamlessly exchange and use health data across organizational and technological boundaries. Achieving interoperability is crucial for creating a connected healthcare ecosystem that facilitates efficient and secure sharing of patient information among various stakeholders, including healthcare providers, patients, insurance companies, and researchers [6]. Data security and privacy are crucial to achieving interoperability. In order to guarantee the safe transmission of health information and the protection of patient privacy in compliance with legal requirements, resilient standards, and procedures are crucial. Also, to guarantee smooth data transfer across various systems, standardizing data formats and implementing interoperability standards like FHIR (Fast Healthcare Interoperability Resources) and HL7 (Health Level Seven) is essential. This entails not just technological issues, but also careful consideration of privacy and security, as well as the creation and implementation of interoperability standards across the healthcare ecosystem [7].

6.3 WIRELESS SENSORS IN HEALTHCARE: INTELLIGENT NETWORKS

Wireless sensors have ushered in a new era in healthcare, enabling unparalleled levels of patient monitoring, data-collecting, and total healthcare management. These sensors, which are frequently included in wearable devices or medical equipment, use wireless technologies like Bluetooth, Wi-Fi, or cellular networks to communicate real-time health data, altering the way healthcare is given and experienced [8].

Vital indicators like as heart rate, blood pressure, and oxygen levels may be tracked in real time using wireless sensors, offering a comprehensive and continuous picture of a patient's health state. This is particularly valuable for patients who live in remote areas or those who require post-operative care but prefer the comfort of their homes. Remote patient monitoring not only improves access to healthcare but also reduces the burden on healthcare facilities and resources [9]. This data-driven approach facilitates evidence-based decision-making, enables early detection of disease outbreaks, and contributes to population health management. As these sensors handle sensitive health information, robust security measures and compliance with privacy regulations such as the Health Insurance Portability and Accountability Act (HIPAA) are paramount to ensure the confidentiality and integrity of patient data. Wireless sensors in healthcare practices have not only transformed patient care but have also catalyzed a shift toward more personalized, proactive, and patient-centric healthcare delivery. As technology continues to advance, the ongoing evolution of wireless sensor capabilities holds the promise of further innovations, ultimately contributing to a more connected, efficient, and effective healthcare ecosystem [10].

6.3.1 Wireless sensor technology: introduction

Wireless sensor technology stands at the forefront of innovation, revolutionizing various industries, with healthcare being a notable beneficiary. These sensors, untethered from traditional wired constraints, leverage wireless communication protocols to collect and transmit data in real time. In the healthcare domain, they are integrated into wearable devices and medical equipment, facilitating continuous patient monitoring, remote healthcare management, and the generation of valuable health data. The advent of wireless sensor technology marks a paradigm shift in healthcare, promising enhanced efficiency, improved patient outcomes, and a more interconnected and proactive approach to medical care [11].

6.3.2 Challenges and considerations in deploying wireless sensors in healthcare

The deployment of wireless sensors in healthcare, while offering significant advantages, comes with a set of challenges and considerations that necessitate careful attention. The foremost concern is data security and privacy, as the sensitive nature of health information demands robust security measures to prevent unauthorized access and protect patient confidentiality. Therefore, adherence to regulatory standards is imperative to ensure compliance and instill trust in patients regarding the protection of their personal health data [12].

The diverse landscape of healthcare systems and devices may lead to compatibility issues, hindering seamless communication between different platforms. The sheer volume of data generated by wireless sensors raises concerns about data overload. Advanced analytics and artificial intelligence solutions are increasingly being employed to manage and interpret the data effectively [13]. In healthcare, reliability and accuracy are critical considerations, where precision is paramount and where any lapses in the accuracy of sensor data could lead to erroneous diagnoses or treatment decisions. The cost of implementation can be a barrier, especially for smaller healthcare facilities. The initial investment in deploying wireless sensor networks, acquiring compatible devices, and establishing necessary infrastructure may be significant. Finally, ethical considerations related to informed consent and patient autonomy must be addressed. Patients should be adequately informed about the purpose of sensor deployment, the data collected, and how it will be utilized. Addressing these considerations not only ensures the successful deployment of wireless sensors in healthcare but also lays the foundation for an ethically sound and technologically advanced healthcare landscape [14].

6.4 IOMT AND ITS SIGNIFICANCE IN MOBILE HEALTHCARE APPLICATIONS

The IoMT stands at the forefront of the technological revolution in healthcare, transforming the landscape of mobile healthcare applications and ushering in a new era of patient-centered and data-driven healthcare. IoMT refers to the interconnected network of medical devices, wearable sensors, and healthcare systems that exchange health-related data through the internet. This interconnectedness brings about a seamless flow of information, creating a dynamic ecosystem where patients, healthcare providers, and devices are interconnected, fostering an intelligent and responsive healthcare environment. Despite these challenges, the significance of IoMT in mobile healthcare applications remains pivotal, promising a future where healthcare is not only more connected but also more accessible, personalized, and responsive to the dynamic needs of patients and healthcare providers alike [15].

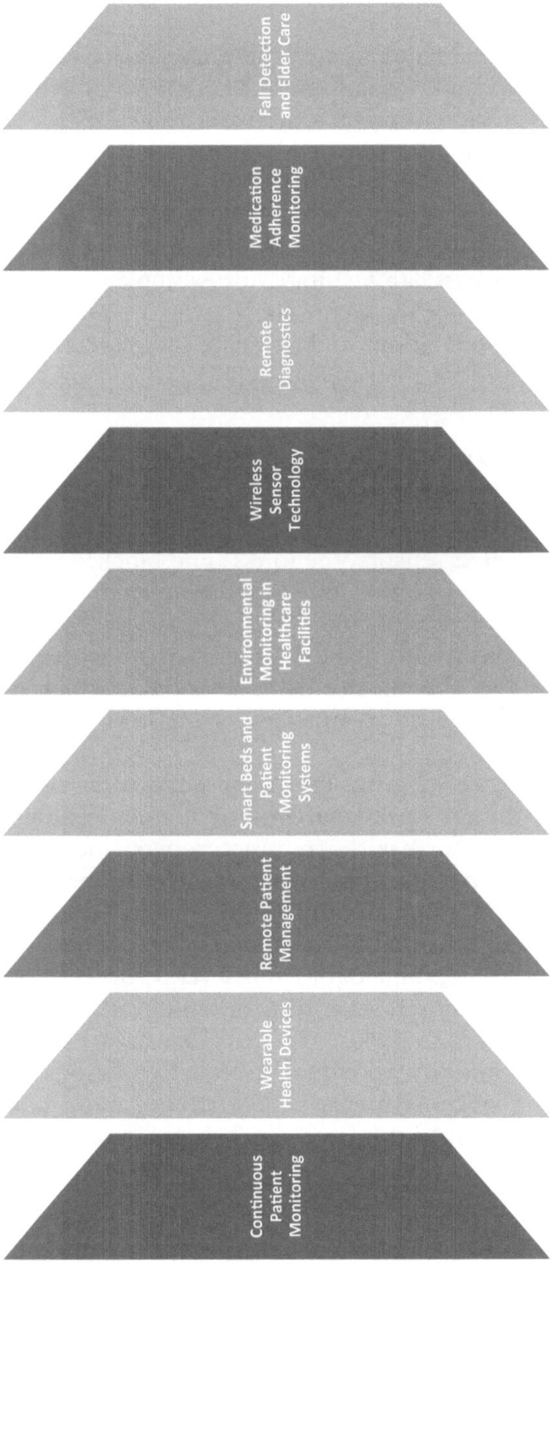

Figure 6.1 Applications of wireless sensors in healthcare.

6.5 BLOCKCHAIN TECHNOLOGY IN HEALTHCARE: INTELLIGENT NETWORKS FOR PATIENT PRIVACY, SECURITY, AND RESILIENCE

Blockchain technology, heralded as a revolutionary force in various industries, is making significant inroads into healthcare, offering intelligent networks that redefine patient privacy, security, and resilience. In the context of healthcare, the application of blockchain introduces a paradigm shift, addressing critical challenges and redefining the way patient data is managed [16]. Healthcare operations are made more automated by incorporating self-executing agreements known as smart contracts into the blockchain. These contracts can facilitate, for example, the automatic execution of insurance claims or the secure sharing of patient data for research purposes, streamlining operations and ensuring that predefined conditions are met without the need for intermediaries [17].

6.6 INTEGRATION OF WIRELESS SENSORS, IOMT, AND BLOCKCHAIN

The integration of wireless sensors, the IoMT, and blockchain marks a significant milestone in the evolution of healthcare technology, promising a transformative impact on patient care and data management. Wireless sensors implanted strategically in medical equipment or worn as wearables record and transmit real-time health data. This data, which ranges from vital signs to particular health parameters, is the IoMT ecosystem's lifeblood [18].

Wireless sensors serve as front-line data collectors, collecting vital health information directly from patients or medical equipment. They are extremely important for healthcare practitioners since they allow for continuous and remote patient monitoring. Healthcare practitioners now have instant access to a complete and up-to-date patient health profile, enabling rapid interventions and individualized treatment strategies [19]. The IoMT's capacity to combine data from several sources improves care coordination by speeding processes and encouraging a comprehensive view of patient health. The constant flow of real-time data enables the early diagnosis of health conditions, allowing for proactive actions and lowering the chance of consequences. Patients benefit from a more individualized and proactive approach to healthcare, which contributes to better health outcomes [20].

6.6.1 Building intelligent networks for seamless integration

Developing intelligent networks for seamless integration is a critical undertaking in many industries, notably healthcare. To construct networked systems that perform cohesively, the integration of varied components such as wireless sensors, the IoMT, and blockchain necessitates a systematic approach. Engaging healthcare providers, IT professionals, regulators, and end-users in the design and implementation phases ensures that the network meets the diverse needs of the ecosystems [21].

6.7 SECURITY MEASURES: INTELLIGENT NETWORKS AND SEMANTIC SCRUTINY OF WIRELESS SENSORS, IOMT, AND BLOCKCHAIN IN MOBILE HEALTHCARE APPLICATIONS

With the integration of intelligent networks and semantic monitoring of wireless sensors, the IoMT, and blockchain, the security environment in mobile healthcare apps has undergone a paradigm change. This convergence results in a comprehensive approach to protecting patient data, maintaining privacy, and strengthening the resilience of healthcare institutions [22]. The deployment of semantic inspection, a process that entails a detailed evaluation of the meaning and context of data, is one of the basic pillars of security in this setting. Semantic inspection in the context of wireless sensors and IoMT enables a more comprehensive interpretation of the health data being delivered. This not only improves data interpretation accuracy but also provides a tool for spotting abnormalities or potential security concerns [23].

Semantic inspection goes beyond standard approaches by contextualizing data in a way that catches the nuances of medical problems and minor changes in patient health. This semantic understanding adds intelligence to the network, allowing it to distinguish between regular patterns and aberrant changes, which might indicate security breaches or data integrity concerns [24]. Because blockchain is decentralized, patient records are not just resistant to unwanted access but also immune to alteration. Each block in the chain is cryptographically connected to the preceding one and has a unique identifier, resulting in an immutable ledger of transactions. This transparency and immutability not only strengthen the security of patient data but also contribute to the overall resilience of the healthcare network. Intelligent networks, powered by semantic scrutiny, facilitate more sophisticated threat detection mechanisms. Anomalies in health data patterns that may indicate security threats, such as unauthorized access or data manipulation, can be identified in real time [25].

Semantic scrutiny with wireless sensors, IoMT, and blockchain in mobile healthcare applications represents a pioneering approach to security. By imbuing the network with semantic intelligence, healthcare systems can achieve a more profound understanding of patient data, enhancing threat detection, ensuring data integrity, and fortifying patient privacy. This intersection of technologies marks a pivotal moment in the evolution of secure and intelligent healthcare ecosystems [26].

6.7.1 Threats and vulnerabilities in healthcare networks

The healthcare industry, with its increasing reliance on digital technologies and interconnected systems, is confronted with a myriad of threats and vulnerabilities that pose significant risks to patient data, operational continuity, and overall system integrity. As healthcare organizations store vast amounts of valuable data, they become attractive targets for malicious actors seeking to exploit vulnerabilities [27]. The proliferation of "connected medical devices" introduces new vulnerabilities. These devices, ranging from insulin pumps to MRI machines, are often networked for real-time monitoring and data collection. Employees, contractors, or third-party service providers with access to healthcare systems can, knowingly or unknowingly, compromise data integrity or security. Insider threats may involve unauthorized access, accidental data leaks, or actions that disrupt system operations. The "proliferation of legacy systems" in healthcare networks poses a challenge. Older systems may lack the security features of modern technologies, making them susceptible to exploitation. As healthcare continues to digitize and innovate, the proactive identification and mitigation of these risks are paramount for ensuring the integrity, privacy, and resilience of healthcare networks [28].

6.7.2 Security protocols and encryption techniques

Security protocols and encryption techniques play a pivotal role in fortifying the integrity and confidentiality of digital information. In the context of securing data at rest, file or disk-level encryption provides an additional layer of protection [29]. The security protocols and encryption techniques create robust defense mechanisms against unauthorized access, data breaches, and cyber threats. They are fundamental components in the

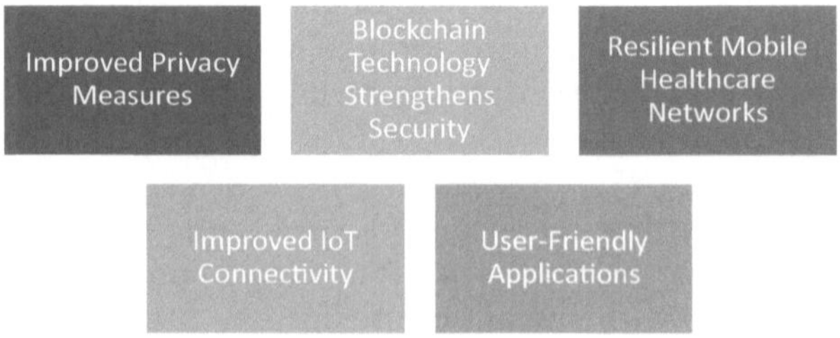

Figure 6.2 Outcome of the present study.

toolkit of cybersecurity measures, playing a critical role in securing sensitive information across digital platforms [30].

6.7.3 Incident response and resilience planning

Incident response and resilience planning are integral components of a proactive cybersecurity strategy, crucial for mitigating the impact of security incidents and ensuring the continuity of operations. A robust resilience plan encompasses not only technological aspects but also factors in personnel, processes, and communication strategies. It ensures that critical systems can recover quickly and that the organization can adapt to changing circumstances [31].

6.8 CHALLENGES AND FUTURE DIRECTIONS

Wireless sensors, IoMT, and blockchain in mobile healthcare applications herald a transformative era, yet it is not without its challenges. Striking the right balance between data accessibility for healthcare professionals and robust patient privacy protection remains a delicate challenge, particularly in an era of evolving regulations and heightened awareness about data breaches [32]. In terms of future directions, the advancement of edge computing holds significant promise. The ability to process data closer to the source, at the edge of the network, can enhance real-time analytics and reduce latency in mobile healthcare applications [33]. These technologies can analyze vast datasets generated by wireless sensors and IoMT devices, providing actionable insights for personalized patient care while contributing to the early detection of health issues. The use of smart contracts in blockchain technology for automating healthcare processes, such as insurance claims and patient consent management, could streamline operations and reduce administrative overhead [34].

6.9 CONCLUSION AND FUTURE SCOPE

The combination of wireless sensors, the IoMT, and blockchain in mobile healthcare applications has set the stage for a revolutionary transformation in patient care. The future trajectory of these intelligent networks points toward a healthcare landscape that is not only technologically advanced but also more resilient, secure, and respectful of patient privacy. The ongoing refinement of these technologies, coupled with collaborative efforts among stakeholders, will shape a future where mobile healthcare applications seamlessly leverage wireless sensors, IoMT, and blockchain for the betterment of patient outcomes and the overall efficiency of healthcare delivery.

REFERENCES

1. Alqaralleh, B. A., Vaiyapuri, T., Parvathy, V. S., Gupta, D., Khanna, A., & Shankar, K. (2021). Blockchain-assisted secure image transmission and diagnosis model on Internet of medical things environment. *Personal and Ubiquitous Computing*, 1–11. https://doi.org/10.1007/s00779-021-01543-2
2. Lakhan, A., Mohammed, M. A., Elhoseny, M., Alshehri, M. D., & Abdulkareem, K. H. (2022). Blockchain multi-objective optimization approach-enabled secure and cost-efficient scheduling for the Internet of Medical Things (IoMT) in fog-cloud system. *Soft Computing*, *26*(13), 6429–6442.
3. Singh, B. (2024). Legal dynamics lensing metaverse crafted for videogame industry and E-sports: Phenomenological exploration catalyst complexity and future. *Journal of Intellectual Property Rights Law*, *7*(1), 8–14.
4. Vinothkumar, J., & Karunamurthy, A. (2022). Recent advancements in artificial intelligence technology: Trends and implications. *Quing: International Journal of Multidisciplinary Scientific Research and Development*, 2(1), 1–11.
5. Parida, P. K., Dora, L., Swain, M., Agrawal, S., & Panda, R. (2022). Data science methodologies in smart healthcare: A review. *Health and Technology*, *12*(2), 329–344.
6. Xu, Y., Liu, X., Cao, X., Huang, C., Liu, E., Qian, S., & Zhang, J. (2021). Artificial intelligence: A powerful paradigm for scientific research. *The Innovation*, 2(4), 100179.
7. Singh, B. (2023). Blockchain technology in renovating healthcare: Legal and future perspectives. In K. Kaushik, S. Dahiya, S. Aggarwal, & A. Dwivedi (Eds.), *Revolutionizing Healthcare Through Artificial Intelligence and Internet of Things Applications* (pp. 177–186). IGI Global. . https://doi.org/10.4018/978-1-6684-5422-0.ch012
8. Kumar, M., Nguyen, T. N., Kaur, J., Singh, T. G., Soni, D., Singh, R., & Kumar, P. (2023). Opportunities and challenges in application of artificial intelligence in pharmacology. *Pharmacological Reports*, *75*(1), 3–18.
9. Al-Turjman, F., Nawaz, M. H., & Ulusar, U. D. (2020). Intelligence in the Internet of medical things era: A systematic review of current and future trends. *Computer Communications*, *150*, 644–660.
10. Taherdoost, H. (2023). Blockchain-based internet of medical things. *Applied Sciences*, *13*(3), 1287.
11. Lin, W., Xu, M., He, J., & Zhang, W. (2023). Privacy, security and resilience in mobile healthcare applications. *Enterprise Information Systems*, *17*(3), 1939896.
12. Kelli, V., Sarigiannidis, P., Argyriou, V., Lagkas, T., & Vitsas, V. (2021, June). A cyber resilience framework for NG-IoT healthcare using machine learning and blockchain. In *ICC 2021-IEEE International Conference on Communications* (pp. 1–6). IEEE.
13. Letafati, M., & Otoum, S. (2023). Digital healthcare in the metaverse: Insights into privacy and security. *arXiv preprint arXiv:2308.04438*.
14. Singh, B. (2023). Federated Learning for Envision Future Trajectory Smart Transport System for Climate Preservation and Smart Green Planet: Insights

into Global Governance and SDG-9 (Industry, Innovation and Infrastructure). *National Journal of Environmental Law*, 6(2), 6–17.
15. Blobel, B., Oemig, F., Ruotsalainen, P., & Lopez, D. M. (2022). Transformation of health and social care systems—an interdisciplinary approach toward a foundational architecture. *Frontiers in Medicine*, *9*, 802487.
16. Sharma, A., & Singh, B. (2022). Measuring impact of E-commerce on small scale business: A systematic review. *Journal of Corporate Governance and International Business Law*, *5*(1). https://doi.org/10.1108/cg-10-2020-0470
17. Chen, J., Yi, C., Okegbile, S. D., Cai, J., & Shen, X. S. (2023). Networking architecture and key supporting technologies for human digital twin in personalized healthcare: A comprehensive survey. *IEEE Communications Surveys & Tutorials*. https://doi.org/10.1109/comst.2023.3308717
18. Ahamad, S. S., Al-Shehri, M., & Keshta, I. (2022). A secure and resilient scheme for Telecare medical information systems with threat modeling and formal verification. *IEEE Access*, *10*, 120227–120244.
19. Bughio, K. S., & Sikos, L. F. (2023). Knowledge organization systems to support cyber-resilience in medical smart home environments. In M. Ahmed & P. Haskell-Dowland (Eds.), *Cybersecurity for Smart Cities: Practices and Challenges* (pp. 61–69). Cham: Springer International Publishing.
20. Abie, H., & Balasingham, I. (2012, February). Risk-based adaptive security for smart IoT in eHealth. In *Proceedings of the 7th International Conference on Body Area Networks* (pp. 269–275).
21. Singh, B. (2022). Understanding legal frameworks concerning transgender healthcare in the age of dynamism. *Electronic Journal of Social and Strategic Studies*, *3*, 56–65.
22. Batista, E., Lopez-Aguilar, P., & Solanas, A. (2023). Smart health in the 6g era: Bringing security to future smart health services. *IEEE Communications Magazine*.
23. Deebak, B. D., Al-Turjman, F., Aloqaily, M., & Alfandi, O. (2019). An authentic-based privacy preservation protocol for smart e-healthcare systems in IoT. *IEEE Access*, *7*, 135632–135649.
24. Singh, B. (2022). Relevance of agriculture-nutrition linkage for human healthcare: A conceptual legal framework of implication and pathways. *Justice and Law Bulletin*, *1*(1), 44–49.
25. Sapra, V., Hasan, M. K., Ghazal, T. M., Bhadrdwaj, A., Bharany, S., Ahmad, M., & Mohamed, T. (2022, October). Privacy-based framework for cyber resilience of healthcare based data for use with machine learning algorithms. In *2022 International Conference on Cyber Resilience (ICCR)* (pp. 1–6). IEEE.
26. Abouzakhar, N. S., Jones, A., & Angelopoulou, O. (2017, June). Internet of things security: A review of risks and threats to healthcare sector. In *2017 IEEE international conference on Internet of Things (iThings) and IEEE green computing and communications (GreenCom) and IEEE cyber, physical and social computing (CPSCom) and IEEE smart data (SmartData)* (pp. 373–378). IEEE.
27. Singh, B. (2022). COVID-19 pandemic and public healthcare: Endless downward spiral or solution via rapid legal and health services implementation with patient monitoring program. *Justice and Law Bulletin*, *1*(1), 1–7.

28. Safavi, S., Meer, A. M., Melanie, E. K. J., & Shukur, Z. (2018, November). Cyber vulnerabilities on smart healthcare, review and solutions. In *2018 Cyber Resilience Conference (CRC)* (pp. 1–5). IEEE.
29. Le Fevre, E. M., & Heleg, G. (2022). Artificial intelligence in medicine: Laws, regulations, and privacy. In *Artificial Intelligence in Cardiothoracic Imaging* (pp. 527–534). Springer International Publishing.
30. Singh, B. (2020). Global science and jurisprudential approach concerning healthcare and illness. *Indian Journal of Health and Medical Law*, *3*(1), 7–13.
31. Xu, M. (2019). *A System Perspective to Privacy, Security and Resilience in Mobile Applications* (Doctoral dissertation). University of Saskatchewan.
32. Hossein, K. M., Esmaeili, M. E., Dargahi, T., Khonsari, A., & Conti, M. (2021). BCHealth: A novel blockchain-based privacy-preserving architecture for IoT healthcare applications. *Computer Communications*, *180*, 31–47.
33. Singh, B. (2019). Profiling public healthcare: A comparative analysis based on the multidimensional healthcare management and legal approach. *Indian Journal of Health and Medical Law*, 2(2), 1–5.
34. Algarni, A. (2019). A survey and classification of security and privacy research in smart healthcare systems. *IEEE Access*, *7*, 101879–101894.

Chapter 7

Energy-efficient data collection scheme for mobile sink-based wireless sensor networks

Vaibhav Agarwal and Shashikala Tapaswi

7.1 INTRODUCTION

Wireless sensor networks (WSNs) have become a crucial component in developing real-time monitoring systems, which include environmental monitoring (Montori, Bedogni, & Bononi, 2017), healthcare monitoring (Y. Zhang, Sun, Song, & Cao, 2014), industrial monitoring (Palattella et al., 2013), military surveillance (Mahamuni, 2016), agricultural monitoring (Hamami & Nassereddine, 2020), disaster management (Chaudhry, Tapaswi, & Kumar, 2019), and emergency evacuation systems (Agarwal, Tapaswi, Chanak, & Kumar, 2023). In a typical WSN, various sensor nodes (SNs) gather real-time values from the monitoring field and send them to the local server, known as a base station (BS). The values can be transmitted directly or in a hop-to-hop fashion as per the application requirement. However, the SNs used in WSNs are battery-operated and have energy constraints. The SNs that are near the BS utilize more energy than far-away nodes. This imbalance in the SNs' average energy consumption rate creates energy holes and partitions within the network. To mitigate these issues, clustering mechanisms are used. Clustering is used to stabilize the energy utilization of SNs. In clustering mechanisms, the network is disjointed into small areas called clusters in which any node becomes a cluster head (CH) based on specific parameters (Agarwal, Tapaswi, & Chanak, 2022). The responsibility of a CH is to gather and accumulate values from the member SNs and disseminate them to the BS. However, a CH faces faster energy depletion because it transmits more data than its members, creating an imbalance in the energy consumption rate. Moreover, in cluster-based techniques, CH nearest nodes transmit more packets than far-away nodes, making the SNs non-functional relatively early and degrading the overall network performance. Therefore, to overcome the limitations of CH-based approaches, an optimal number of rendezvous points (RPs) have been identified in the monitoring field for data-gathering purposes. An RP is a location that is identified based on the SNs' location in a particular region so that the SNs' energy utilization will be minimal.

DOI: 10.1201/9781003541363-7

Recently, various researchers have utilized mobile sinks (MSs) for practical information gathering by optimizing their path length. The MS is capable of moving in the monitoring environment to collect data from the SNs. Various moving objects, such as robots, vehicles, and people, can carry the MS during data collection. For example, in the case of air pollution monitoring, SNs are placed on the top of the buildings, and wireless communication devices, or MSs, are used in public transport, such as buses, for sensed data collection. In forest monitoring applications, animals are fitted with wireless communicating devices to gather sensed data from different areas of forests. In military and battlefield applications, soldiers in the field can use the wireless device to acquire information about hostile objects from the positioned SNs. The movement of the MS is divided into three broad categories (Chatzigiannakis, Kinalis, & Nikoletseas, 2006) described below:

- **Random mobility:** In this scheme, the device is attached to mobile units, such as animals, known as data mules. Animals have unpredictable speeds with undefined paths and directions. An animal's next step is uncertain. The researchers used these patterns in delay-sensitive applications and conditions with random SN deployment. In these situations, the performance of the random sink mobility pattern shows better performance, as the network lifetime is enhanced.
- **Fixed path mobility:** In this pattern, the MS will follow a predefined path for data collection. The MS injects its route information into the network, which predicts the sink's future location. The MS does not change its trajectory on the difference in data generation rate or emergency events happening. In this scenario, a sink is attached to a vehicle, let's say a bus, to gather the data from the WSNs with non-uniform node distribution. Such WSNs may be present in delay-tolerant applications like scheduled air pollution monitoring or any industrial area. Another example of a fixed path mobility pattern is a city traffic management system where public transport, i.e., MS, follows the roads to collect data from SNs positioned on the building.
- **Controlled mobility:** In this scenario pattern, the MS adopts a dynamic path based on some network parameters, such as the remaining energy of a node or an event happening at some location. In these cases, the MS is attached to a robot that considers various parameters while making decisions. The MSs prepare their movement (both speed and trajectory) to enhance the overall performance based on the system's demand. This also decreases the data collection latency compared to random and fixed path mobility patterns.

The MS selects the mobility patterns depending on the application requirement. Further, data-gathering is classified into two classes. The first is direct data-gathering from every node. The second is indirect data-gathering in which the MS visits some locations (RPs) or some nodes (CHs) and gathers the data only from a set of nodes based on its configuration. Direct data-gathering is easy to implement and prevents sensor node energy consumption, leading to longer delays and buffer overflows. On the other hand, indirect data-gathering from points or nodes manages fair energy consumption and data latency at far ends. Therefore, choosing the optimal number of points is crucial for data collection. If the RPs are larger in number, then the path length of the MS is much larger, leading to buffer overflows and increasing data latency. On the other hand, if the number of RPs is significantly lower, there is multi-hop communication between SNs, prompting an irregularity in the energy utilization rate. That is why a trade-off should be maintained while optimizing RPs and MS paths. This chapter proposes an optimal number of RP selection schemes, and an MS is utilized for information gathering to upgrade the network's performance.

The primary highlights of this work are presented below:

- The elbow method computes the value of *k*.
- This *k* is passed as an input for the cluster formation.
- The centroids of these clusters are chosen as RPs.
- After selecting RPs, the MS gathers the information by visiting them.
- The simulations are performed to depict the credibility of this work regarding various performance metrics.

The chapter is further arranged as follows: Section 7.2 describes the related work. Section 7.3 reveals the details of the system model. Section 7.4 explains the methodology of this work in detail. Section 7.5 shows performance analysis and result comparison. Finally, Section 7.6 concludes the chapter.

7.2 RELATED STUDIES

Recently, various literature has discussed the optimization of RPs and route of MS. This section discusses various existing works that perform optimization for the RP and MS tour.

Latiff, Latiff, and Ahmad (2011) proposed a particle swarm optimization-based mobile base station (PSO-MSB) algorithm for WSNs. This scheme selects the CHs depending on the residual energy and minimal path using PSO, and a mobile base station is used for data collection. However, making a whole base station mobile is not feasible in practical scenarios. Feng, Qi, and Tang (2011) proposed an improved energy-efficient PEGASIS protocol by addressing the CH selection issue. In this approach, the CH is selected using a weight value function, which calculates the leftover energy in the SN and its distance from the BS. However, a longer chain in the network is the major issue with this algorithm. Jafri, Javaid, Javaid, and Khan

(2013) presented an MS-based IEEP protocol by maximizing the network's performance using multiple chains. The authors divided their network into four equal regions, and each region had a single chain by taking 25 nodes. This approach minimizes the data traveling path from a normal node to the leader. The authors also used RP-based data collection by using MS. In this approach, the MS stays at predefined locations for a temporary amount of time, collects the leader's data, and transmits that data to the BS.

Salarian, Chin, and Naghdy (2013) presented a weighted rendezvous planning (WRP) mechanism for computing the optimal path for an MS. This approach calculates a weight value based on the distance and packets transmitted. However, this approach selects more RPs, leading to a longer path length of the MS. Due to the increase in MS tour, a buffer overflow occurs at some SNs. Also, because of various computations, the time complexity is very high. Sitanayah, Brown, and Sreenan (2014) tried to increase the WSNs' reliability by deploying some additional *k* nodes. In this scheme, the value of *k* is identified using their dynamic programming variant. In addition, a centralized offline algorithm is used to minimize the number of deployed relays. The main advantage of this technique is that if any path becomes faulty, the data delivery is possible because the data is transmitted with more paths. This increases the reliability of the network by ensuring data delivery. However, with the rise in relay nodes, the number of data packets also rises. Since the data is transmitted through several paths, more copies with the same data are created, which degrades the network's overall performance. The multiple packets are transmitted through various paths and nodes; the energy of intermediate nodes is also wasted. This paper uses a static sink approach, which creates an energy hole near the sink region. R. Zhang, Pan, Xie, and Wang (2015) described a clustering and mobile collection scheme for data acquisition from the deployed SNs. Initially, hierarchical clustering and routing algorithms were performed for RP selection. Further, the MS gathered the data from these RPs. However, this algorithm ignores the re-selection of RPs, which leads to network partitioning.

Hu, Ding, Ren, Hao, and Han (2015) proposed an endocrine cooperative particle swarm optimization algorithm (ECPSOA) for WSNs, where the authors used PSO for RP selection. In this approach, weight is calculated for all the SNs in the WSNs based on the total data they have received from other SNs. The MS gathers data from those RPs, which have massive amounts of data. However, ECPSOA has the drawback of not selecting the CH efficiently because it does not consider important CH selection parameters like residual energy and number of cluster members. Chang and Shen (2016) proposed a tree-based mechanism in which data is forwarded via the root node. First, a cluster is formed as per tree properties, and then path planning is done using the dynamic sorting approach. However, the tree-based structure overburdened the root node, leading to the early death of that node. Wang et al. (2017) presented a PSO using variable dimensions (VD-PSO) to select RPs efficiently. It identifies the shortest route between pairs of RPs. The main shortcoming of this approach is that it is unsuitable for large-scale WSNs because of high running time complexity.

Kaswan, Nitesh, and Jana (2017) proposed an efficient scheme for reducing the energy dissipation of nodes. Initially, this algorithm selected the optimal positions for the RPs based on the SN deployment. Furthermore, this algorithm optimizes the number of points chosen to ensure single-hop communication. Finally, the MS visits these RPs for data acquisition. However, this algorithm takes exponential time complexity to identify the number of RPs and MS path formation. Sharma, Puthal, Jena, Zomaya, and Ranjan (2017) constructed a data-gathering region in the center of the monitoring field. Multiple trees are created in different regions from the center. The child nodes pass their data through intermediate nodes, and these nodes send it to the parent node. Due to this, the parent becomes overburdened with the data, leading to its early death. Renold and Ganesh (2019) proposed a secure data routing protocol (secure MRL). This approach is divided into three parts: identification of data collection points, MS path planning, and secured data transmission. The data collection points cover all the SNs by forming a boundary region. An MS visits each point and forms a closed region. Since all the data collection points are in the boundary region, the center nodes face multi-hop routing, creating an imbalance in energy consumption.

Chauhan and Soni (2020) presented a nature-inspired firefly optimization algorithm for cluster head selection. Initially, this scheme partitioned the monitoring field into rectangular regions and selected a CH for each region. After CHs are selected, the centroids of these CHs are chosen as RP. However, this algorithm ignores important parameters like node centrality, residual energy, and hop count while selecting the CHs. Fu and He (2020) proposed an energy-saving mechanism by balancing the clusters' energy (BIIE) to proportionate energy utilization within each cluster. In addition, some RPs are identified based on CH locations. Finally, an MS is utilized for data collection from the CHs by visiting RPs. However, fewer RPs are chosen against more CHs, causing buffer overflow at some network points. Gupta and Saha (2020) used a hybrid meta-heuristic algorithm in which the artificial bee colony algorithm is utilized along with differential evolution to proportionate the energy utilization of CHs. In addition, data was collected from CHs using an MS. However, a low convergence rate and a bad stability period are the major drawbacks of this scheme. In addition, the imbalance in energy dissipation leads to network performance degradation.

Donta, Amgoth, and Annavarapu (2021) proposed an extended ant colony optimization (EACO) for WSNs. This scheme identifies RPs on the basis of data generated by the SNs. Later, new RPs are identified when the data generation rate varies. Additionally, the MS tour length is bounded by a threshold value, and the proposed scheme chooses the RPs based on this value. However, this algorithm requires a massive amount of time to identify the RPs. Rathee et al. (2024) proposed a quantum-inspired genetic algorithm (Q-GEMS) for WSNs using multiple MSs. In this work, the authors used a genetic algorithm to optimize the energy utilization among the SNs to enhance the network lifetime. In addition, multiple MSs were utilized for fair consumption of energy within the networks. The proposed scheme also adapts to any change in network topology due to sensor failures and external influences. However, managing multiple MSs introduces new challenges, such as synchronization, even network distribution, and optimal path length of each MS, which are not covered in detail in the proposed scheme. In Gowda and Jayasree (2021), a routing protocol using group teaching and the bald eagle search mechanism (GTA-BE) for WSNs was presented. The authors utilized the mean shift clustering algorithm for clustering and the new bald eagle search mechanism for CH selection. Further, based on the data packets and hop distance, RPs were identified. Finally, the ideal route between SNs and RPs was computed using the group teaching algorithm. However, high message overhead and transmission delays were challenges.

All the above-mentioned approaches focused on selecting RPs by ignoring the data acquisition latency and message overhead. Many algorithms do not consider the re-selection of RPs in a sufficient amount of time. Therefore, to overcome these limitations, this chapter proposes an optimal selection of RPs considering SN deployment and using the MS for efficient data collection by focusing on improving the network's performance.

7.3 SYSTEM MODEL

This section depicts the system model. First, the network model is described, then the energy model is defined, and finally, some assumptions are considered in the proposed scheme.

7.3.1 Network model

This work considers a monitoring field M in which n SNs $S = \{s_1, s_2, \ldots, s_n\}$ are spread randomly in the monitoring environment, which is considered in the form of a 2D graph G(S, E) in which S and E are the set of SNs and links among them. A link is said to be present between (s_i, s_j) if they fall within the transmission range R. It can also be expressed as $E = \{(s_i, s_j) || D_{ij} \leq R \; \& \; s_i, s_j \in S\}$, where D_{ij} is the optimal distance between two nodes on an x-y axis. Further, an MS traverses the monitoring field and collects the data from these SNs. The network model is shown in Figure 7.1, in which black dots show the SNs and blue points show the RPs. The BS is the starting and ending point for the MS. In between the MS visits, each RP collects the data from the SNs. Finally, when the MS completes one round and returns to the BS, it offloads the collected data there. The BS then transfers that data to the control center via the Internet.

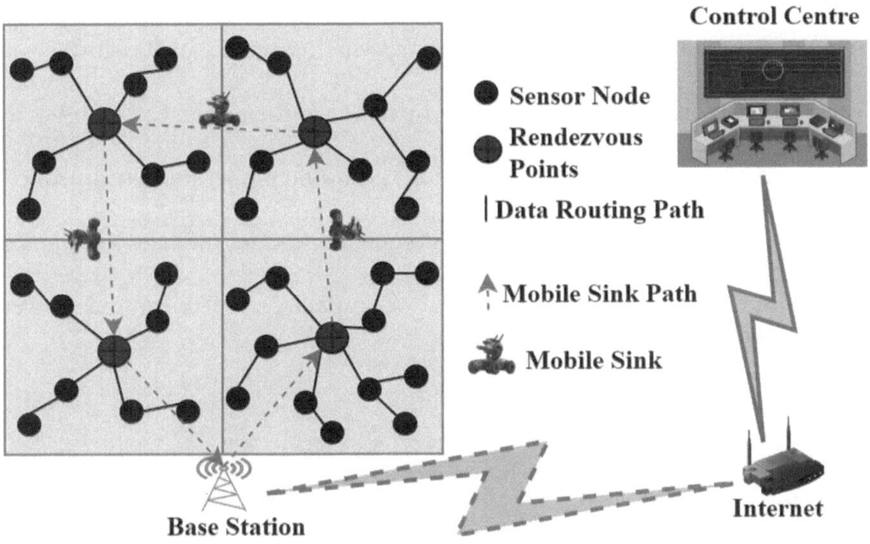

Figure 7.1 A basic architecture of a rendezvous point-based data collection scheme using a mobile sink. In this figure, sensor nodes are randomly deployed, and data is collected using MS by visiting RPs and transmitting the data to the control center via the base station.

7.3.2 Energy model

The radio model utilized to calculate the energy dissipation of this work is the same as that in Jindal, Agarwal, and Chanak (2021). To transmit k-bit data to a distance d, the energy consumption is calculated as follows:

$$E_{tx} = \begin{cases} (E_{el} + E_{free} * d^2) * k, if\ d < d_0 \\ (E_{el} + E_{multi} * d^2) * k, if\ d \geq d_0 \end{cases} \tag{7.1}$$

where E_{el} energy is consumed by transmitter circuitry, and E_{free} and E_{multi} are energies consumed in the single-hop and multi-hop transmission, respectively. The threshold distance d_0 is computed as follows:

$$d_0 = \sqrt{\frac{E_{free}}{E_{multi}}} \tag{7.2}$$

In receiving k-bit data, the energy consumption is calculated as:

$$E_{rx} = k * E_{el} \tag{7.3}$$

7.3.3 Assumptions

This work considers various assumptions, which are described below:

- SNs are spread indiscriminately in the monitoring field of the A × A $[m^2]$.
- The SNs are homogeneous, static, and battery-operated with limited energy.
- SNs have a global positioning system built into them, which helps identify their location.
- BS and MS have high computation capacity and unlimited power supply.

7.4 PROPOSED SCHEME

This work considers two phases: cluster formation and MS-based data-gathering. In the first phase, the optimal value of *k* is identified using the elbow method based on the SNs' deployment. After finding the value of *k*, the *k*-means clustering algorithm is executed to form optimal-size clusters. Furthermore, the centroids of these clusters are chosen as RPs. In the last, the MS visits these RPs to gather the data from all the cluster members belonging to that RP and forwards it to the BS.

7.4.1 Cluster formation

All the cluster formation algorithms pre-require the value of the number of clusters (k). However, knowing the optimal value of clusters before executing the algorithm is very hard. Therefore, an elbow method is used in this work to pre-identify the optimal *k* value, which is the input for the *k*-means clustering algorithm. The elbow method computes the within-cluster sum of squares (WCSS) value, which identifies the optimal value of clusters concerning the average distance between all the nodes. The output of the elbow method shows a sharp bend near the optimal value based on the SNs' deployment. Figure 7.2 shows the output of the elbow method when executed with 100 SNs randomly deployed in the 100 × 100 $[m^2]$ network field.

After obtaining the optimal value of *k* from the elbow method, it forms clusters within the network. *k*-means clustering is one of the famous unsupervised machine learning techniques. The main idea behind its procedure is to find at least one center for each point with minimum distance. This algorithm places the centers as far as possible from each other. When the loop gets iterated, all the *k* centers change their location step by step. Let us assume, $n = \{n_1, n_2, \ldots n_n\}$ is the set of data points and $k = \{k_1, k_2, \ldots, k_k\}$ is the set centers. The primary goal is to optimize the objective function, which is given by:

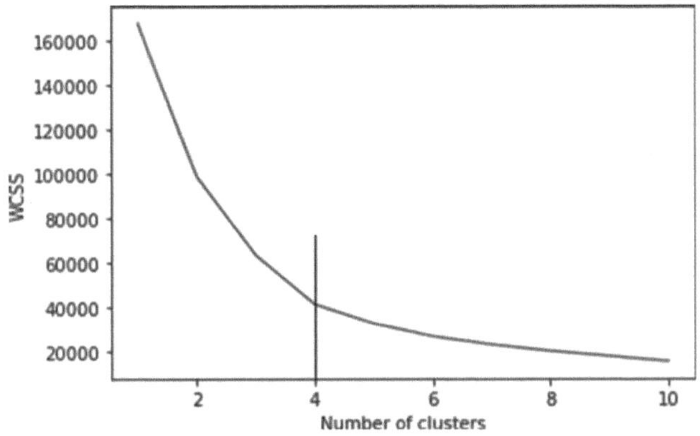

Figure 7.2 This figure shows the identification of an optimal number of clusters using the elbow method based on the sensor nodes' deployment and the minimal distance from each RP.

$$V(\mathrm{k}) = \sum_{i=1}^{n} \sum_{j=1}^{k} \left(\left\| n_i - k_j \right\| \right)^2 \tag{7.4}$$

Figure 7.3 shows the cluster formation with different colors and their centroid in black color. These centroids are chosen as RPs for the MS-based data collection.

7.4.2 Mobile sink-based data-gathering

After the cluster formation and identification of optimal locations for RPs, an MS is used to visit these RPs and gather the information saved by the SNs. For efficient data-gathering, optimal path planning for the MS is needed. The BS has all the locations of RPs, and it computes the shortest path using the traveling salesman problem with Christofides heuristics (Johnson & McGeoch, 2007). Since all the RPs are fixed, there is no need to reconstruct the path for MS again and again. In the proposed work, MS uses a fixed path mobility pattern, which helps to avoid the unnecessary burden of MS path construction in each round. After the path formed by the BS, the MS visits each RP in the order identified by the BS. After reaching an RP, the MS broadcasts a request message with $\mathrm{R_{id}}$ for data uploading, which all the SNs belonging to that particular RP receive. After receiving the *request* message, the SNs check the $\mathrm{R_{id}}$ along with their buffer condition. If the SNs belonging to the $\mathrm{R_{id}}$ have any data to upload, then they send a *reply* message to the MS with their ID, amount of data, and residual energy. The MS designates a time frame for each SN based on the data and receives the data from all the SNs belonging to that RP. After collecting the data from one

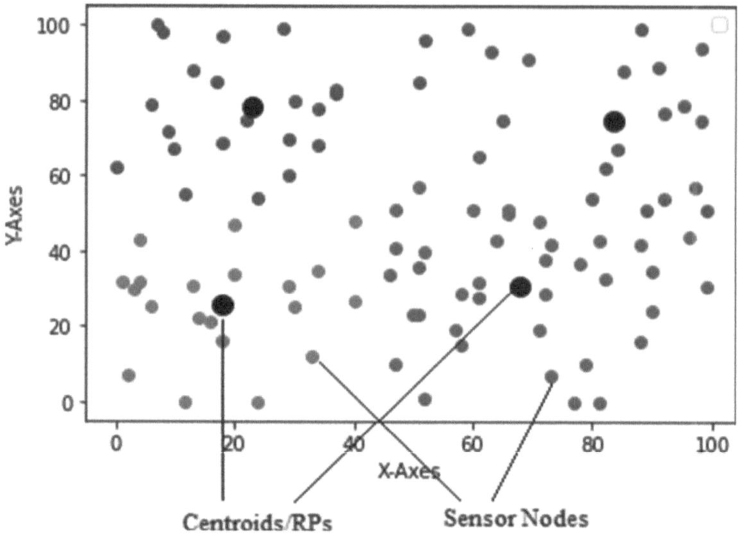

Figure 7.3 This figure shows the selection of coordinates for the RPs using the *k*-means clustering algorithm. The optimal value of *k* is identified using the elbow method in the previous step.

RP, the MS moves forward and collects the data from the remaining RPs. Finally, the MS uploads all the data to the BS.

7.5 PERFORMANCE ANALYSIS AND RESULT COMPARISON

This section presents the performance analysis of this work by comparing it with secure MRL, RkM, and WRP in terms of lifetime, energy utilization, and reliability. NS-3.36 is used as simulation software. The comparison is done by taking a variable number of SNs ranging from 50 to 200. The SNs are randomly spread in the monitoring field of $100 \times 100[m^2]$. Table 7.1 shows the remaining simulation parameters used in this work.

7.5.1 Stability period

The stability period refers to the number of rounds for becoming the first node dead. A node is said to be dead if its residual energy becomes zero. The network is said to be stable if all the nodes are alive. For efficient working, the stability period should be very high.

Figure 7.4 shows the efficiency of this work in terms of the stability period. On average, it improved by 26.73%, 31.74%, and 45.5% compared with secure MRL, RkM, and WRP, respectively. The improvement is due to identifying the optimal number of clusters using the elbow method and

Table 7.1 Simulation parameters

Parameters	*Values*
Monitoring area	100 × 100[m²]
Number of SNs	50 to 200
Radio propagation model	Two-way ground
MAC layer	802.11
Physical medium	Wireless
Initial energy of sensor node	0.5 J
E_{el}	50 nJ/bit
E_{free}	10 pJ/bit/m^2
E_{multi}	0.0013 pJ/bit/m^4
E_{agg}	5 nJ/bit
Size of data packet	512 Bytes
Size of control packet	64 Bytes
BS coordinates	(0,50)

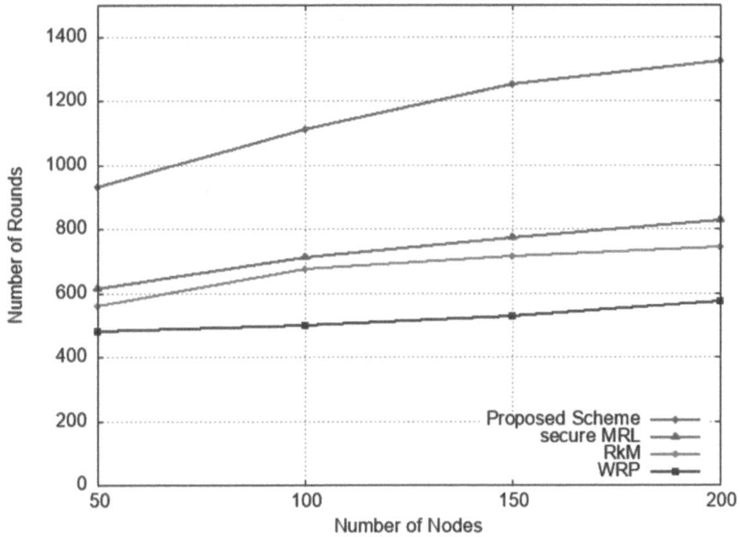

Figure 7.4 This figure shows the comparison of all four schemes regarding the stability period of the network based on the number of rounds.

clustering the SNs with minimal distance from the centroid, which becomes the RP for MS-based data collection.

7.5.2 Network lifetime

Network lifetime means the number of rounds to which 100% of the nodes become dead. For better network performance, the network lifetime should be high so that the data can be collected from the network for more rounds.

Figure 7.5 shows the comparison of network lifetime. Here, the network lifetime is improved by 23.42%, 30.33%, and 42.32% compared with secure MRL, RkM, and WRP, respectively. The improvement is due to the balance of the energy consumption among the nodes and the efficient use of MS-based data-gathering.

7.5.3 Average energy consumption

This is computed as the energy utilization by each SN up to some rounds. This parameter should be significantly lower so that the energy is consumed in a balanced way and the network remains alive for a longer time span.

Figure 7.6 depicts the comparison between all four schemes regarding average energy consumption, which is reduced by 22.19%, 37.26%, and 48.21% in comparison to secure MRL, RkM, and WRP, respectively. This improvement in the performance is because of the optimal distribution of SNs in each cluster depending on the locations and efficient data aggregation by the MS from them by visiting the nearest RP.

7.5.4 Throughput

Throughput is calculated in terms of data packets reaching the BS with packets transmitted by the SNs. For better reliability of the network, it should be as high as possible.

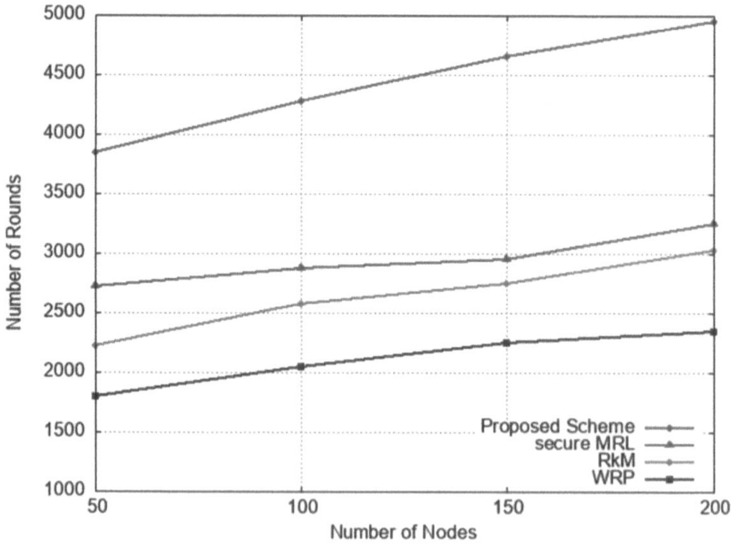

Figure 7.5 This figure shows the comparison of all four schemes regarding network lifetime based on the number of rounds.

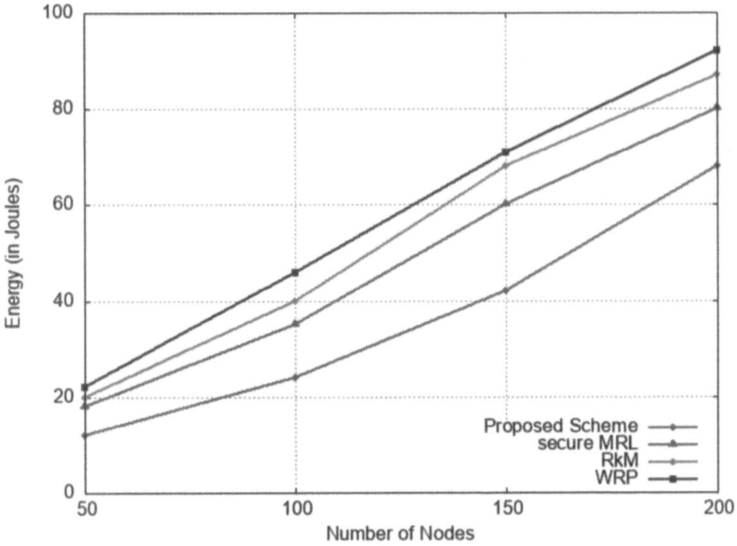

Figure 7.6 This figure shows the comparison of all four schemes regarding the average energy consumption of the network in Joules.

Figure 7.7 shows the comparison of all four algorithms regarding throughput. The results show that the network throughput is improved by 21.78%, 30.59%, and 41.07% when compared with secure MRL, RkM, and WRP, respectively. This improvement in throughput is because of the timely collection of the data from the SNs, which reduces packet losses and helps improve reliability.

7.6 CONCLUSION AND FUTURE OPPORTUNITIES

This chapter proposes an energy-saving mechanism for MS-based WSNs. First, the optimal clusters (k) are identified using the elbow method as per the SNs' deployment. The computed k value becomes the input for the k-means algorithm. Further, centroids of these clusters become the optimal locations for the RPs. The MS visits these RPs and aggregates the information from the members belonging to that RP. After visiting all the RPs, the MS transmits the acquired data to the BS. This work contrasts with cutting-edge algorithms like secure MRL, RkM, and WRP. The simulation results reveal an outperformance regarding various network parameters such as lifetime, energy utilization, and reliability.

This study can be improved by using query processing within the BS in the future. This study will have many practical applications, such as healthcare monitoring, the medical field, forest fire detection, and industrial monitoring.

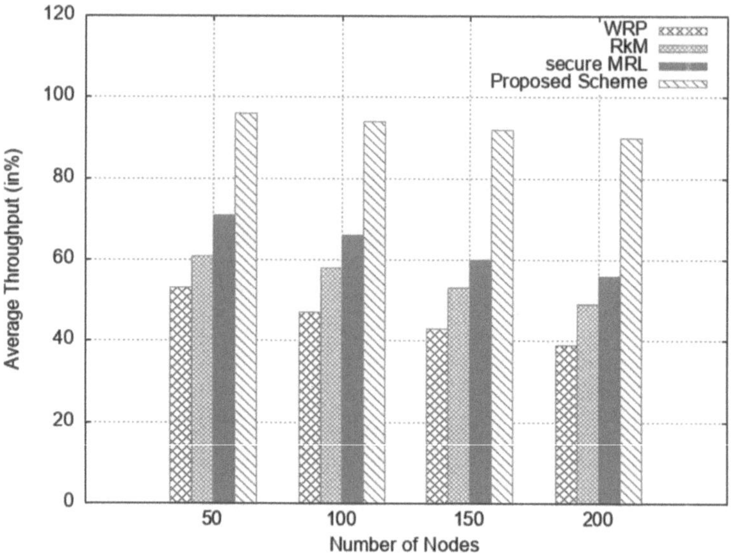

Figure 7.7 This figure shows the comparison of all four schemes regarding network throughput in percentages.

ACKNOWLEDGMENT(S)

The authors thank NIT Kurukshetra and ABV-IIITM, Gwalior, for providing sufficient resources to implement this research.

DISCLOSURE STATEMENT

No relevant interest is involved in this paper.

Funding

No funding is involved in this paper.

REFERENCES

Agarwal, V., Tapaswi, S., & Chanak, P. (2022). Energy-efficient mobile sink-based intelligent data routing scheme for wireless sensor networks. *IEEE Sensors Journal*, 22 (10), 9881–9891.

Agarwal, V., Tapaswi, S., Chanak, P., & Kumar, N. (2023). Intelligent emergency evacuation system for industrial environments using iot-enabled wsns. *IEEE Transactions on Instrumentation and Measurement*, 72, 1–12.

Chang, J.-Y., & Shen, T.-H. (2016). An efficient tree-based power saving scheme for wireless sensor networks with mobile sink. *IEEE Sensors Journal*, 16 (20), 7545–7557.

Chatzigiannakis, I., Kinalis, A., & Nikoletseas, S. (2006). Sink mobility protocols for data collection in wireless sensor networks. *In Proceedings of the 4th ACM international workshop on mobility management and wireless access* (pp. 52–59).

Chaudhry, R., Tapaswi, S., & Kumar, N. (2019). A green multicast routing algorithm for smart sensor networks in disaster management. *IEEE Transactions on Green Communications and Networking*, 3 (1), 215–226.

Chauhan, V., & Soni, S. (2020). Mobile sink-based energy efficient cluster head selection strategy for wireless sensor networks. *Journal of Ambient Intelligence and Humanized Computing*, 11 (11), 4453–4466.

Donta, P. K., Amgoth, T., & Annavarapu, C. S. R. (2021). An extended aco-based mobile sink path determination in wireless sensor networks. *Journal of Ambient Intelligence and Humanized Computing*, 12 (10), 8991–9006.

Feng, S., Qi, B., & Tang, L. (2011). An improved energy-efficient pegasis-based protocol in wireless sensor networks. In *2011 eighth international conference on fuzzy systems and knowledge discovery (fskd)* (Vol. 4, pp. 2230–2233). IEEE.

Fu, X., & He, X. (2020). Energy-balanced data collection with path-constrained mobile sink in wireless sensor networks. *AEU-International Journal of Electronics and Communications*, 127, 153504.

Gowda, C. S., & Jayasree, P. (2021). Rendezvous points-based energy-aware routing using hybrid neural network for mobile sink in wireless sensor networks. *Wireless Networks*, 27 (4), 2961–2976.

Gupta, G. P., & Saha, B. (2020). Load balanced clustering scheme using hybrid metaheuristic technique for mobile sink based wireless sensor networks. *Journal of Ambient Intelligence and Humanized Computing*, 1–12. https://doi.org/10.1007/s12652-020-01909-z

Hamami, L., & Nassereddine, B. (2020). Application of wireless sensor networks in the field of irrigation: A review. *Computers and Electronics in Agriculture*, 179, 105782.

Hu, Y.-F., Ding, Y.-S., Ren, L.-H., Hao, K.-R., & Han, H. (2015). An endocrine cooperative particle swarm optimization algorithm for routing recovery problem of wireless sensor networks with multiple mobile sinks. *Information Sciences*, 300, 100–113.

Jafri, M. R., Javaid, N., Javaid, A., & Khan, Z. A. (2013). Maximizing the lifetime of multichain pegasis using sink mobility. *arXiv preprint*, arXiv:1303.4347.

Jindal, A., Agarwal, V., & Chanak, P. (2021). Emergency evacuation system for clogging-free and shortest-safe path navigation with iot-enabled wsns. *IEEE Internet of Things Journal*, 9 (13), 10424–10433.

Johnson, D. S., & McGeoch, L. A. (2007). Experimental analysis of heuristics for the STSP. In *The traveling salesman problem and its variations* (pp. 369–443). Boston: Springer.

Kaswan, A., Nitesh, K., & Jana, P. K. (2017). Energy efficient path selection for mobile sink and data gathering in wireless sensor networks. *AEU-International Journal of Electronics and Communications*, 73, 110–118.

Latiff, N. A. A., Latiff, N. M. A., & Ahmad, R. B. (2011). Prolonging lifetime of wireless sensor networks with mobile base station using particle swarm optimization. *In 2011 fourth international conference on modeling, simulation and applied optimization* (pp. 1–6).

Mahamuni, C. V. (2016). A military surveillance system based on wireless sensor networks with extended coverage life. *In 2016 international conference on global trends in signal processing, information computing and communication (icgtspicc)* (pp. 375–381).

Montori, F., Bedogni, L., & Bononi, L. (2017). A collaborative internet of things architecture for smart cities and environmental monitoring. *IEEE Internet of Things Journal*, 5 (2), 592–605.

Palattella, M. R., Accettura, N., Grieco, L. A., Boggia, G., Dohler, M., & Engel, T. (2013). On optimal scheduling in duty-cycled industrial iot applications using ieee802.15.4e tsch. *IEEE Sensors Journal*, 13 (10), 3655–3666.

Rathee, M., Kumar, S., Dilip, K., Dohare, U., Aanchal, & Parveen. (2024). Towards energy balancing optimization in wireless sensor networks: A novel quantum inspired genetic algorithm based sinks deployment approach. *Ad Hoc Networks*, 153, 103350. https://www.sciencedirect.com/science/article/pii/S1570870523002706

Renold, A. P., & Ganesh, A. B. (2019). Energy efficient, secure data collection with path-constrained mobile sink in a duty-cycled unattended wireless sensor network. *Pervasive and Mobile Computing*, 55, 1–12.

Salarian, H., Chin, K.-W., & Naghdy, F. (2013). An energy-efficient mobile-sink path selection strategy for wireless sensor networks. *IEEE Transactions on Vehicular Technology*, 63 (5), 2407–2419.

Sharma, S., Puthal, D., Jena, S. K., Zomaya, A. Y., & Ranjan, R. (2017). Rendezvous-based routing protocol for wireless sensor networks with mobile sink. *The Journal of Supercomputing*, 73 (3), 1168–1188.

Sitanayah, L., Brown, K. N., & Sreenan, C. J. (2014). A fault-tolerant relay placement algorithm for ensuring k vertex-disjoint shortest paths in wireless sensor networks. *Ad Hoc Networks*, 23, 145–162.

Wang, W., Shi, H., Wu, D., Huang, P., Gao, B., Wu, F., & Chen, X. (2017). Vd-pso: An Efficient mobile sink routing algorithm in wireless sensor networks. *Peer-to-Peer Networking and Applications*, 10 (3), 537–546.

Zhang, R., Pan, J., Xie, D., & Wang, F. (2015). Ndcmc: A hybrid data collection approach for large-scale wsns using mobile element and hierarchical clustering. *IEEE Internet of Things Journal*, 3 (4), 533–543.

Zhang, Y., Sun, L., Song, H., & Cao, X. (2014). Ubiquitous WSN for healthcare: Recent advances and future prospects. *IEEE Internet of Things Journal*, 1 (4), 311–318.

Chapter 8

IoT-enabled smart green cities using blockchain technology

Sustainable Applications

Amrit Suman, Preetam Suman, Sana Zeba, Sasmita Padhy, and Vishal Jain

8.1 INTRODUCTION

As an outcome of the government's campaign for cities that are intelligent and intuitive with large expenditures, it is projected that the Internet of Things (IoT) will play a crucial role in managing the oversight of the bulk of scarce commodities like water, gasoline, and food. However, an elevated level of communication between numerous large IoT devices is necessary for the real-time deployment of such applications. Thus, the Internet of Things has gained popularity among individuals. According to Dr. Judith Dayhoff, it has turned our inanimate physical world into "a digital nervous system." IoT gadgets behave very much like an extension of the human body. Figure 8.1 demonstrates how IoT devices gather data at the perception layer, move toward an IoT hub or IoT gateway for data collection and transfer, and then analyze the collected data before taking any further action.

It is practically impossible for anyone to replace existing data records with blockchain encryption. A further layer of protection is added to the network to stop nefarious intruders from accessing it by using blockchain to store IoT data. Energy management systems have undergone tremendous improvements in recent years in order to handle the rising amount of integrated renewable energy output.

Within this chapter, we research the role of cutting-edge blockchain technology within the context of an environmentally conscious IoT. This research endeavor has successfully illuminated the fundamental components, prospects, and challenges integral to the development of a sustainable "green" IoT ecosystem.

The key aspects covered in this chapter are:

1. Considering the relevance of the IoT ecosystem: This chapter emphasizes the pivotal role of the IoT ecosystem in a society that prioritizes environmental consciousness. It articulates the crucial factors that necessitate attention to enable a sustainable green IoT environment.

DOI: 10.1201/9781003541363-8

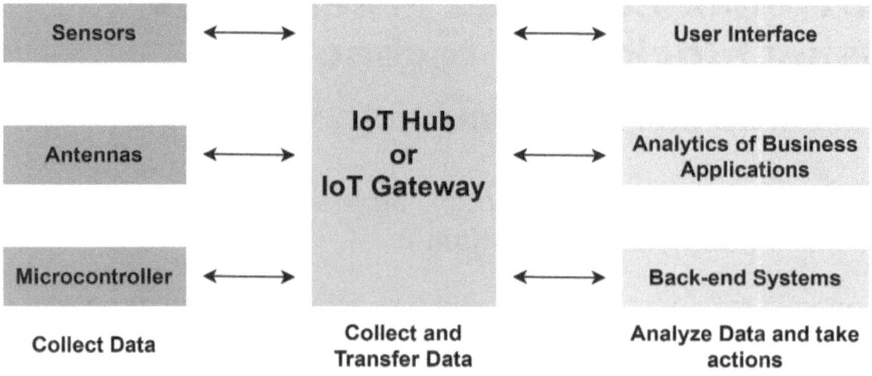

Figure 8.1 Example of an IoT system.

2. Eco-friendly IoT with blockchain: This chapter showcases how contemporary blockchain technology can contribute to rendering the IoT ecosystem more ecologically sustainable.
3. Unearthing pending challenges and research avenues: This chapter outlines unresolved issues and potential research directions for the realization of a long-lasting, eco-friendly IoT ecosystem underpinned by blockchain technology.

The chapter's structure is divided as follows: Section 8.3 reviews existing literature on IoT and blockchain. Section 8.4 delves into the initial considerations related to IoT, green IoT, and blockchain technology. Section 8.5 comprehensively examines diverse applications of green IoT, while Section 8.6 focuses on the applications of blockchain in green smart IoT. Section 8.7 engages in an insightful discussion about pertinent issues and delineates future research avenues. Finally, Section 8.8 culminates the research work with a comprehensive conclusion.

8.2 OUTCOME

In the future, the world will be a greener place due to significant improvements in green IoT. The intelligent communication between machines, drones, sensors, and other equipment is witnessed every day as they work together to preserve a sustainable and green environment. Green IoT can have a direct impact on the IoT industry, in addition to aiding other companies in sustainable development and environmental protection. The dynamics of creating an environmentally friendly and sustainable Internet of Things ecosystem will be greatly impacted by blockchain technology's limitless application possibilities and problem-solving capabilities.

It is crucial for researchers to evaluate the applicability of existing theories to the phenomena that arise at the intersection of sustainable IoT ecosystems and blockchain-based technologies. The book chapter provides an explanation of the concept of green IoT, which is predicted to utilize eco-friendly technology and design in the future.

8.3 LITERATURE REVIEW

As cities expand and evolve, it is possible to enhance the quality of life of urban residents by incorporating safe, interoperable, and integrated internet services. Government organizations are cognizant of the potential of digital currency to solve social problems. Smart cities can benefit from the new organizational paradigm of blockchain technology and a more transparent economic framework for resource management. In its beginnings, it was linked to the virtual currency bitcoin. Smart contracts between companies, citizens, and government agencies are proposed as an ecosystem model for smart cities. Additionally, it provides a summary of the potential application areas for the technology [1].

Prince Waqas Khan et al. [2] introduced a technique for recording films to help identify fake videos and assure the validity of stored video recordings in blockchain ledgers. Through copyright, this blockchain-based leger lowers forgery dangers. To address future data loss, extending the sensor chain with a blockchain contract was suggested [3]. MobiChain was recommended, an app for mobile devices that makes use of blockchain to effectively carry out mining operations in order to secure the transactions of m-commerce applications. To tailor the design of the healthcare application, which involves numerous parties like doctors, insurance providers, health service providers, and pharmacies, a multi-layer IoT/blockchain architecture was proposed [4]. By generating a hash value for each piece of data, a security foundation for blockchain-based healthcare multimedia IoT data applications was established, making data change or tampering impossible [5]. In order to protect all candidates in the IoT by eliminating third-party dependency, the secure classifiers KNN were suggested to offer the privacy of trained IoT data. Blockchain plus partial homomorphism known as Paillier was also proposed [6, 7]. With the goal statement "because we feel that technology can drive improved health, we exist to harness all sector vices to build a healthier future for all," IET Healthcare was created for the future of healthcare [8, 9]. It was discovered that deep learning was a superior method for identifying intrusions in an IoT context [10]. Healthcare security vulnerabilities were also examined, and a "Smart Medical System" (SMS) secure architecture for the municipal system was suggested. For computations, this system took advantage of the secured mobile edge computing idea [11]. A ground-breaking framework called BlockID was created, which uses trusted computer technology and biometric verification or authentication in order to authenticate people's identities. Additionally, the features, security procedures, and current and experimental uses of BlockID were considered [12]. This security framework provides a creative framing platform for secure communication. A compact smart greenhouse design built on blockchain technology that improves security and privacy by using the blockchain ledger was also suggested [13]. To build IAM resilience for the global risk in the IoT, the blockchain approaches IOTA and DAG were evaluated. This evaluation used the distributed ledger of the public permission-less blockchain [14, 15]. A thorough discussion of combining blockchain technology with IoT applications is also necessary, and a study looked at comparative studies and provided a thorough literature assessment of these topics [16]. Another study has offered a thorough analysis of developing blockchain-based IoT healthcare technologies and other relevant applications [17]. For home quarantine, Zhang and Wu [18] suggested a blockchain-based system and offered a use case for an Internet of Things (IoT) system. A fresh viewpoint on trust for evaluating the reliability of IoT data in two different contexts, system trust and user trust, was also provided [19]. Privacy concerns of the proposed framework based on verification, identification, and signing have also been discussed [20]. The Hyperledger blockchain framework was discussed, including a proposed security architecture based on this framework [21]. K. P. Satamraju et al. [22] described a model to use the blockchain to solve IoT security and authentication problems. C. Varghese et al. [23] suggested a solution that creates a decentralized network that offers security and transparency. Surveys have been conducted to find solutions to the problems of anonymity, flexibility, and integrity for a successful integration of blockchain with IoT devices [24, 25]. Haowen Tan et al. [26] created a system based on homomorphism authentication for VANETs that performs monitoring of all ambulances. Additionally, VC and edge units assisted in the development of another decentralized vehicle based on a blockchain recording system. It is recommended that the suggested design be used in future work as a new COVID-19 pandemic regulator.

8.4 BACKGROUND STUDY

8.4.1 Internet of Things (IoT)

The IoT is the connection of devices. This applied to not only traditional computers or smartphones but everyday objects equipped with embedded microprocessors, sensors, and actuators. What makes IoT truly remarkable is that these objects can communicate with each other without the need for direct human-to-human or computer-mediated interaction. The statistics are awe-inspiring, to sum it up. An astonishing 4.4 zettabytes of data were generated by over 50 billion IoT devices worldwide in 2020. The growth is truly remarkable when you realize that IoT devices only produced 100 billion megabytes of data in 2013, which was just seven years before. The world of IoT is revealed in an intriguing way by IoT Analytics Research. Their estimates suggest that there will be a staggering 27 billion active IoT connections by 2025, and that number excludes PCs, laptops, and tablets. Although this projection is positive, there are concerns about the impact of the global chip shortage on IoT device connectivity, particularly beyond 2023. System security and cost-effectiveness are of utmost importance due to the widespread use of IoT in diverse applications. Furthermore, IoT devices must be capable of carrying out a variety of tasks [27].

The architecture is divided into four layers:

1. Perception layer: Information about the environment is gathered by this top layer, sometimes called the sensing layer. It has RFID readers, temperature sensors, pressure sensors, proximity measurement devices, and GPS for data collection.
2. Network layer: This layer is responsible for data transmission between various applications.
3. Middleware layer: This layer is responsible for connections between the network and application layer.
4. Application layer: This is the top layer, which can communicate with the other interfaces.

All four layers provide a secure structure for the IoT environment, as shown in Figure 8.2.

8.4.2 Green Internet of Things

The greatest need of the environment is to reduce CO2 emissions. The green IoT proposed in this chapter is an initiative to reduce CO2 emissions. The IoT is growing all over the world, as IoT provides intelligent support for many tasks. Green IoT is the integration of energy-efficient techniques with an IoT infrastructure. The ultimate goal is to reduce carbon from the environment. The interconnected phases and the importance of environmental responsibility from design to disposal are illustrated in Figure 8.3, which depicts the life cycle of a green IoT. The ecological footprint of IoT technologies can be minimized and sustainability promoted by taking a comprehensive approach.

These are some of the essential aspects of the green IoT and its applications:

1. Green RFID tags are designed with radio-frequency identification (RFID) technology that is energy-efficient, leading to decreased power consumption and more sustainable RFID applications.
2. The achievement of green IoT objectives requires the use of green wireless sensor networks (WSNs). Sensors that operate with limited

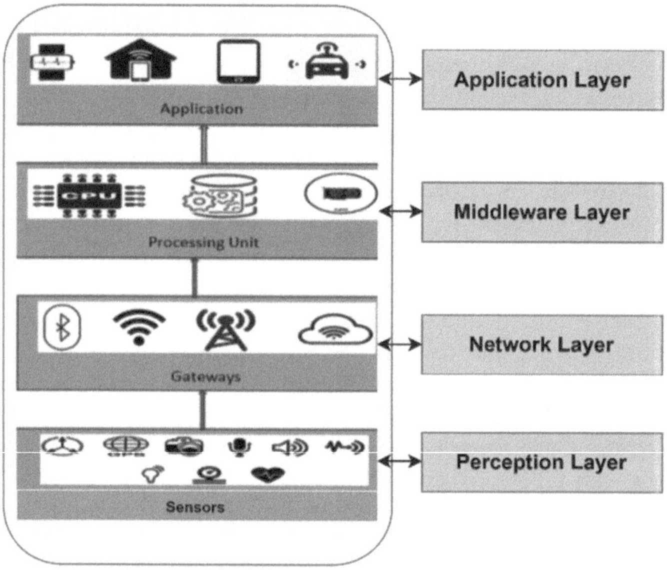

Figure 8.2 Layered Internet of Things (IoT) architecture.

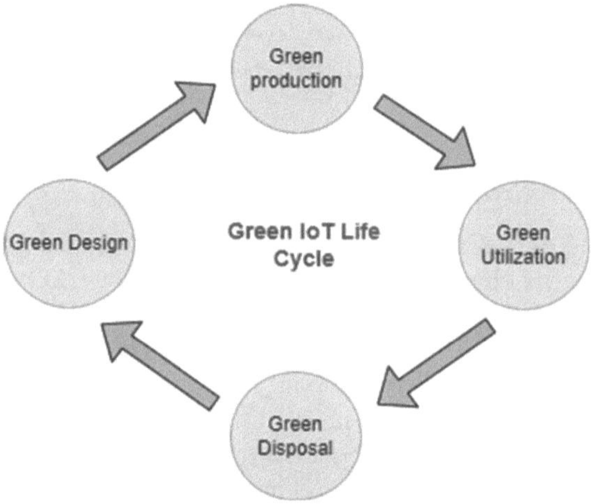

Figure 8.3 Green IoT life cycle.

power and storage are used in wireless sensor networks, which are made up of nodes.

3. Green IoT optimizes the energy required for machine-to-machine communication.
4. Green data centers allow most data to be saved in an eco-friendly environment.

5. Green IoT can establish more responsible and sustainable IoT for the future.

8.4.3 Blockchain technology

Blockchain technology has many potential applications. These include recording virtual trade items. The important components are shown in Figure 8.4, which are vital in its functionality. Blockchain's reputation as a transformational technology with the potential to revolutionize asset management is established by its characteristics and components. In various areas, such as business, finance, and intellectual property, transactions are conducted and trust is established [28, 29]. In turn, efficiency, transparency, and trust in the overall business environment can significantly increase with the implementation of this technology.

Several key features and components that make it a powerful and versatile tool for a wide range of applications, such as IoT systems, distinguish blockchain technology. These features include:

1. Distributed ledger technology, which allows all network users to access the distributed ledger that powers blockchain. The elimination of redundancy and the recording of all activities in one, unalterable record is achieved by recording transactions only once in this shared ledger.
2. Once a transaction is recorded on the blockchain, it is immutable and cannot be altered or tampered with. In order to correct an error, a new transaction must be created and added to the ledger.

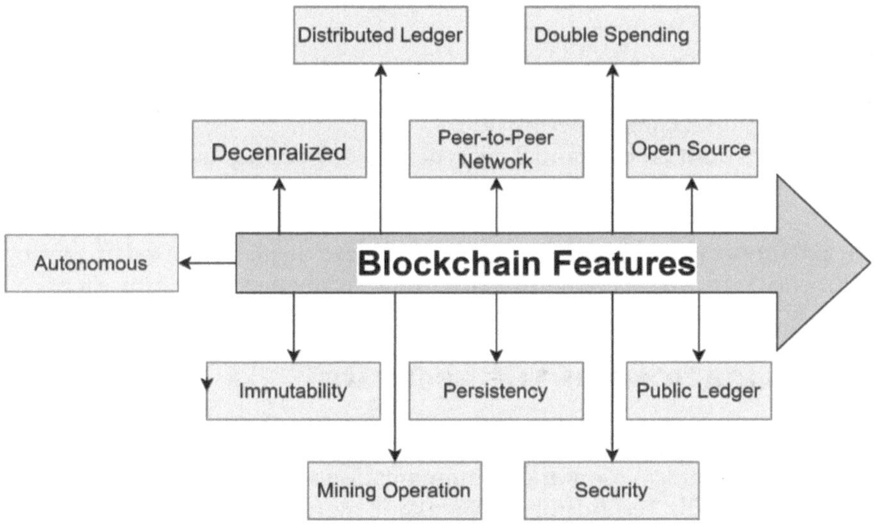

Figure 8.4 Important components and key features of blockchain.

3. Predefined instructions that are stored on the blockchain are what smart contracts are made of. Transactions are executed automatically if certain conditions are met. Many processes, from company bond transfers to travel insurance requirements, can be simplified with the help of this feature.
4. Due to the decentralized nature of blockchain, transaction validation can be done without the need for a central authority. The system is maintained in working order by users who are a part of the network.
5. Interacting directly with the blockchain allows users to maintain anonymity.
6. Participating nodes in the IoT network can independently verify all data revisions due to the immutability of the distributed ledger.
7. Blockchain security is enhanced through cryptographic technologies like public and private keys and blockchain hashing.
8. Blockchain technology effectively solves the problem of double spending, preventing the same digital asset from being spent twice on the network.

Blockchain technology is a robust solution for various applications, particularly in IoT systems, due to the combination of these features. Ensured security, transparency, and efficiency can be achieved through transaction handling. There are a variety of ways to build a blockchain network. They can be created by a consortium, granted permission, public, or private [28]. Blockchain networks can accommodate a variety of use cases and requirements. Blockchain networks come in different types.

- **Public blockchain:** This is open to all and used by everyone. It is open for use and membership.
- **Private blockchain:** This type of blockchain is limited to specific users.
- **Permitted blockchain:** This type of blockchain is open for participation, but it requires permission.
- **Consortium blockchain:** This type of blockchain is used for organizations at various levels.

Blockchain provides various use cases to achieve the desired level of privacy and security. It can be configured according to the needs of the users.

8.5 APPLICATIONS OF THE GREEN IOT

"Green computing" is the practice of using computers, servers, peripherals, and other information and communication technologies in a way that is environmentally responsible. To make it more sustainable, the IoT has included some green computing techniques. Storage links and the virtualization of resources like servers mean that fewer overall pieces of equipment are required, and as a result, less energy is required to run everything. Fewer devices translate into lower costs, lower greenhouse gas emissions, and faster application and storage space management. Other tried-and-trusted approaches to integrating green technology into the IoT include virtualizing desktops, reducing consumables, and using recycling and reuse (Figure 8.5).

The green IoT has broadened to encompass diverse technologies and advancements to curb energy usage, decrease carbon discharges, and advance sustainability. Adaptations to radio-frequency identification (RFID) technology are being made with the goal of enhancing its eco-friendliness and energy efficiency, also known as green RFID. Green wireless sensor networks (WSNs) are specifically engineered for energy efficiency and sustainability purposes, making them perfectly suited for use in IoT applications. Data centers have started green initiatives that play a crucial role in IoT infrastructure. The goal is to reduce their energy consumption and carbon emissions. The goal of green communication is to achieve energy efficiency and save the environment. IoT applications utilize internet technologies that prioritize sustainability and energy conservation. These are also known as green internet technologies. The green IoT is currently being improved. A few ways to improve it are as follows:

1. The first step is the energy harvesting using RF. This enables IoT devices to conserve more energy. This also promotes green energy by reducing CO2 emissions.
2. The green IoT has the potential to enhance environmental preservation and energy efficiency. The key factors are enhanced bandwidth, enhanced throughput, and a decrease in delay.
3. The green IoT can improve environmental conservation and energy efficiency by incorporating sustainable technology.

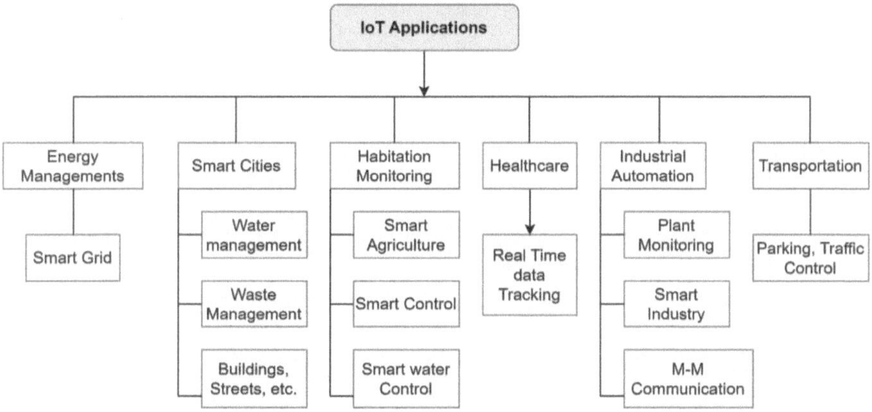

Figure 8.5 Applications of IoT.

The growth of the green IoT could increase with the following:

1. The migration of wireless sensor networks (WSN) into cloud technology. This could secure more energy and lead to lower carbon emissions. Normally, WSNs use a multi-hop concept to send the data. If each node of the sensor node is connected to a cloud setup, then the transmission energy will reduce. Additionally, data loss due to distance and other factors could be minimized. Data could be analyzed in the cloud itself. The only hurdle would be network availability and signal strength.
2. Green Social Network as a Service (SNaaS): Green SNaaS focuses on improving the energy efficiency of the entire ecosystem, including systems, services, WSNs, and cloud administration. The objective is to make social networks more environmentally friendly and sustainable.
3. UAVs (unmanned aerial vehicles): Future UAVs have the potential to replace IoT devices, leading to a significant decrease in power consumption and environmental pollution. The potential exists for green IoT devices to be low-cost and highly effective.
4. Ease of installation and usage: The ease of installation and usage for end users is a crucial consideration for IoT devices and applications in the future. Devices should prioritize user-friendliness while preserving energy efficiency.
5. Energy-efficient sensor techniques: Sensors often experience rapid energy depletion due to continuous detection. To conserve energy, various sleeping techniques have been proposed. Continued research is necessary to find innovative ways to enhance sensor energy efficiency.
6. Data context preservation: Gathering data without context can lead to unnecessary resource consumption. Preserving data context is a future challenge in IoT to ensure that data processing makes more sense and reduces resource waste.
7. Uninterrupted connectivity: As the IoT device market expands, there will be increased competition for available spectrum and potential interference issues. Future IoT devices must be designed to operate in interference-prone environments and effectively reduce interference from other IoT devices while maintaining energy efficiency.

These advancements reflect the growing importance of energy efficiency, sustainability, and user-friendliness in shaping the future of IoT. Green IoT will play a critical role in addressing environmental concerns and achieving efficient, eco-friendly solutions. The following is a transcript from the June 15, 2021, meeting of the Hyperledger Climate Action and Accounting Special Interest Group, which featured Kamlesh Nagware, CTO of Snapper Future Tech, and Veronica Garcia-Heller, founder and CEO of Bitlumens:

Using a blockchain and an older smart metre, Bitlumens makes it easier to scale renewable solutions. In order to do this, we collaborate with PV small grids in rural areas with sporadic or non-existent electrical availability. India is our key market because of how quickly and with government support the energy sector is being digitalized there. This offers an excellent testing ground for the use of carbon budgeting and trading communities. With the help of our platform, which integrates the Internet of Things (IoT) with blockchain, we offer data protection, verification, and traceability. We provide each family a credit score and a carbon credit using survey and IoT data. These calculations are kept on the blockchain for authorized stakeholders to access. We think that including stakeholders in carbon reduction right from the start of a project encourages investment flows into renewable energy in rural areas.

8.6 APPLICATIONS OF BLOCKCHAIN TECHNOLOGY IN GREEN SMART IOT

Thanks to the innovation of blockchain, it has been used to generate electricity to power IoT devices by a range of industries, including financial institutions, manufacturers, agricultural organizations, and various businesses. These efforts to be environmentally conscious reflect the shift toward technologies that are beneficial to the environment rather than harmful. Mini-power plants are a significant application of blockchain in the deployment of green IoT. Traditional centralized power plants are unnecessary for long-distance power transmission when decentralized power sources can operate independently and be located in various settings. Increased energy efficiency and sustainability are among the benefits of this approach. Numerous businesses have utilized blockchain technology to engage customers and investors in the advancement of the green fuel economy. Investors can participate in environmentally friendly initiatives by the issuance of tokens to qualified investors.

The visual representation in Figure 8.6 demonstrates the contribution of blockchain technology to the creation of a robust green IoT environment. The various ways in which blockchain technology promotes a sustainable and eco-friendly IoT ecosystem are highlighted in this section. Blockchain that drives environmental conservation and energy efficiency will have potential for future generations as technology continues to evolve.

1. **Green computing:** The field of green computing, which is often referred to as environmentally responsible computing, is on the rise. By using resources more effectively, efficiency is enhanced. By outlining the benefits and drawbacks of implementing this method as well as the metrics that businesses may use to gauge their performance, it reduces energy use and lowers environmental dissipation. It improves energy

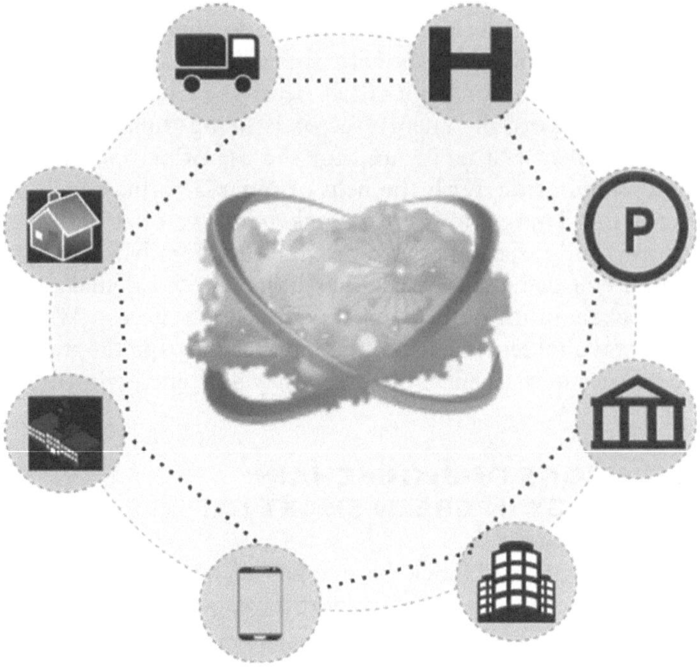

Figure 8.6 Demonstration of blockchain in green IoT applications.

efficiency, organizes information, and gives businesses the essential analytical tools so they can more successfully manage their environmental impact. The concept of "green computing" is crucial because it has the potential to protect the environment and improve the ecosystem's greenness. It is now inevitable that the creation of cloud computing for a future that is environmentally friendly will move in the direction of a successful, safe, open, and affordable resource exchange mechanism. Distributed data storage has grown in popularity.

2. **Green energy:** Renewable sources of energy have recently gained relevance and now need to be taken into consideration due to their significantly less harmful environmental effects when compared to petroleum and coal. Energy is created by utilizing renewable resources like wind, geothermal warmth, sunlight, and rain. The smart grid is the key piece of technology in this scenario for using these widely available and abundant renewable sources of energy. The phrase "smart grid" describes a modernized electrical grid that uses analog or digital communication as well as information technologies. Smart grid safety, connection, data interchange, authorization management, etc., are among a number of significant challenges. Renewable energy sources' inconsistent and unpredictable nature makes their integration

more difficult, which stresses the necessity to incorporate cutting-edge technologies into the current network.

3. **Green finance:** Green finance is able to get around the barriers preventing the shift to a green IoT environment and enhance financial choices because of its unrivaled capacity to make data more easily accessible, quicker, more accurate, and cost-effective. Blockchain technology has an opportunity to reshape electronic government and green finance, bridging the gap between cutting-edge technology and international initiatives to mitigate global warming.
4. **Green smart city:** In order to develop livable urban settings, smart green cities serve as a collaborative hub that links business, academia, government, and the general public. By their very nature, cities are dynamic and complicated. Energy systems are transitioning to a more decentralized paradigm that accepts a variety of competing energy sources and energy storage technologies (ESS).
5. **Green safe transportation:** Truck fleets, traffic issues, and associated operations can be monitored using technology. It aids in enhancing security and transparency, and decreasing potential fraud or threats. Environmental conservation can benefit greatly from the implementation of blockchain technology.

The blockchain can be applied to many social and economic tasks, for example organizing conservation efforts and implementing effective wildlife protection initiatives. It can be utilized in supply chains that prioritize environmental accountability and sustainability. It is also helping to reduce the gas emissions due to greenhouse effects.

8.7 ISSUES AND FUTURE WORK

The green IoT will be very useful in reducing the carbon emissions of electronic devices. However, there are still several issues and challenges in implementing green IoT. There are two major factors that can affect the growth of green IoT. The first issue is the use of sensor networks, and the second is the amount of data generated by them. As sensor networks are in a phase of development, there is a need for more connecting devices. More connecting sensor nodes can consume more energy. This then causes more carbon emissions. Large sensor networks also require more communication devices, which can generate the signals in the environment.

8.8 CONCLUSION

Future green IoT will bring about major improvements that will contribute to a greener world. We witness intelligent communication between sensors, machines, drones, and other devices every day as they work to maintain a green and sustainable environment. Green IoT has a direct impact on the IoT sector in addition to assisting other businesses in environmental protection and sustainable development. The dynamics of building a green, ecological IoT ecosystem will be dramatically changed by blockchain technology's endless application options and its abilities for problem-solving. Researchers are urged to carefully assess how applicable current theories are to the phenomena that are emerging at the intersection of blockchain-based technology and sustainable IoT ecosystems. Research that encompasses this will lay the groundwork for more investigation in this area. In the future, green IoT will be an environmentally friendly design and technology.

REFERENCES

1. Davlyatov, S. (2023). "Smart-City Ecosystem Using Block-Chain Technology," 2023 3rd International Conference on Advance Computing and Innovative Technologies in Engineering (ICACITE), Greater Noida, India, pp. 1077–1080, https://doi.org/10.1109/ICACITE57410.2023.10183248
2. Khan, P. W., Byun, Y. C., & Park, N. (2020). A data verification system for cctv surveillance cameras using blockchain technology in smart cities. *Electronics (Switzerland)*, 9(3). https://doi.org/10.3390/electronics9030484
3. Shahid, A. R., Pissinou, N., Staier, C., & Kwan, R. (2019). Sensor-chain: A lightweight scalable blockchain framework for internet of things. Proceedings - 2019 IEEE International Congress on Cybermatics: 12th IEEE International Conference on Internet of Things, 15th IEEE International Conference on Green Computing and Communications, 12th IEEE International Conference on Cyber, Physical and So, 1154–1161. https://doi.org/10.1109/iThings/GreenCom/CPSCom/SmartData.2019.00195
4. Rifi, N., Agoulmine, N., Taher, N. C., & Rachkidi, E. (2018). Blockchain technology: Is it a good candidate for securing IoT sensitive medical data ? *Wireless Communications and Mobile Computing, 2018*(1), 9763937.
5. Rathee, G., Sharma, A., Saini, H., Kumar, R., & Iqbal, R. (2019). A hybrid framework for multimedia data processing in IoT-healthcare using blockchain technology. *Multimedia Tools and Applications*, *79*(15), 9711–9733.
6. Ul Haque, R., Hasan, A. S. M. T., Jiang, Q., & Qu, Q. (2020). Privacy-preserving k-nearest neighbors training over blockchain-based encrypted health data. *Electronics (Switzerland)*, 9(12), 1–20. https://doi.org/10.3390/electronics9122096
7. Kim, J., McCourt, M., You, T., Kim, S., & Choi, S. (2021). Bayesian optimization with approximate set kernels. *Machine Learning*, September 2020. https://doi.org/10.1007/s10994-021-05949-0
8. Ilesh Dattani. (n.d.). *Blockchain in Healthcare.* In Report by The Institution Engineering and Technology (IET).
9. Anna Roy, Dr. Rajiv Kumar, A. K. (2020). *Blockchain: The India Stratefy.* Draft Discussion Paper Blockchain: The India Strategy.
10. Liang, C., Shanmugam, B., Azam, S., Karim, A., Islam, A., Zamani, M., Kavianpour, S., & Idris, N. B. (2020). Intrusion detection system for the

internet of things based on blockchain and multi-agent systems. *Electronics (Switzerland)*, 9(7), 1–27. https://doi.org/10.3390/electronics9071120

11. Tripathi, R. K. (2020). A Novel Algorithm for Salient Region Detection. In *Communications in Computer and Information Science: Vol. 1241 CCIS*. https://doi.org/10.1007/978-981-15-6318-8_33
12. Gao, Z., Xu, L., Turner, G., Patel, B., Diallo, N., Chen, L., & Shi, W. (2018). Blockchain-based identity management with mobile device blockchain-based identity management with mobile device. September. https://doi.org/10.1145/3211933.3211945
13. Patil, A. S., Tama, B. A., Park, Y., & Rhee, K. (2018). *A framework for blockchain based secure smart green*. Springer Nature Singapore Pte Ltd. [J. J. Park et Al. (Eds.), Advances in Computer Science and Ubiquitous Computing. *Lecture Notes in Electrical Engineering*, 1162–1167. https://doi.org/10.1007/978-981-10-7605-3_185]
14. Partida, A., Criado, R., & Romance, M. (2021). Identity and access management resilience against intentional risk for blockchain-based IOT platforms. 1–26. www.mdpi.com/journal/electronics
15. Oktian, Y. E., Lee, S. G., & Lee, H. J. (2020). Hierarchical multi-blockchain architecture for scalable internet of things environment. *Electronics (Switzerland)*, 9(6), 1–27. https://doi.org/10.3390/electronics9061050
16. Atlam, H. F., Azad, M. A., Alzahrani, A. G., & Wills, G. (2020). A review of blockchain in internet of things and AI. *Big Data and Cognitive Computing*, 4(4), 28.
17. Khezr, S., Moniruzzaman, M., Yassine, A., & Benlamri, R. (2019). Blockchain technology in healthcare: A comprehensive review and directions for future research. *Applied Sciences (Switzerland)*, 9(9), 1–28. https://doi.org/10.3390/app9091736
18. Zhang, J., & Wu, M. (2020). Blockchain use in iot for privacy-preserving anti-pandemic home quarantine. *Electronics (Switzerland)*, 9(10), 1–16. https://doi.org/10.3390/electronics9101746
19. Byabazaire, J., O'hare, G., & Delaney, D. (2020). Data quality and trust: Review of challenges and opportunities for data sharing in IoT. *Electronics (Switzerland)*, 9(12), 1–22. https://doi.org/10.3390/electronics9122083
20. Alam, T. (2020). mHealth Communication Framework using Blockchain and IoT Technologies. July. https://doi.org/10.20944/preprints202006.0180.v1
21. Attia, O., & Adjih, C. (2019). An IoT-Blockchain architecture based on hyperledger framework for healthcare monitoring application. 2019 10th IFIP International Conference on New Technologies, Mobility and Security (NTMS), 1–5.
22. Satamraju, K. P., & Malarkodi, B. (2019). A secured and authenticated internet of things model using blockchain architecture. Proceedings of the 2019 TEQIP - III Sponsored International Conference on Microwave Integrated Circuits, Photonics and Wireless Networks, IMICPW 2019, 19–23. https://doi.org/10.1109/IMICPW.2019.8933275
23. Varghese, C., & Jose, J. (2019). IoT device management using blockchain. *International Journal of Science, Engineering and Technology Research (IJSETR)*, 8(3), 79–84.
24. Nyamtiga, B. W., Sicato, J. C. S., Rathore, S., Sung, Y., & Park, J. H. (2019). Blockchain-based secure storage management with edge computing

for IoT. *Electronics (Switzerland)*, 8(8), 1–22. https://doi.org/10.3390/electronics8080828

25. Noh, J., Jeon, S., & Cho, S. (2020). Distributed blockchain-based message authentication scheme for connected vehicles. *Electronics (Switzerland)*, 9(1). https://doi.org/10.3390/electronics9010074
26. Tan, H., Kim, P., & Chung, I. (2020). Practical homomorphic authentication in cloud-assisted vanets with blockchain-based healthcare monitoring for pandemic control. *Electronics (Switzerland)*, 9(10), 1–21. https://doi.org/10.3390/electronics9101683
27. Zeba, S., & Amjad, M. (2020). Sustainable and effective blockchain-based solutions for the security and privacy of healthcare data. In *Blockchain for Healthcare Systems*, pp. 159–176. CRC Press.
28. Sana Zeba, Shameemul Haque, M. A. H. (2021). *Revolution in IoT with Blockchain Technology*. Lambert Academic Publishing, Book (ISBN-13: 978-620-3-91142-8), 1–137.
29. Zeba, S., & Amjad, M. (2022). *Sustainable and Secure IoT-Chain Architecture Integrated with Blockchain Technology BT - Proceedings of International Conference on Intelligent Cyber-Physical Systems* (B. Agarwal, A. Rahman, S. Patnaik, & R. C. Poonia (eds.); pp. 51–64). Springer Nature Singapore.

Chapter 9

Leveraging wireless technology and IoT in developing a smart judiciary system with smart dust sensors

Kanishka Vaish, Mousam Sharma, Samta Kathuria, Shweta Pandey, Rajesh Singh, Anita Gehlot, Shaik Vaseem Akram, and Praveen Kumar Malik

9.1 INTRODUCTION

The IoT is at the vanguard of a technological revolution that will change the way we interact with our environment. Its influence goes well beyond mere convenience, ushering in a fundamental revolution in how we live, work, and interact with technology. At its core, the IoT is a network of networked devices, each equipped with sensors, processors, and communication modules to gather and share data easily over the internet. While cell phones, laptop computers, and household appliances have become widespread, they still mostly function as stand-alone devices, generally with rudimentary Wi-Fi or Bluetooth connectivity. These devices have specialized functions but often operate alone, unable to collect and send data in real time or communicate with other systems. IoT devices, on the other hand, are designed to actively acquire and send data, permitting connection not only between devices but also with larger linked systems.

The specialized architecture of IoT devices allows them to gather data from their surroundings and send it over networks, opening a world of possibilities for real-time analysis and application. These devices, which are equipped with sensors, collect information about their surroundings, such as temperature, movement, light, or other factors, and process it inside their internal systems. Following this, the processed data is sent across the internet, encouraging interconnection, and opening the way for a plethora of applications.

Consider smart home devices as an example of a practical application of IoT. Smart thermostats, which are coupled with sensors and internet connectivity, continually collect data about the temperature and occupancy patterns of a home. This data is then communicated and analyzed in real time, allowing the thermostat to automatically modify settings for maximum comfort and energy savings. In industrial settings, IoT-enabled

DOI: 10.1201/9781003541363-9

machinery may identify performance irregularities, allowing for predictive maintenance and reducing downtime, boosting productivity and cost-effectiveness. The potential of IoT goes beyond simply convenience; it has far-reaching consequences for many aspects of our lives. Wearable devices with sensors can monitor vital signs in healthcare, allowing for remote patient monitoring and prompt treatments. IoT-enabled infrastructure in urban planning may optimize traffic flow and energy usage, resulting in more sustainable and efficient cities. Furthermore, the quantity of data provided by IoT devices provides chances for in-depth analysis, resulting in informed decision-making and innovation. Businesses may get useful insights, improve operational efficiency, and customize services to changing customer requirements by leveraging this data.

However, as IoT becomes more prevalent in everyday life, questions regarding data privacy, security, and ethical implications have surfaced. To prevent possible hazards connected with IoT growth, sensitive information must be safeguarded and comprehensive security measures implemented. In essence, the Internet of Things is a paradigm shift that combines technology and connection to boost productivity, optimize resource utilization, and improve quality of life across several areas. Its incorporation into our daily lives has the potential to open up new opportunities and shape a more linked and efficient society [1].

The IoT network of physical objects like cars, appliances, and other household products allows devices to connect and share data because they are equipped with software, sensors, electronics, and connectivity. It alludes to the notion of utilizing the internet to link to any device having an on/off switch [2].

IoT devices can be put strategically in areas with elevated crime to monitor activities and give law enforcement access to real-time data. This can serve as a deterrent to crime and as proof in the event of a crime. For instance, smart cameras with software that recognizes faces can be used to identify suspects and follow their whereabouts, making it simpler for law enforcement to capture them [3].

Small, wireless sensors employed in smart dust technology can be utilized to track criminal activities in real time and deliver an early warning about possible hazards. To detect anomalous behavior and warn law enforcement, for instance, smart dust detectors could be installed in high-crime areas. Law enforcement would then be able to react to the situation promptly and effectively.

Due to its distinct characteristics, including its small size, cost-effectiveness, and covert nature, smart dust is evolving and its possible applications are receiving a lot of attention. Robotics, environmental sensing, local and national security, military surveillance, healthcare, farming, transportation, management of parking, research into biology, planetary exploration, machinery monitoring, and geographic data collection are just a few of the industries that smart dust has the potential to revolutionize. These characteristics make the smart dust an extremely flexible technology with a variety of impacts [4].

IoT and smart dust technologies may be able to make a major difference in reducing crime, but it's important to consider any potential ethical and privacy concerns [5]. When considering whether to implement such technology, the right to privacy must be addressed, and it is critical to ensure that the data acquired is only utilized for authorized law enforcement objectives. IoT and smart dust technologies can be valuable tools in the fight against crime, but their usage must be properly scrutinized and overseen to ensure that they are utilized ethically and morally. The following are the study's primary objectives:

1. To look at the viability of developing a smart judicial system using wireless and IoT technology and intelligent dust sensors for better evidence collecting.
2. To understand the potential and constraints of smart dust sensors in terms of collecting and communicating environmental facts. The objective is to remove obstacles to IoT and smart dust implementation like privacy worries, challenging terms in technology, and operating in sensitive regions like high-crime areas.

We discuss the many enabling technologies necessary for the development of a smart court system in Section 9.3 of this chapter. The relationship between these technologies and the development of an intelligent judicial system will be thoroughly discussed in this section. The use of wireless technologies and IoT in the judicial system is further described in Section 9.7. The specific function of smart dust sensors in a smart judicial system is explored in Section 9.6 in more detail. It will examine how these sensors can be used to track various facets of the legal system in real time, such as court hearings, and offer insightful data. In Section 9.9, the difficulties encountered in implementing this technology and solutions to them are covered. Section 9.8 discusses ideas and suggestions for future study in a smart court system utilizing wireless technology and IoT with smart dust sensors. The study is brought to a close in Section 9.10 by stressing the potential effects of a smart court system that uses wireless technology, the Internet of Things, and intelligent dust sensors on the overall efficacy and efficiency of the legal system.

9.2 RESEARCH METHODOLOGY

By utilizing smart dust as a method to identify significant topics, trends, and gaps in the existing information, the authors have undertaken non-empirical research with the purpose of exploring the corpus of literature on wireless technology and IoT in developing smart judicial. In order to create a smart court system, the authors will address study topics that focus on identifying the potential benefits and drawbacks of integrating wireless technology plus IoT with smart dust sensors. This chapter's goal is to shed light on how these technologies can be used effectively to enhance the speed and accuracy of the legal system.

Using keywords like smart dust, IoT, smart judiciary, wireless technology, and many more, a thorough search strategy was designed to find pertinent research articles on this subject. In order to identify the uses of wireless technology and IoT to create a smart judiciary system, which aims to facilitate the collection of evidence, provide assistance, detect crime, and carry out other related functions, researchers searched academic databases like PubMed, JSTOR, and Google Scholar as well as manually searching the reference lists of pertinent articles. All articles found using the search method had their titles and abstracts checked to see if they were pertinent to the study issue. The data from each study was extracted and synthesized to identify key themes and trends. Based on the results of the literature review, conclusions are drawn and recommendations are made for future research. This will include suggestions for areas where further research is needed and recommendations for the design and conduct of future studies on the topic.

9.3 ENABLING TECHNOLOGY – IOT AND ITS BENEFITS

9.3.1 IoT

The combination of hardware, software, and network connectivity comprise IoT technology. These elements work together to give IoT devices the ability to gather and communicate data, make decisions using that data, and collaborate with other devices and systems to accomplish a particular task. The IoT encompasses a symbiotic fusion of hardware, software, and network infrastructure. This convergence enables IoT devices to collect, transmit, and process data, fostering autonomous decision-making based on gathered information. Furthermore, IoT facilitates interoperability among devices, enabling seamless collaboration to execute tasks and enhancing efficiency across various domains like healthcare, agriculture, and smart cities.

9.3.2 Sensors and actuators

These are the actual hardware devices that are affixed to things, equipment, or products and gather information about their surroundings or carry out specified tasks. A refrigerator's temperature sensor or an actuator that opens and closes the door are two examples of such devices. Sensors and actuators form the foundational hardware of IoT, capturing data and performing actions. They encompass various types such as temperature, motion, and proximity sensors, as well as actuators like motors or valves. Evolving technologies include smart sensors equipped with AI capabilities, enhancing data accuracy, while actuators enable remote control for automated tasks in diverse IoT applications.

9.3.3 Connectivity

To share data and communicate with other IoT devices, devices must be connected to a network. Wi-Fi, Bluetooth, Zigbee, and cellular networks (such as 4G or 5G) are the most popular connectivity alternatives [6]. Connectivity is crucial for IoT devices to exchange data and communicate effectively. Apart from Wi-Fi, Bluetooth, Zigbee, and cellular networks (4G/5G), emerging technologies like LPWAN (low power wide area network) enable long-range communication with minimal power consumption, ideal for applications requiring low data rates and extended device battery life in remote or challenging environments.

9.3.4 Hardware

In order to complete the necessary functions, IoT devices need specific hardware. This could include memory, RF modules, power management chips, and microcontrollers, among other components [7]. IoT devices rely on a tailored amalgamation of hardware components to execute their multifaceted functions effectively. These devices necessitate diverse hardware elements such as memory for data storage, RF (radio frequency) modules for wireless communication, power management chips for energy optimization, and microcontrollers to process and control data flow, ensuring seamless operations and connectivity.

9.3.5 Cloud computing

IoT devices generate a lot of data, which is frequently stored and analyzed in the cloud. The massive amounts of data generated by IoT devices may be managed by platforms for cloud computing because they provide scalable and adaptable computing resources [8]. Cloud computing plays a pivotal role in managing the substantial volume of data generated by IoT devices. These platforms offer scalable and flexible computing resources, enabling efficient storage, processing, and analysis of vast datasets. Leveraging cloud infrastructure allows seamless handling of IoT-generated data, ensuring adaptability and scalability to meet evolving demands.

9.3.6 Data analytics

IoT data is examined to draw forth valuable facts and insights. Advanced algorithms like machine learning, deep learning, and natural language processing are used for this [9]. In IoT ecosystems, data analytics involves scrutinizing the vast array of information generated by connected devices to extract valuable insights. Utilizing sophisticated algorithms like machine learning, deep learning, and natural language processing, data analysis identifies patterns, trends, and anomalies. These advanced techniques empower businesses to derive actionable intelligence, make informed decisions, predict future outcomes, and optimize processes within various sectors, enhancing efficiency and innovation.

9.3.7 Application development

Software apps are created to communicate with Internet of Things (IoT) devices and to give users a way to monitor and operate their own devices. Numerous platforms, including mobile devices, computers, and web browsers, are compatible with these programs [10]. Application development for IoT involves crafting software that interfaces with IoT devices, enabling users to monitor and control devices seamlessly across multiple platforms.

9.3.8 Security control devices

IoT security is a significant issue since connected devices are vulnerable to hackers. To ensure the security of IoT systems, several security measures, such as encryption, authentication, and access control, have been implemented [11]. Many security control devices can be used to protect IoT systems. Examples include firewalls, IDS/IPS, antivirus software, and security information and event management (SIEM) solutions. Figure 9.1 indicates how to search through a vast volume of legal papers using smart dust sensors to discover relevant documents and analyze them.

Figure 9.1 illustrates the process of user-initiated legal file search utilizing smart dust sensors. The user triggers the search through an interface connected to smart dust sensors dispersed in the environment. These sensors detect, gather, and transmit data related to file locations. The information is relayed to a centralized system, which processes the data and retrieves the requested legal files. The smart dust sensors act as discreet data collectors, enabling efficient and covert search operations, and enhancing the accessibility and retrieval of legal documents for the user. In the depicted scenario, the smart dust sensors are strategically deployed within a legal environment, such as a law office or a legal department within an organization. These sensors are tiny, wireless devices equipped with various sensors like RFID (radio-frequency identification), motion detectors, and perhaps even cameras or environmental sensors.

When a user initiates a search for specific legal files or documents, they interact with a user interface, possibly a computer application or a dedicated platform. This interface communicates with the smart dust network, triggering the sensors to start scanning their surroundings for the requested

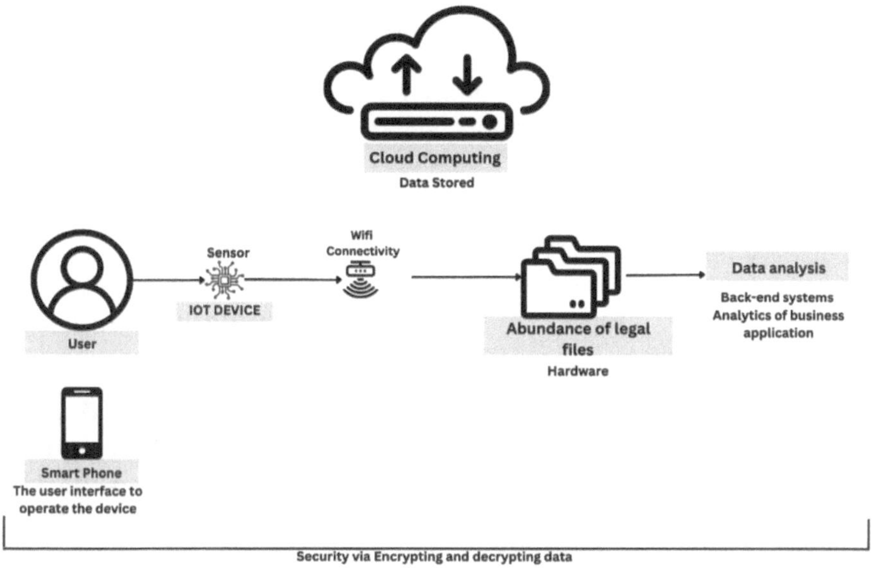

Figure 9.1 The process of searching legal files by the user using smart dust sensors.

files. The sensors' RFID capabilities might detect tags or signals embedded in physical files or document storage units, providing location data. Simultaneously, motion detectors and environmental sensors may contribute to refining the search, detecting movement or changes in the environment that could indicate file usage or recent access.

Once the smart dust sensors collect relevant data, they transmit this information wirelessly to a central processing system or a cloud-based server. This centralized system employs algorithms and data processing techniques to analyze the incoming data, interpret the location clues provided by the sensors, and identify the precise whereabouts of the requested legal files. Finally, the system retrieves the identified files or documents from their location within the legal environment, allowing the user to access them through the same interface they used to initiate the search. This innovative use of smart dust sensors streamlines and enhances the process of locating specific legal materials, offering efficient and discreet file retrieval within a legal workspace.

9.4 CONCEPT OF SMART DUST

The ideology behind smart dust originated in the 1990s. The concept of smart dust was developed in 1998 by Dr. Kris Pister of the University of California, Berkeley [12]. One of the pioneers in developing smart dust technology is Dr. Pister, a well-known electrical engineer and computer scientist. In order to acquire data and send it back to a hub, he recommended using small, wireless sensor units that could be spread across an environment. This proposal can be recognized as the foundation stone for the development of smart dust technology and generated a significant deal of interest and research in the field. A number of tiny MEMS sensors known as "smart dust" may communicate with other MEMS sensors as well as with a centralized system. The term "smart dust" refers to these tiny, wireless, battery-free sensors that operate on an external power source, such as radio waves or solar energy. The sensor is powered by an external power source, which also transmits data to a central point for analysis. Because of its compact size and lack of batteries, smart dust is highly energy-efficient when powered externally.

Smart dust can sense, compute, and communicate despite being so small that it could be eaten or inhaled [13]. Each smart dust "mote," or sensor, consists of a power source, a radio transmitter, sensors, and a CPU. The sensors may gather data on elements such as temperature, light, pressure, and acceleration before initiating the transfer of information to other mops or a centralized system. Smart dust has several applications including environmental monitoring, industrial process control, surveillance, and safety [14]. Since they are minuscule in size and have low power consumption, smart dust motes can be utilized in large quantities and put in a vast system of sensors that can cover a large area. Smart dust has a wide range of possible applications, from tracking animal migration as part of conservation efforts to tracking crop health and soil moisture in agriculture.

Component miniaturization, reduced power consumption, and data protection are only a few of the issues that must be resolved because the technology is still in its infancy [15]. Smart dust's use in the legal system would, however, bring up a number of ethical and privacy issues, including the potential for surveillance and inappropriate use of the data collected, the veracity and accuracy of the data, and the requirement for adequate security measures to prevent unauthorized access to the data [16]. In order to protect people's rights and privacy, any application of smart dust in the judicial system requires diligent consideration to analyze the merits and demerits.

9.5 CASE STUDY: THE NEED FOR A SMART JUDICIARY SYSTEM

A functioning judicial system is required in every democratic society. Justice means treating everyone fairly and evenly, regardless of social or economic background. Because of the rapid expansion in population, there is a backlog of court cases, as well as an increase in the number of legal concerns. The backlog of cases not only makes it difficult to deliver justice, but it also has a substantial impact on the lives of people affected. An intelligent legal system might be beneficial in this situation.

9.5.1 Crime scene investigation

Crime scenes can have smart dust sensors installed to collect information about humidity, temperature, and air quality. The crime scene can be recreated using this data, and suspects could be found. By giving investigators precise and timely data, smart dust sensors might save time and resources. Crime scene investigation smart dust sensors offer a compact, high-tech solution for forensic analysis. These miniature sensors, dispersed over crime scenes, collect diverse data on environmental factors, DNA traces, chemical residues, and movement patterns. Through real-time monitoring and data collation, they provide crucial insights for investigators, aiding in reconstructing events and identifying key evidence. These sensors facilitate rapid and precise analysis, enhancing forensic capabilities by offering a comprehensive understanding of crime scenes. Their use significantly expedites investigations, enabling law enforcement to gather critical information swiftly, contributing to more efficient and thorough crime scene examinations.

9.5.2 Courtroom monitoring

Courtrooms can be equipped with intelligent dust sensors to keep an eye on the judges, attorneys, and witnesses. This information can be utilized to make sure that everyone follows the rules and that the procedures are reasonable and fair. In addition to providing data that could be used to discover trends or patterns in courtroom conduct, smart dust sensors might additionally offer information that could be utilized to enhance the legal system as a whole. Courtroom monitoring smart dust sensors offer discreet and sophisticated surveillance within legal proceedings. Strategically placed throughout courtrooms, these compact sensors capture audio, visual, and environmental data. They ensure transparency by monitoring proceedings, verifying the integrity of testimonies, and detecting any anomalies or disruptions during trials. These sensors enhance courtroom security, enabling real-time monitoring of interactions and maintaining an accurate record of events. Their unobtrusive nature aids in preserving the sanctity of legal procedures while providing invaluable data for analysis, ensuring fairness and accountability within the judicial process.

9.5.3 Effective case management

The traditional legal system prioritizes paperwork heavily, which causes case processing to lag. A smart judicial system can assist in the efficient handling of cases by digitizing each step of the process. Digital technologies like e-filing, online case management, and conference calls can reduce the backlog of cases by saving time and effort. Effective case management smart dust sensors offer a dynamic tool for optimizing legal proceedings. Integrated into case management systems, these sensors track document movements, monitor case file access, and analyze workflow patterns. By collecting real-time data on file interactions, resource allocation, and time management, they streamline case progression. These sensors facilitate efficient allocation of resources, identify potential bottlenecks, and enhance task prioritization. Their data-driven insights enable legal professionals to make informed decisions, expedite processes, and ensure timely case resolutions. Ultimately, they bolster the efficiency of case management systems, enabling a more organized and responsive approach to handling legal matters.

9.5.4 Intelligent data analysis

A smart judiciary system can help judges make well-informed judgments by providing them with intelligent data analysis. Using machine learning and artificial intelligence algorithms, the system can assess previous cases and identify trends and patterns that could be used to make better decisions. Intelligent data analysis empowered by smart dust sensors revolutionizes the legal landscape. These sensors collect intricate data points from diverse sources, facilitating comprehensive analysis through advanced algorithms and AI [17]. They decipher complex patterns, trends, and correlations within legal databases, offering predictive insights and informed decision-making for legal professionals. By processing large volumes of information swiftly, they aid in identifying crucial evidence, potential case outcomes, and precedents, streamlining legal research and case strategy formulation. These sensors serve as a cornerstone for legal analytics, empowering the legal community with invaluable tools to navigate complexities and derive actionable intelligence from vast data repositories, ensuring a more strategic and informed approach to legal proceedings.

9.5.5 Simplified communication

Communication is a crucial aspect of the legal system. The outdated method, on the other hand, relies on manual intervention, which might result in errors and delays. Communication between the court, attorneys, and litigants, as well as between all other parties engaged in a case, can be accelerated and improved by an intelligent judicial system. Smart dust sensors enable simplified communication within legal frameworks by facilitating seamless connectivity among stakeholders. These compact sensors, integrated into communication systems, offer real-time data transmission and reception, ensuring instant and secure exchanges between legal professionals, clients, and judicial authorities. They streamline communication channels, providing efficient sharing of critical information, case updates, and documentation. By bridging geographical barriers and enhancing accessibility, these sensors promote collaborative efforts, enabling swift decision-making and timely responses. Their role in fostering efficient and secure communication networks contributes to a more cohesive and responsive legal ecosystem, optimizing interactions and facilitating smoother proceedings within the legal domain.

9.5.6 Greater accessibility

The traditional court system can be frightening and intimidating to the average member of the public, preventing them from seeking justice. A savvy court system can close this gap by providing easy access to legal aid, legal materials, and online conflict resolution methods. People might be able to receive justice in this way without having to invest time and money in protracted court disputes. Smart dust sensors contribute to greater accessibility within the legal sphere by enabling inclusive and remote access to legal resources and services. These sensors, when integrated into legal systems, offer enhanced connectivity and data retrieval options. They facilitate remote access to legal databases, case files, and resources, ensuring lawyers, clients, and stakeholders can retrieve information irrespective of their physical location. This accessibility promotes inclusivity, enabling individuals from diverse backgrounds and regions to engage with legal services more easily. By breaking down geographical barriers, smart dust sensors foster a more accessible legal environment, empowering a broader spectrum of individuals to participate in legal processes and benefit from legal resources and expertise.

The world right now requires a complex legal framework. With the use of digital technologies, the judicial system could be streamlined and made more effective, with pending cases reduced, and with a better evidence-gathering system. It could also serve to bridge the gap between the legal system and the general public, ensuring that everyone has a right to justice. The justice system could change if smart dust is used to gather evidence, as it will be more accurate, powerful, and effective. Data could be gathered utilizing smart dust sensors at criminal scenes, courtrooms, and accident scenes to make certain every person involved receives justice.

9.6 THE ROLE OF SMART DUST SENSORS IN A SMART JUDICIARY SYSTEM

An intelligent judicial system could use smart dust to gather information and provide quick knowledge about various aspects of the legal system, such as courtroom conditions, the presence and movement of witnesses, or even the feelings and reactions of those present. By collecting and analyzing this data, a smart judicial system may improve the efficiency and openness of the legal system, provide new views on decision-making, and assist the fair and equitable administration of justice [18].

As seen in Figure 9.2, the purpose of this chapter is to go into further detail on how smart dust technology might be used to identify crime in society. Law enforcement organizations may be able to track person movement and spot suspect patterns of behavior by placing smart dust sensors around an urban area. This necessitates the utilization of smart dust technology for crime identification within society. By strategically situating these miniature sensors across urban areas, law enforcement gains the ability to monitor and analyze person-to-person movement and behavior patterns. These sensors, equipped with various detection capabilities, including motion, sound, or even chemical sensors, discreetly gather data on activities and behavioral patterns. Analyzing this data enables law enforcement agencies to identify suspicious behavior, track suspect movements, and potentially predict or prevent criminal activities, bolstering community safety and crime prevention efforts.

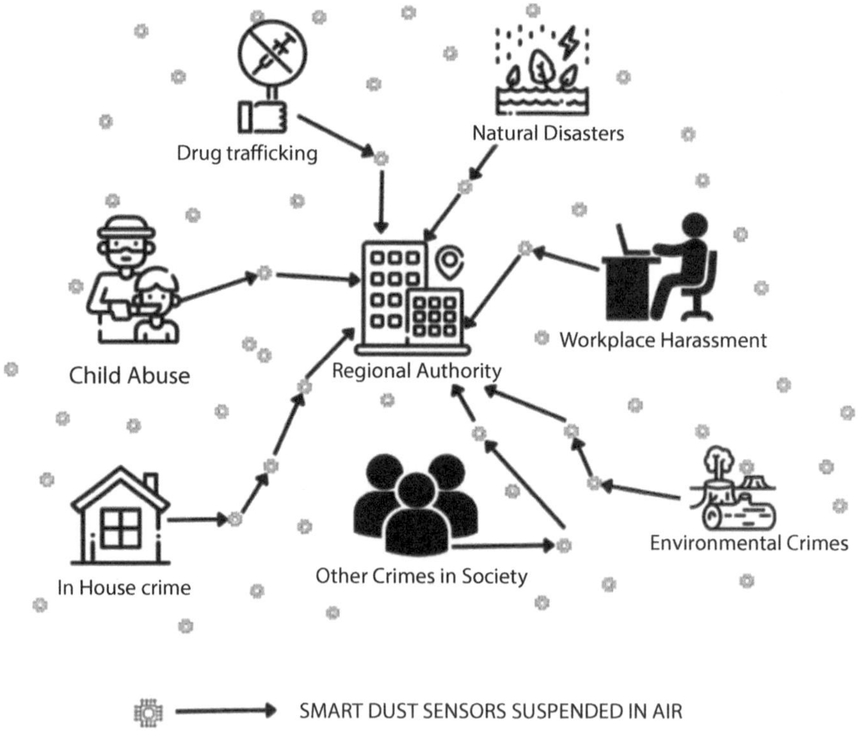

Figure 9.2 Use of smart dust sensors to detect crime in society.

The notion of smart dust, which consists of small, wireless sensors with impressive capabilities, has the potential to greatly transform crime detection and prevention, especially in locations where it is difficult to properly identify or monitor criminal activity. These tiny sensors, which are often smaller than a grain of sand, possess diverse capabilities such as sensing, computation, and wireless communication. This allows them to collect and send a broad range of data. Smart dust is an essential instrument in the field of crime prevention because it may silently gather data in situations where conventional surveillance techniques may be restricted or unfeasible. These sensors may be strategically positioned in metropolitan locations, isolated places, or even within structures to observe and record vital data about suspicious actions or possible threats [19].

Smart dust, which is equipped with a variety of sensors including motion detectors, cameras, microphones, and chemical sensors, could identify abnormalities or trends that may indicate illegal behavior. For example, in metropolitan environments, these sensors could identify abrupt movements or unusual crowd behavior that might indicate possible criminal actions such as vandalism, theft, or violence. In regions that are far or have limited surveillance, smart dust has the capability to identify atypical changes in the environment or unauthorized entry into restricted areas, therefore initiating prompt warnings or notifications to law enforcement or security officials.

The inconspicuous characteristics of smart dust provide a notable benefit in clandestine surveillance, allowing law enforcement authorities to collect crucial information without arousing suspicion. These sensors may be discreetly distributed, incorporated into the surroundings, or hidden inside ordinary items, enabling effortless data gathering without sacrificing their visibility.

Moreover, the capacity of smart dust to transmit data in real time allows for swift reaction and intervention. Smart dust may expedite decision-making by consistently transmitting data to centralized systems or authorities, enabling law enforcement to take preemptive measures in addressing possible threats or illegal actions before they escalate. Nevertheless, it is essential to meticulously tackle the ethical implications and privacy issues that arise from the extensive use of smart dust in monitoring and data gathering. To prioritize public safety while respecting people's private rights, it is crucial to implement strong privacy protections, use data encryption, and enforce stringent legislation limiting the usage of smart dust.

To summarize, the use of smart dust offers a great opportunity to improve crime detection and prevention in difficult or inaccessible areas. Smart dust, with its unobtrusive nature, powerful sensor capabilities, and real-time data transfer, has the potential to greatly enhance security measures and assist law enforcement agencies in efficiently combating crime. Nevertheless, ensuring responsible and ethical execution is vital to strike a balance between the advantages of new technology and the concerns around privacy and social ramifications [20]. These devices can be used to monitor illicit activities like trafficking in drugs, theft, and homicide in high-crime regions [21]. Law enforcement organizations can use the data collected by these devices to create strategies for lowering crime and enhancing public safety.

9.7 OVERVIEW OF WIRELESS TECHNOLOGY AND IOT IN THE JUDICIARY SYSTEM

The Internet of Things and wireless technology have changed several businesses, including the legal sector. One of the main applications of wireless technology in the judicial system is remote communication between attorneys, judges, and other legal professionals [22]. Through the provision of precise and current information on their concentrations and toxicity, technology can help identify the causes of pollution and the individuals responsible. To make the polluter comply with environmental standards, this can require imposing a fine or pursuing legal action. Zigbee is a wireless communication protocol with a low data rate and power consumption. Medical equipment, smart home devices, and industrial measurement and control systems are just a few of the devices and systems that use it [23].

The integration of wireless technology and the Internet of Things (IoT) has significantly impacted various sectors, including the legal domain, by revolutionizing communication and data management. Remote communication among legal professionals, facilitated by wireless connectivity, has enhanced collaboration, enabling attorneys, judges, and other stakeholders to interact seamlessly regardless of their physical locations.

Moreover, wireless technology plays a pivotal role in addressing environmental concerns related to pollution. By providing real-time data on the concentration and toxicity of pollutants, technological advancements aid in identifying sources of pollution and holding responsible individuals or entities accountable. This involves leveraging accurate information to enforce environmental standards through legal means, such as imposing fines or initiating legal actions against polluters.

Among the prominent wireless communication protocols, Zigbee stands out due to its low data rate and energy-efficient characteristics, finding applications in diverse sectors like medical equipment, smart home systems, and industrial control systems. However, Zigbee networks are not immune to security vulnerabilities. The susceptibility to threats, particularly the risk of a man-in-the-middle (MITM) attack, poses a significant concern. In an MITM attack, hackers intercept and manipulate communication between devices, potentially gaining unauthorized access to sensitive data or seizing control of targeted systems.

Preventing such security breaches requires robust protective measures to safeguard Zigbee networks. Implementing encryption, authentication mechanisms, regular security updates, and stringent access controls are essential steps to mitigate the risks associated with these attacks. Looking ahead, the continuous evolution of IoT and wireless technology is poised to further transform the legal landscape, offering increased efficiency, effectiveness, and security. As these technologies advance, they are likely to play an even more substantial role in reshaping how legal professionals operate and interact, ensuring a more secure and interconnected judicial system.

However, Zigbee networks are susceptible to security threats and attacks just like any other technology. A MITM attack is one of the deadliest attacks that can happen on a Zigbee network. An MITM attack entails a hacker intercepting and altering the traffic between two networked devices to get access to confidential data or seize control of the targets. Make sure Zigbee networks are well protected if you want to stop this kind of attack. Overall, wireless technology and IoT are helping to transform the judiciary system, making it more efficient, effective, and secure. As these technologies continue to evolve, they will likely play an even larger role in the legal profession in the years to come.

9.8 POTENTIAL BENEFITS OF SMART JUDICIARY SYSTEM

A smart judiciary system offers numerous potential benefits, including streamlined processes through digitization, enhancing accessibility to legal services for citizens regardless of location, and facilitating remote hearings for increased efficiency. Utilizing data analytics can expedite case management, improve decision-making, and reduce backlog. Smart technologies enable secure storage and retrieval of legal documents, ensuring data integrity and confidentiality. Automation of routine tasks minimizes human errors and frees up time for legal professionals to focus on complex issues. Overall, a smart judiciary system promises increased transparency, accessibility, and efficiency, ultimately advancing fairness and effectiveness in delivering justice.

The study of employing smart dust to recognize criminals has drawn more attention in recent years. Smart dust detectors have the capacity to offer an unprecedented amount of data and accuracy in detecting and identifying criminal conduct sed to monitor high-crime areas, identify unusual or suspicious conduct, or track the movements of people who are suspected of being involved in unlawful conduct. The sensors can additionally be used to gather data on ambient variables, including temperatures, light, and sound levels, in order to further aid in an inquiry into illegal acts [24].

Figure 9.3 serves as an example of how the smart judicial system can make use of smart dust and IoT. The process of a crime being committed,

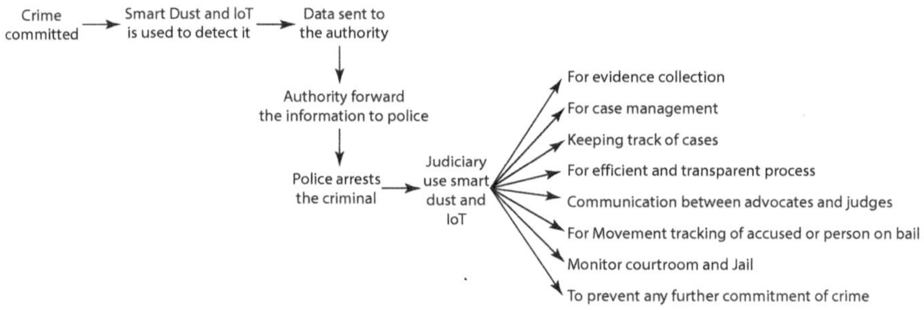

Figure 9.3 Uses of smart dust and IoT by smart judiciary system in various sectors.

information being sent to the authorities, and the authorities being made aware of the crime are all depicted in the diagram. This also emphasizes the numerous uses of smart dust in the judicial system, such as acquiring evidence, managing cases, tracking cases, ensuring efficient and open procedures, keeping track of the whereabouts of suspects, keeping an eye on courts and jails, and stopping crime.

Additionally, by giving law enforcement agencies real-time data and alarms, smart dust sensors might be utilized to improve already-existing criminal detection structures, such as security cameras. With the aid of these technologies, parents and other responsible adults can monitor and track children's whereabouts and activities in order to protect them. They can also be used to spot and stop cyberbullying, online abuse, and other forms of harassment [25]. The implementation of a smart judiciary system promises a myriad of potential benefits that could profoundly reshape the legal landscape. Through the infusion of cutting-edge technology, such a system would revolutionize traditional legal processes, leading to enhanced efficiency, accessibility, and transparency within the judicial framework. Automating routine tasks and digitizing documentation processes significantly expedites case proceedings, document management, and administrative operations, thereby streamlining the overall workflow. Additionally, the system's ability to facilitate remote access to legal services and resources ensures inclusivity, enabling individuals from diverse locations to engage with the legal system. Leveraging data analytics, this innovative framework enables data-driven decision-making, empowering legal professionals with insights derived from extensive datasets for more informed judgments. Furthermore, the system's real-time monitoring and data collection mechanisms promote transparency, ensuring accurate record-keeping and fostering trust in the judiciary. Ultimately, these advancements not only offer the potential to reduce backlog cases but also bolster security measures to safeguard sensitive legal information, ensuring confidentiality and integrity within the legal domain.

9.9 CHALLENGES AND SOLUTIONS IN ADOPTING TECHNOLOGIES IN THE JUDICIARY SYSTEM

A fast-expanding topic with many potential advantages is the incorporation of IoT and smart dust technology into many systems and industries. Before widespread adoption can take place, a number of problems and challenges must be resolved. Challenges in adopting technologies in the judiciary include data security risks and resistance to change. Solutions entail robust cybersecurity measures, comprehensive training for stakeholders, and a phased implementation strategy to overcome resistance, ensuring a seamless integration of technology while safeguarding sensitive information. These difficulties include, among others:

9.9.1 Data security and privacy

Data security and privacy are issues that are brought up by the growing deployment of IoT and smart home technology. The vast amounts of data that these devices gather raise the possibility of unauthorized access, theft, and misuse of personal information [26]. These devices are made to gather an immense amount of information about us and our surroundings, information that can be utilized in a variety of different ways to make our lives better. However, as IoT and smart home devices proliferate, serious worries regarding the security and privacy of personal information have surfaced. This information could always be obtained by unauthorized people or groups.

9.9.2 Interoperability

Technology can be used to improve the efficiency and precision of the judicial system, from case administration to the collection and processing of evidence. The integration of IoT and smart dust devices into different systems and sectors requires compatibility and interoperability across various devices, networks, and platforms. Without this interoperability, the potential benefits of these technologies cannot be fully realized [27]. For example, IoT devices can be used to track and monitor offenders who are placed under home arrest, eliminating the need for probation officers to visit the offender in person. Smart dust monitors can be used to check the cleanliness and usability of courtrooms and other spaces. In order to help judges and attorneys make better decisions, these tools can also be used to acquire and analyze data regarding court procedures. To be effective in the judicial system, these instruments must be interoperable. For real-time data and information transfers between different devices and systems, smooth communication between them is required.

9.9.3 Cost

The cost of establishing and maintaining large IoT and smart grid networks could be a significant barrier to their widespread implementation and utilization from a legal perspective. Especially in underdeveloped countries, the court system typically has a limited budget, making it challenging to invest in innovative technologies. In order to integrate IoT and smart dust systems in tribunals, prisons, and other justice facilities, it may be costly to buy the necessary infrastructure, hardware, and software. Additionally, regular upkeep and updates are required to keep these systems functional and current. Another factor to take into account is the cost of educating staff and workers on how to use this novel piece of equipment. These technologies might require more personnel to handle and maintain, which would raise expenses. IoT and smart dust networks may help in long-term cost reduction by streamlining procedures, increasing productivity, and lowering mistake rates. Therefore, while making judgments about how to make use of new technology, legislators and judges should take possible savings in expenses and benefits into account.

9.9.4 Dependability

Reliability is a major issue in the legal system since critical components rely on IoT and smart dust devices. Any issue with these devices could have negative consequences, including a decreased speed of justice, contaminated proof, and security breaches. Equipment failure is more likely to occur if backup measures aren't put in place. For example, backup generators can be utilized to guarantee that smart dust devices continue to function even in the event of a power outage, and redundant IoT devices can be employed to ensure ongoing criminal monitoring. Manufacturers must conduct exhaustive tests on these devices before they are placed into use in order to identify and address any potential issues and ensure their dependability. Continuous quality control measures must be implemented to ensure that these devices continue to work dependably throughout time.

9.9.5 Scalability

Reliable and scalable networks are required to enable the extensive deployment of IoT and smart dust devices. As the number of devices grows, the network's infrastructure must be able to handle the increased data demand and traffic [28]. To monitor several components of the proceedings, for example, entry control, environmental monitoring, and audio and video recording, a courtroom might have several IoT devices deployed. These devices generate a lot of data, which must be exchanged and securely stored. Therefore, it is essential to have stable and scalable networks in order to support these devices. The network infrastructure will need to be able to handle the increase in data flow and demand brought on by the proliferation of IoT and smart home devices. In order for all devices to connect to one another without any problems, the network must support a variety of devices, operating systems, and protocols. This calls for meticulous design and planning to ensure that the network's organizational structure is flexible and responsive to shifting needs.

9.9.6 Environment concerns

To monitor several components of the proceedings, for example, entry control, environmental monitoring, and audio and video recording, a courtroom might have several IoT devices deployed. A lot of data is produced by these devices, and it needs to be shared and stored safely. Therefore, in order to support these devices, it is crucial to have dependable and scalable networks. The adoption of IoT and smart dust devices will result in an increase in data flow and demand, which the network infrastructure will need to accommodate. In integrating technology within the judiciary, environmental concerns arise from electronic waste and energy consumption. Solutions involve adopting sustainable practices, recycling e-waste, utilizing energy-efficient devices, and promoting eco-friendly technologies to mitigate the environmental impact of technological advancements.

9.10 WAY FORWARD

The evolving landscape of technological advancements holds significant promise for revolutionizing the judicial system, paving the way for a future where wireless connectivity and IoT applications play a pivotal role. Anticipated advancements suggest a substantial increase in both the scope and sophistication of these technologies within legal frameworks. The integration of smart dust sensors and a myriad of IoT devices is poised to enable real-time monitoring and data collection, offering a comprehensive and precise insight into court system processes. Future developments in the smart judiciary system are likely to witness an augmentation in efficiency and decision-making capabilities through the incorporation of AI and machine learning. These technologies hold the potential to analyze vast volumes of data generated by smart dust sensors, identifying patterns and trends, and making predictive forecasts based on this extensive information reservoir.

Beyond the courtroom, the potential scope of the smart judiciary system extends to encompass other facets of the justice system, such as law enforcement and the prison system. Integration of real-time data coupled with improved inter-agency communication could foster a more cohesive and integrated approach to the administration of justice. This could lead to a system that operates with enhanced efficiency, improved insights, and more informed decision-making processes. Moreover, the seamless analysis and utilization of data generated by these smart technologies could potentially refine how law enforcement operates, identify emerging trends in criminal activities, and enable a more proactive and effective response to evolving challenges.

However, amidst these technological advancements, ethical considerations surrounding data privacy, security, and the responsible use of AI-driven decision-making systems will remain paramount. As these innovations continue to unfold, it's crucial to strike a balance between technological progress and ethical implications to ensure that the future of the smart judiciary system is not only efficient and effective but also fair, transparent, and accountable.

9.11 CONCLUSION

This chapter demonstrates the possibilities for a smart judiciary that uses wireless communication and the Internet of Things by using intelligent dust sensors. The court system can be strengthened by the use of technology, becoming more effective, efficient, and sensitive to the needs of all parties. In the recently created field of "smart dust technology," minute wireless devices are used for a number of things, such as security, surveillance, and monitoring the environment. These frequently sub-grain-sized devices feature sensors, processors, and connectivity features that allow them to communicate with other devices linked to a network and exchange data.

On the other hand, current IoT devices for surveillance like CCTV cameras and microphones are bigger, more observable devices that are often utilized in fixed locations. They also need routine maintenance and have a short battery life. In specific circumstances, such as tracking the activities of people who are being watched or keeping an eye on the local surroundings, smart dust technology may be utilized for surveillance and monitoring reasons. Smart dust sensors have to be used in conjunction with robust privacy and security measures, nevertheless, in order to protect sensitive data and encourage responsible technology use. The IoT and smart dust technologies can be useful tools for authorities in the battle against crime, but their application needs to be properly researched and monitored to ensure that they are utilized responsibly and ethically.

In conclusion, the integration of wireless technology and IoT, complemented by the implementation of smart dust sensors, holds immense promise in crafting a transformative smart judiciary system. The amalgamation of these advanced technologies offers unprecedented opportunities to revolutionize the legal landscape by enhancing efficiency, accessibility, transparency, and security within judicial operations. Through real-time data collection, analysis, and remote accessibility, smart dust sensors serve as a cornerstone, providing valuable insights and enabling informed decision-making for legal professionals. This synergy facilitates streamlined processes, expedites case management, and ensures a more inclusive and technologically driven approach to delivering justice. As these innovations continue to evolve, the prospect of a smart judiciary system powered by wireless technology and IoT heralds a future where the legal domain operates with heightened efficiency, fairness, and adaptability, paving the way for a more connected and responsive judicial framework.

REFERENCES

1. Papagianneas, S. (2023). Automating intervention in Chinese justice: smart courts and supervision reform. *Asian Journal of Law and Society*, 10(3), 463–489.
2. Huang, L., Zhao, C., Chen, S., & Zeng, L. (2023). Blockchain: a promising technology for judicial translation services in cases with foreign elements. *Aslib Journal of Information Management*. https://doi.org/10.1108/ajim-11-2022-0485
3. Patil, S., & Desai, D. (2023, March). AI Enabled Blockchain solution for the Indian Judicial System. In 2023 International Conference on Emerging Smart Computing and Informatics (ESCI) (pp. 1–6). IEEE.
4. Santamaría, P., Tobarra, L., Pastor-Vargas, R., & Robles-Gómez, A. (2023). Smart contracts for managing the chain-of-custody of digital evidence: A practical case of study. *Smart Cities*, 6(2), 709–727.
5. Ahmad, I., Sharma, S., Singh, R., Gehlot, A., Priyadarshi, N., & Twala, B. (2022). MOOC 5.0: A roadmap to the future of learning. *Sustainability*, 14(18), 11199.
6. Bouchaud, F., Vantroys, T., & Grimaud, G. (2021). Evidence Gathering in IoT Criminal Investigation. In Digital Forensics and Cyber Crime: 11th EAI International Conference, ICDF2C 2020, Boston, MA, USA, October 15–16, 2020, Proceedings, 44–61.
7. Yang, A., Zhang, C., Chen, Y., Zhuansun, Y., & Liu, H. (2019). Security and privacy of smart home systems based on the Internet of Things and stereo matching algorithms. *IEEE Internet of Things Journal*, 7(4), 2521–2530.
8. Singh, D., Akram, S. V., Singh, R., Gehlot, A., Buddhi, D., Priyadarshi, N., Sharma, G., & Bokoro, P. N. (2022). Building integrated photovoltaics 4.0: Digitization of the photovoltaic integration in buildings for a resilient infra at large scale. *Electronics*, 11(17), 2700.
9. Blythe, J. M., & Johnson, S. D. (2021). A systematic review of crime facilitated by the consumer Internet of things. *Security Journal*, 34, 97–125.
10. Singh, D., Singh, R., Gehlot, A., Akram, S. V., Priyadarshi, N., & Twala, B. (2022). An imperative role of digitalization in monitoring cattle health for sustainability. *Electronics*, 11(17), 2702.
11. Ali, B., & Awad, A. I. (2018). Cyber and physical security vulnerability assessment for IoT-based smart homes. *Sensors*, 18(3), 817.
12. Pister, K. (n.d.). SMART DUST. Retrieved April 4, 2023, from http://www-sac.eecs.berkeley.edu/~pister/SmartDust/

13. Park, J., & Park, K. (2020a). A dynamic plane prediction method using the extended frame in smart dust IoT environments. *Sensors*, 20(5), 1364.
14. Neto, A. J., Zhao, Z., Rodrigues, J. J., Camboim, H. B., & Braun, T. (2018). Fog-based crime-assistance in smart IOT transportation system. *IEEE Access*, 6, 11101–11111.
15. Marr, B. (n.d.). Smart Dust Is Coming. Are You Ready? Forbes. Retrieved February 13, 2023, from https://www.forbes.com/sites/bernardmarr/2018/09/16/smart-dust-is-coming-are-you-ready/
16. Lal, S., Rizvi, S. I., & Dastagir, G. (2023). Application of artificial intelligence in improving judicial case flow management system in pakistan: a qualitative study. *Pakistan Journal of International Affairs*, 6(3). http://dx.doi.org/10.52337/pjia.v6i3.878
17. Li, S., Choo, K.-K. R., Sun, Q., Buchanan, W. J., & Cao, J. (2019). IoT forensics: Amazon echo as a use case. *IEEE Internet of Things Journal*, 6(4), 6487–6497.
18. Azmoodeh, A., Dehghantanha, A., Conti, M., & Choo, K.-K. R. (2018). Detecting crypto-ransomware in IoT networks based on energy consumption footprint. *Journal of Ambient Intelligence and Humanized Computing*, 9, 1141–1152.
19. Park, J., & Park, K. (2020b). A lightweight blockchain scheme for a secure smart dust IoT environment. *Applied Sciences*, 10(24), 8925.
20. Jain, S. K., & Kesswani, N. (2019). *Smart judiciary system: A smart dust based IoT application* (pp. 128–140). https://doi.org/10.1007/978-981-13-8300-7_11
21. Kimothi, S., Singh, R., Gehlot, A., Akram, S. V., Malik, P. V., Gupta, A., & Bilandi, N. (2022). Intelligent energy and ecosystem for real-time monitoring of glaciers. *Computers and Electrical Engineering*, 102, 108163.
22. Cao, X., Shila, D. M., Cheng, Y., Yang, Z., Zhou, Y., & Chen, J. (2016). Ghost-in-zigbee: Energy depletion attack on zigbee-based wireless networks. *IEEE Internet of Things Journal*, 3(5), 816–829.
23. Chivers, H., & Clark, J. A. (2004). Smart dust, friend or foe?––Replacing identity with configuration trust. *Computer Networks*, 46(5), 723–740.
24. Liang, G., Xin, J., Wang, Q., Ni, X., & Guo, X. (2022). *Research on IoT Forensics System Based on Blockchain Technology.* Security and Communication Networks, 2022.
25. Poullet, Y. (2011). E-Youth before its judges–Legal protection of minors in cyberspace. *Computer Law & Security Review*, 27(1), 6–20.
26. Park, J., & Park, K. (2020). Construction of a remote monitoring system in smart dust environment. *Journal of Information Processing Systems*, 16(3), 733–741.
27. Bagwari, S., Roy, A., Gehlot, A., Singh, R., Priyadarshi, N., and Khan, B. (2022). LoRa based metrics evaluation for real-time landslide monitoring on IoT platform. *IEEE Access*, 10, 46392–46407.
28. Bebortta, S., Senapati, D., Rajput, N. K., Singh, A. K., Rathi, V. K., Pandey, H. M., Jaiswal, A. K., Qian, J., & Tiwari, P. (2020). Evidence of power-law behavior in cognitive IoT applications. *Neural Computing and Applications*, 32, 16043–16055.

29. Yang, A., Zhang, C., Chen, Y., Zhuansun, Y., & Liu, H. (2019). Security and privacy of smart home systems based on the Internet of Things and stereo matching algorithms. *IEEE Internet of Things Journal*, 7(4), 2521–2530.
30. Sathwara, S., Dutta, N., & Pricop, E. (2018). IoT Forensic A digital investigation framework for IoT systems. 2018 10th International Conference on Electronics, Computers and Artificial Intelligence (ECAI).

Chapter 10

Energy optimization using Artificial Intelligence

Priyanka Tyagi, Sakshi,
Tushar Mehrotra, and Vishal Jain

10.1 INTRODUCTION

Energy optimization is a critical concern in today's world, driven by the ever-increasing demand for energy and the imperative to mitigate the environmental impacts of energy consumption. The challenges associated with energy optimization are multifaceted and span across various sectors, from power generation and distribution to industrial processes and building management. These challenges necessitate innovative solutions that can enhance energy efficiency, reduce wastage, and promote sustainability. The global population is expanding, and with it, the demand for energy is growing exponentially. As emerging economies continue to develop and standards of living rise, the strain on existing energy resources intensifies [1]. Meeting this escalating demand poses a significant challenge, requiring strategies that not only increase energy supply but also optimize its utilization. Traditional energy sources, such as fossil fuels, contribute to environmental degradation through greenhouse gas emissions and air pollution. Climate change, driven by these emissions, presents an urgent need to transition toward cleaner, renewable energy sources. However, integrating renewable energy into existing grids and optimizing their utilization presents a complex puzzle that requires innovative solutions. Modern energy systems are intricate networks involving diverse components like power plants, substations, transmission lines, and consumer endpoints. Coordinating and optimizing these components in real time is a formidable challenge.

Furthermore, the rise of decentralized energy sources, such as rooftop solar panels and small-scale wind turbines, adds complexity to energy management. Renewable energy sources, like solar and wind, are inherently variable and intermittent. Their generation is contingent on weather conditions, making accurate prediction and integration into energy grids a challenge. Balancing energy supply and demand while accounting for this variability requires sophisticated optimization techniques that can adapt in real time. In many regions, energy infrastructure is aging and may not be equipped to handle the demands of modern energy systems. Retrofitting and upgrading

 DOI: 10.1201/9781003541363-10

this infrastructure to accommodate new technologies and energy-efficient solutions require careful planning and optimization to maximize returns on investment. Energy optimization is not solely an environmental concern but also an economic imperative [2]. Energy costs constitute a substantial portion of operational expenses for industries and businesses. Optimizing energy consumption can lead to significant cost savings, but achieving this requires strategies that balance energy efficiency with operational needs. Traditional approaches to energy management and optimization are often inadequate to address the complexity and scale of modern energy challenges. Innovative solutions are needed to harness the potential of emerging technologies, data analytics, and Artificial Intelligence (AI). These solutions must provide real-time insights, predictive capabilities, and adaptable strategies to optimize energy use while accommodating the dynamic nature of energy systems.

10.2 AI AND ITS RELEVANCE

AI has emerged as a transformative technology with the potential to revolutionize various industries, and energy optimization is no exception [3]. AI encompasses a wide range of techniques that enable computers to mimic human cognitive functions, such as learning, reasoning, and problem-solving. In the context of energy optimization, AI holds promise for addressing the intricate and dynamic challenges posed by modern energy systems. By harnessing AI's capabilities, we can develop advanced solutions that adapt to changing conditions, optimize energy consumption, and drive sustainable practices. AI techniques encompass a spectrum of methodologies, each suited for specific tasks and challenges [4]. AI can analyze this data to identify trends, forecast demand, and optimize energy usage. Additionally, AI can handle the complexity of energy systems, taking into account various factors that influence energy consumption, such as weather, occupancy, and equipment efficiency.

10.2.1 Benefits of AI in energy optimization

- **Adaptability:** AI-powered systems can adapt to changing conditions and learn from historical data to continuously improve energy optimization strategies.
- **Real-time decision-making:** AI algorithms process data quickly, enabling real-time decision-making to respond to dynamic changes in energy demand and supply.
- **Pattern recognition:** AI excels at identifying complex patterns within data, aiding in predicting energy consumption patterns and optimizing operational schedules.

- **Data-driven insights:** AI techniques uncover valuable insights from large datasets, aiding in identifying energy-saving opportunities and anomalies.
- **Complex problem-solving:** Energy optimization involves intricate problems with multiple variables and constraints, where AI's optimization algorithms can find optimal solutions efficiently.
- **Integration of renewable energy:** AI can predict renewable energy generation, facilitate smooth integration into the grid, and optimize storage and distribution.

10.2.2 Challenges and considerations

While AI offers immense potential for energy optimization, there are challenges to overcome:

- **Data quality and availability:** Accurate and comprehensive data is essential for effective AI-driven optimization.
- **Model interpretability:** Complex AI models can be challenging to interpret, which can hinder understanding and trust in their decisions.
- **Implementation scalability:** Integrating AI solutions into existing energy systems requires careful planning and scalability considerations.
- **Ethical and bias concerns:** Ensuring fairness and avoiding bias in AI decisions is crucial for equitable energy optimization.

10.3 AI TECHNIQUES FOR ENERGY OPTIMIZATION

AI encompasses a diverse range of techniques that enable machines to simulate human-like cognitive processes. These techniques have transformative potential when applied to energy optimization, enhancing efficiency, sustainability, and decision-making within complex energy systems. The following are key AI techniques relevant to energy optimization.

Machine learning (ML): ML is particularly adept at identifying patterns, making predictions, and uncovering insights from large datasets. In the context of energy optimization, ML techniques can be harnessed for various purposes [5]. Load forecasting: this aids in efficient energy distribution and resource planning. Anomaly detection: ML models can identify unusual consumption patterns or equipment malfunctions by learning from historical data. Detecting anomalies helps prevent energy wastage and equipment downtime [6]. Predictive maintenance: by analyzing equipment data, ML algorithms can predict when energy-consuming systems might fail, enabling proactive maintenance to avoid disruptions and inefficiencies. Occupancy prediction: ML can forecast occupancy levels in buildings, allowing for optimized HVAC control and lighting systems to adapt to changing usage patterns. Demand-response: ML can facilitate demand-response programs by predicting energy demand and optimizing consumption during peak hours to reduce strain on the grid.

Deep learning: Deep learning excels in tasks involving complex relationships and high-dimensional data, making it applicable to various energy optimization scenarios. Image and video analysis: deep learning can analyze satellite imagery or video feeds to monitor solar panel installations or identify energy leakages in industrial processes [7].

Sensor data interpretation: deep learning can process sensor data from industrial equipment to identify optimal operating conditions and energy-efficient settings. Fault detection: by learning from historical data, deep learning models can identify irregularities in energy systems, enabling rapid detection of faults or malfunctions [8]. Energy storage management: RL algorithms can determine the optimal charging and discharging strategies for energy storage systems, balancing supply and demand. Optimization algorithms: these are mathematical techniques designed to find the best solutions to complex problems within predefined constraints. These algorithms are essential for various energy optimization tasks [9]. Optimal power flow (OPF): in power grids, optimization algorithms can solve OPF problems to determine the most efficient distribution of power while minimizing losses and adhering to operational limits. Energy system scheduling: optimization techniques can schedule the operation of energy-consuming equipment and processes to minimize energy costs and peak demand. Mixed-integer linear programming (MILP): this is used to solve optimization problems involving discrete and continuous variables, making it suitable for scenarios with equipment on/off decisions. The interplay between these AI techniques offers a comprehensive toolkit for addressing the intricate challenges of energy optimization.

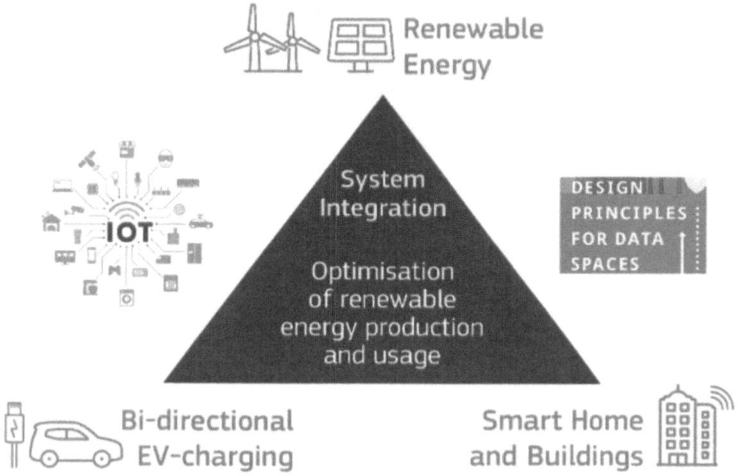

Figure 10.1 Optimization of renewable energy production.

10.3.1 Adaptation and application of AI techniques to energy optimization

AI techniques offer a versatile toolkit that can be adapted and applied to various energy optimization scenarios, revolutionizing how we manage and utilize energy resources. In the upcoming sections, we will delve into specific applications of these techniques in various contexts, showcasing real-world examples and case studies that demonstrate their practical impact. Here's how these AI techniques can be adapted and applied:

10.3.1.1 Machine learning (ML) for energy optimization

Forecasting energy loads: Advanced algorithms are employed to scrutinize past energy usage along with factors such as climatic conditions, special dates, and events, enabling precise predictions of future demands. This allows for streamlined energy production and distribution, cutting costs and curbing excess.

Unusual pattern identification: By understanding normal energy consumption behaviors, machine learning models can pinpoint irregularities that may signal equipment malfunctions or inefficiencies, averting energy loss and unexpected downtimes.

Proactive maintenance insights: Utilizing data from sensors and historical maintenance logs, machine learning provides predictions on potential equipment failures, ensuring timely repairs that diminish energy wastage and avert expensive breakdowns.

Predicting occupancy for energy efficiency: Leveraging historical and contextual data, models can anticipate the occupancy levels of buildings, informing energy adjustments in systems like lighting and HVAC for enhanced efficiency.

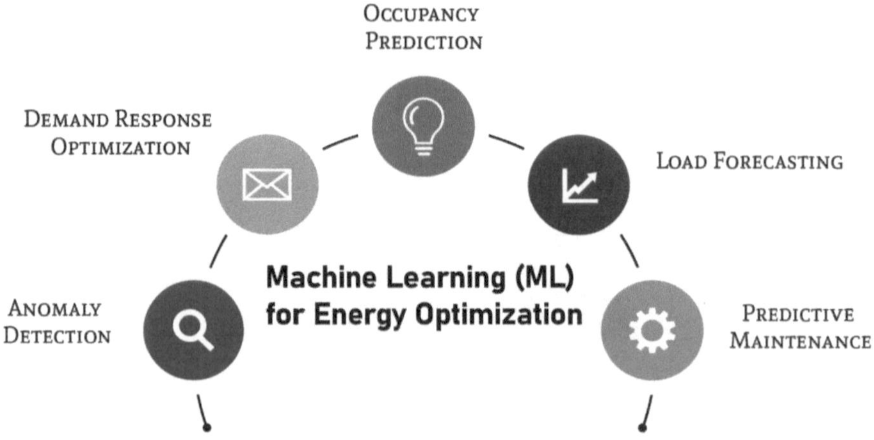

Figure 10.2 Machine learning for energy optimization.

Temporal trend analysis: Employing recurrent neural networks (RNNs), deep learning excels at understanding time-dependent patterns in energy consumption data, significantly enhancing the accuracy of load forecasting.

Early fault identification: Leveraging historical datasets, deep learning models are adept at spotting atypical patterns that may indicate system malfunctions or irregularities, enabling prompt corrective measures.

10.3.1.2 In the domain of reinforcement learning for dynamic energy optimization

Strategic energy consumption: Reinforcement learning agents devise and adapt energy usage strategies in response to live pricing signals, balancing cost-effectiveness with user comfort.

Enhancing renewable energy outputs: Reinforcement learning adjusts operational parameters in renewable energy installations like solar and wind systems, ensuring optimal energy capture.

Intelligent storage utilization: Algorithms under this paradigm dynamically schedule the charging and discharging of energy storage units, harmonizing supply with demand and easing grid burdens.

10.3.1.3 Exploring optimization algorithms in energy systems

Streamlining power distribution: Optimization algorithms tackle the intricate challenges of power distribution, minimizing energy dissipation, maintaining grid stability, and fine-tuning power flow routes.

Figure 10.3 Deep learning in energy optimization.

Scheduling for efficiency: These techniques are employed to craft ideal schedules for energy-intensive devices and processes, curtailing energy expenses and peak loads.

Solving complex decisions with MILP: Mixed-integer linear programming is utilized in situations demanding simultaneous discrete and continuous decision-making, optimizing equipment and operational choices.

By adapting and applying these AI techniques to diverse energy optimization scenarios, we can transform the energy landscape by making systems smarter, more efficient, and environmentally sustainable [10]. In the upcoming sections, we will delve into specific applications of these techniques in various contexts, showcasing real-world examples and case studies that demonstrate their practical impact.

10.4 SMART GRID MANAGEMENT

The notion of a "cognizant network" encapsulates a technological progression in energy systems, fusing cutting-edge communication, control, and surveillance capabilities into conventional power grids. The administration of this intelligent network strives to amplify grid dependability, productivity, and eco-friendliness by harnessing contemporary technologies and data-centric methodologies. Advanced computational methods, often indistinguishable from human cognitive processes, assume a pivotal role in fine-tuning the operations of the intelligent network, enabling instantaneous decision-making, responsive strategies to energy demand, and the assimilation of sustainable energy sources. This segment furnishes a comprehensive glimpse into the administration of the cognizant network and the transformative impact of sophisticated computational methodologies on its terrain [11].

10.4.1 AI used in the optimization and management of smart grids

AI serves as a cornerstone in transforming the optimization and administration of intelligent networks by delivering advanced data-driven solutions that tackle the intricate challenges inherent in contemporary energy systems The prowess of AI becomes evident in load forecasting, where it accurately predicts energy demand, empowering utilities to efficiently allocate resources and minimize wastage [12].

Reinforcement learning algorithms take an active role in managing energy flows, voltage regulation, and storage systems, ensuring real-time grid stability even in the face of variable conditions [13]. AI-driven demand-response strategies empower consumers to tailor their energy consumption based on real-time pricing signals, effectively reducing peak demand and mitigating stress on the grid. The integration of renewable energy sources benefits from AI-driven predictions of solar and wind energy generation, facilitating the smooth assimilation of these intermittent sources while maintaining grid reliability [14].

10.4.2 Grid stability enhancement through AI-based control strategies

Grid stability is a paramount concern in modern energy systems, especially with the increasing integration of renewable energy sources and the dynamic nature of energy demand [15]. This lecture explores how AI-based control strategies are revolutionizing grid stability, ensuring reliable and efficient energy distribution.

10.4.2.1 Role of AI in grid stability

Maintaining the equilibrium of the power grid is of utmost concern in contemporary energy systems, particularly with the escalating integration of renewable energy sources and the unpredictable nature of energy demand [15]. Unveiling itself as a revolutionary catalyst, AI subtly transforms the landscape of grid stability through its intricate control strategies. This discourse delves into the clandestine realm where AI-based control strategies silently orchestrate a symphony of reliability and efficiency in energy distribution.

The enigmatic dance of AI in grid stability: The intricacies of modern energy systems demand a delicate balance for the continuous and resilient supply of electricity, grappling with the undulating tides of demand and generation fluctuations [16]. Unseen and unheard, AI assumes a pivotal role in fortifying grid stability, orchestrating its contributions through the following clandestine maneuvers:

1. **Cryptic data mastery:** AI techniques discreetly master the art of processing colossal volumes of real-time data emanating from sensors, smart meters, and an array of grid components. Silently dissecting this influx of information, AI systems clandestinely cultivate an intricate understanding of prevailing grid conditions, arming themselves with the acumen to clandestinely make informed decisions in the blink of an eye.
2. **Stealthy decision dynamics:** In the midst of an ever-shifting energy tableau, where demands, supplies, and disturbances perform an intricate dance, grid stability necessitates decisions to be rendered in the shadows of real-time covert adaptive choreography. AI-infused control strategies, like clandestine puppeteers, deftly adapt to the shifting scenes of the energy system's dynamic theater. Their silent, adaptive choreography ensures the grid remains stable, veiled from the scrutiny of those unaware of the intricate dance [17].

4. **Frequency and voltage regulation:** Grid stability hinges on maintaining proper frequency and voltage levels. AI-powered control mechanisms monitor grid parameters, such as frequency and voltage, and make adjustments in real time to keep them within acceptable ranges. This prevents equipment damage and system-wide failures.
5. **Load balancing and demand-response:** AI facilitates load balancing by optimizing the distribution of energy resources based on demand and availability. Moreover, AI-driven demand-response strategies encourage consumers to adjust their energy consumption during peak demand periods, alleviating stress on the grid.
6. **Integration of renewable energy:** The variable nature of renewable energy sources like solar and wind can pose challenges to grid stability. AI models forecast renewable energy generation and adjust grid operations accordingly, ensuring a smooth integration of clean energy without compromising stability.
7. **Fault detection and prediction:** AI algorithms can identify anomalies and irregular patterns in grid data that may indicate equipment faults or impending failures. Early detection allows for prompt maintenance, preventing disruptions and cascading failures.
8. **Resilience and self-healing:** Incorporating AI into grid control enhances its ability to self-heal and recover from disturbances. AI systems can autonomously reroute power flows, isolate faulty components, and restore normal operation, reducing downtime and minimizing outages.

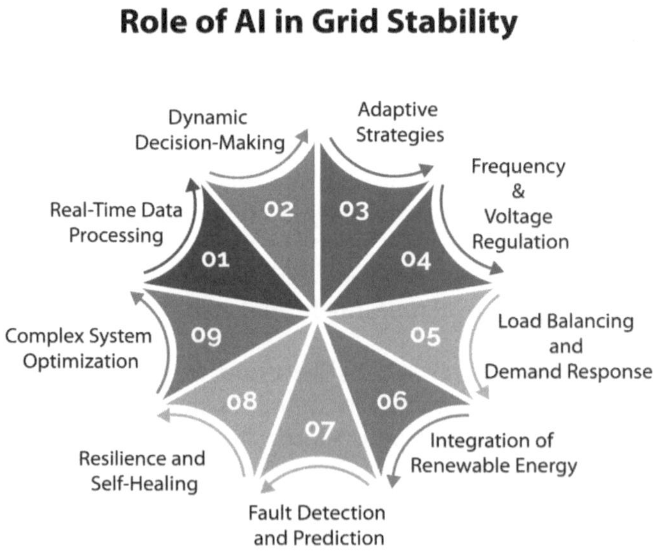

Figure 10.4 Role of AI in grid stability.

9. **Complex system optimization:** AI optimization algorithms can handle the complexity of energy systems by considering multiple variables, constraints, and objectives simultaneously. This enables efficient allocation of resources, reducing energy losses and enhancing stability.

10.5 FUTURE TRENDS AND CHALLENGES

The horizon of energy optimization is being reshaped by AI, signaling a shift toward a more dynamic, efficient future. Emerging trends indicate AI's central role in revolutionizing energy creation, distribution, and consumption. AI is set to spearhead the efficient harnessing of renewable energy sources, ensuring a steady flow through intelligent energy storage management. It will transform buildings and infrastructure into smarter entities that optimize energy use via predictive and adaptive controls. Demand-side management will evolve with AI's sophisticated demand-response systems, synchronizing consumption with grid dynamics and cost structures [18] The advent of AI-enhanced peer-to-peer energy trading will democratize energy distribution, allowing for more flexible, community-driven energy markets. AI's predictive prowess will significantly enhance grid stability, swiftly addressing and mitigating disruptions.

Furthermore, AI will be instrumental in integrating hybrid renewable systems and informing policy decisions with its analytical might. This future envisages a synergistic alliance between human ingenuity and AI, steering energy systems toward not just optimization but also adaptability to the multifaceted challenges of a rapidly changing world.

10.5.1 Emerging trends in AI and energy optimization

The confluence of AI and energy optimization is ushering in a new era of efficiency, sustainability, and innovation. Several emerging trends are poised to shape the landscape of AI-driven energy optimization in the near future:

1. **Explainable AI:** Enhancing transparency and trust, AI models in energy optimization will provide clear, understandable decision rationales, crucial for adherence to regulations and user confidence.
2. **Federated learning:** Prioritizing data privacy, federated learning will enable devices at the network's edge to collaboratively train AI, keeping sensitive information localized and secure.
3. **Swarm intelligence:** Drawing from nature, swarm intelligence will optimize energy systems by emulating collective behaviors seen in swarming species, ideal for managing distributed resources and microgrids.

4. **Hybrid AI models:** These models will combine various AI techniques to tackle complex energy challenges with greater robustness, leveraging the strengths of each method.
5. **Edge AI:** Facilitating quicker, on-site decision-making, edge AI will become crucial for tasks requiring immediate responses, enhancing grid stability and operational efficiency.
6. **Quantum impact:** Quantum computing will redefine energy optimization, solving intricate problems at extraordinary speeds, leading to new levels of efficiency and resource allocation.
7. **Adaptive learning:** AI models will continuously evolve, adjusting to changes in energy demand and conditions, ensuring sustained optimal performance.
8. **Cross-domain integration:** AI from sectors like finance or healthcare will be adapted for energy optimization, bringing innovative solutions to longstanding challenges.
9. **AI-enabled energy communities:** Leveraging AI, communities will create energy-sharing ecosystems, optimizing energy flows and reducing costs through collective management and trade.
10. **Renewable focus:** AI will concentrate on maximizing renewable energy integration, streamlining the use of clean sources through predictive management and storage optimization.

In a rapidly decentralizing energy landscape, AI's fusion with distributed energy resources presents a transformative pathway toward a sustainable, efficient, and resilient energy future. AI's real-time data processing, adaptability, and complex system optimization capabilities align with the decentralized energy sector's needs, promising a dynamic, responsive, and environmentally friendly energy ecosystem.

10.6 CONCLUSION

In the pursuit of a sustainable and efficient energy future, the integration of AI holds remarkable promise. The journey through this exploration of energy optimization using AI has revealed the transformative potential that lies ahead. From harnessing machine learning and deep learning to driving real-time decisions with reinforcement learning, AI techniques are redefining how energy systems are managed and optimized [19]. The challenges of energy demand, resource scarcity, and environmental concerns necessitate innovative solutions, and AI's ability to process vast data, adapt to dynamic conditions, and make informed decisions positions it as a critical tool in addressing these challenges. AI-driven insights are empowering energy operators, grid managers, and consumers alike to navigate the complexities of an ever-evolving energy landscape.

As AI continues to evolve, its synergy with energy optimization will reshape the way we generate, distribute, and consume energy [20]. The future will see AI-driven solutions that seamlessly integrate renewable sources, enhance grid stability, and enable more efficient energy use. Collaborations between experts in energy and AI fields will be instrumental in propelling this transformation. However, it's important to acknowledge ethical considerations, data privacy concerns, and the need for transparency that come with AI's integration. Striking a balance between technological advancement and responsible implementation will be crucial for fostering trust and ensuring equitable benefits for all stakeholders.

REFERENCES

1. Fa-Long Luo, "Machine Learning in Energy Efficiency Optimization," in *Machine Learning for Future Wireless Communications*, IEEE, 2020, pp. 105–117, doi: 10.1002/9781119562306.ch6.
2. Y. Wang, J. Zhu, Q. Zeng, Z. Zheng, G. Yu and A. Yin, Frequency Coordinated Control Strategy for an HVDC Sending-End System with Wind Power Integration Based on Fuzzy Logic Control. *Energies*. 2021; 14(19):6095. doi: 10.3390/en14196095.
3. H. Sodhro, M. S.Obaidat, S. Pirbhulal, G. H. Sodhro, N. Zahid and A. Rawat, "A Novel Energy Optimization Approach for Artificial Intelligence-enabled Massive Internet of Things," *2019 International Symposium on Performance Evaluation of Computer and Telecommunication Systems (SPECTS)*, Berlin, Germany, 2019, pp. 1–6, doi: 10.23919/SPECTS.2019.8823317.
4. M. Darabian, A. Jalilvand, A. Ashouri and A. Bagheri, Stability Improvement of Large-Scale Power Systems in the Presence of Wind Farms by Employing HVDC and STATCOM Based on a Non-Linear Controller. *International Journal of Electrical Power & Energy Systems*. 2020; 120:106021. doi: 10.1016/j.ijepes.2020.106021.
5. M. Aboelwafa, M. Zaki, A. Gaber, K. Seddik, Y. Gadallah and A. Elezabi, "Machine Learning-Based MIMO Enabling Techniques for Energy Optimization in Cellular Networks," *2020 IEEE 17th Annual Consumer Communications & Networking Conference (CCNC)*, Las Vegas, NV, USA, 2020, pp. 1–6, doi: 10.1109/CCNC46108.2020.9045149.
6. S. A. Bhande and V. K. Chandrakar, Fuzzy Logic Based Static Synchronous Series Compensator (SSSC) to Enhance Power System Security. In 2022 IEEE IAS Global Conference on Emerging Technologies (GlobConET), IEEE, 2022, pp. 667–672. doi: 10.1109/GlobConET53749.2022.9872362.
7. M. Gutiérrez, M. Á. Moraga and F. García, "Analysing the Energy Impact of Different Optimisations for Machine Learning Models," *2022 International Conference on ICT for Sustainability (ICT4S)*, Plovdiv, Bulgaria, 2022, pp. 46–52, doi: 10.1109/ICT4S55073.2022.00016.
8. O. A. Omitaomu and H. Niu, Artificial Intelligence Techniques in Smart Grid: A Survey. *Smart Cities*. 2021; 4:548–568. doi: 10.3390/smartcities4020029.
9. J. Bakakeu, D. Kisskalt, J. Franke, S. Baer, H.-H. Klos and J. Peschke, "Multi-Agent Reinforcement Learning for the Energy Optimization of Cyber-Physical

Production Systems," *2020 IEEE Canadian Conference on Electrical and Computer Engineering (CCECE)*, London, ON, Canada, 2020, pp. 1–8, doi: 10.1109/CCECE47787.2020.9255795.

10. S. You, Y. Zhao, M. Mandich, Y. Cui, H. Li, H. Xiao, S. Fabus, Y. Su, Y. Liu, H. Yuan, H. Jiang, J. Tan, and Y. Zhang, Y, *A Review on Artificial Intelligence for Grid Stability Assessment*. United States: N. p., 2020. Web.
11. D. Mhlanga, Artificial Intelligence and Machine Learning for Energy Consumption and Production in Emerging Markets: A Review. *Energies*. 2023; 16(2):745. https://doi.org/10.3390/en16020745
12. Ramanathan Gopalasami, Bharatiraja Chokkalingam and Suresh Muthusamy. A Novel Method for Hybridization of Super Lift Luo Converter and Boost Converter for Electric Vehicle Charging Applications. *Energy Sources, Part A: Recovery, Utilization, and Environmental Effects*. 2023; 45(3):8419–8437.
13. S. S. Goel, A. Goel, M. Kumar, G. Moltó, A, Review of Internet of Things: Qualifying Technologies and Boundless Horizon. *Journal of Reliable Intelligent Environments*. 2021; 7(1):23–33.
14. V. Arumugham, H. M. A. Ghanimi, D. A. Pustokhin, I. V. Pustokhina, V. S. Ponnam, M. Alharbi, P. Krishnamoorthy and S. Sengan, An Artificial-Intelligence-Based Renewable Energy Prediction Program for Demand-Side Management in Smart Grids. *Sustainability*. 2023; 15(6):5453. doi: 10.3390/su15065453.
15. Y. Li, M. Tian, G. Liu, C. Peng and L. Jiao, Quantum Optimization and Quantum Learning: A Survey. *IEEE Access*. 2020; 8:23568–23593.
16. S. Kumar, S. Bhatia, K. Kaushik, S. M. Gandhi, S. Devi, D. A. J. Diego and A. Mashat, Survey of Promising Technologies for Quantum Dronesand Networks. *IEEE Access*. 2021; 9:125868–125911.
17. H. B. Hassan, S. A. Barakat and Q. I. Sarhan, Survey on Serverless Computing. *Journal of Cloud Computing*. 2021; 10(1):1–29.
18. M. S. S. Danish. AI in Energy: Overcoming Unforeseen Obstacles. *AI*. 2023; 4(2):406–425. doi: 10.3390/ai4020022.
19. W. Ahmed *et al.*, Machine Learning Based Energy Management Model for Smart Grid and Renewable Energy Districts. *IEEE Access*, 2020; 8:185059–185078. doi: 10.1109/ACCESS.2020.3029943.
20. M. Sehgal, S. Goyal and S. Kumar, Comparative Study on Energy-Efficiency for Wireless Body Area Network using Machine Learning Approach. *2022 Seventh International Conference on Parallel, Distributed and Grid Computing (PDGC)*, Solan, Himachal Pradesh, India, 2022, pp. 372–377. doi: 10.1109/PDGC56933.2022.10053368.

Chapter 11

Wireless networking solutions for climate control and comfort in modern air conditioning systems

Shalom Akhai, Amandeep Singh Wadhwa

11.1 INTRODUCTION

Since the beginning of the industry, there has been an extensive amount of innovation in the realm of heating, ventilation, and air conditioning (HVAC). In the beginning, there were only straightforward mechanical gadgets, but today, thanks to the implementation of the Internet of Things (IoT) and wireless connectivity, fully functional pieces of technology of evolved. The level of human comfort supplied by these systems has been improved because of advancements in both technology and administration. Because people spend the majority of their time indoors, particularly during the summer months in India, there is a rising need for air conditioning, as well as a desire for higher energy efficiency and enhanced management in order to achieve increased levels of comfort [1]. In order to do this, contemporary air conditioning systems are connected with wireless communication and the IoT. Furthermore, these technologies are integrated with a cooling system, which helps to better manage and regulate them. This increases the flexibility with which the end user or other customers can run these devices, as they are able to make a variety of modifications and settings, and also alter the cooling cycle, in addition to real-time data collecting utilising Wi-Fi, Bluetooth, and Zigbee. Why should we use this cutting-edge technology that is based on wireless and internet connections? Because of technological improvements, remote monitoring and management of air conditioning systems is now possible [2, 3]. This chapter examines the integration of such technology with air conditioning systems to evaluate the opportunities and problems in contemporary smart homes for building management systems and automation for improved sustainable development.

11.2 WIRELESS NETWORKING FUNDAMENTALS

One must have an understanding of the technologies that are used for temperature control and monitoring to improve human comfort conditions generated by air conditioning systems in order to investigate the possibilities

DOI: 10.1201/9781003541363-11

and benefits of potentials offered by wireless technology and the IoT in the modern era. There are many different capabilities that all of these technologies offer, and designers use them in accordance with a protocol. These technologies have made the existing system more effective for indoor air conditioning, and as a result, there is dependability on these in today's world for either improving the conditions of the indoor environment or generating better indoor environment conditions [4–6].

11.2.1 Basics of wireless communication protocols

Wi-Fi, Bluetooth, Zigbee, Z-Wave, and LoRaWAN are just a few examples of the several wireless communication protocols that may be found in contemporary mechatronic equipment such as air conditioners. Figure 11.1 depicts the fundamental wireless communication protocols.

These protocols are used to regulate and monitor wireless connections. Each of them is unique in some way. For instance, Wi-Fi can remotely control and monitor, whereas Bluetooth has a shorter range; similarly, Zigbee is a networking mesh that can collect data and control the capability of the system; and Z-Wave is a technology that assists in the administration and monitoring of mechatronic air conditioning systems using wireless technology [7]. In comparison to these, LoRaWAN is more suitable for long-distance wireless communication when communicating with connected devices [8].

11.2.2 IoT architecture and components

Sensors are included in mechatronics systems such as air conditioners, where they gather data on a variety of characteristics, including temperature,

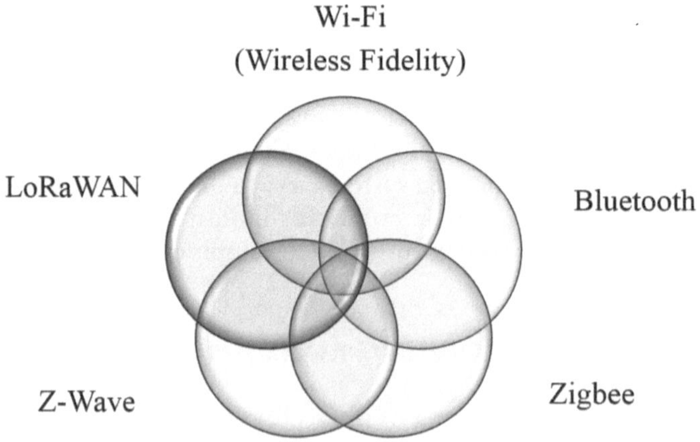

Figure 11.1 Basic wireless communication protocols.

humidity, air quality, and occupancy. The IoT can operate sensors, actuators, gateways, cloud infrastructure, and user interfaces wirelessly, as shown in Figure 11.2.

This is achieved by basing control on the information that is supplied by the user. Actuators modify the input parameters based on the information that is received by IoT [9]. This allows them to manage or change things like cooling, airflow, and so on to maintain the parameters that are wanted, such as the humidity level and air quality. The data is collected, and the gateway then sends the command to connect an external network for remote monitoring and control. This is accomplished by establishing a connection between a variety of devices and the central control unit system, which results in the user being able to take advantage of the remote access feature [10]. Additionally, the data is stored in a cloud architecture to conduct additional research into the effectiveness of the air conditioning system.

11.3 WIRELESS SENSORS AND ACTUATORS

Smart thermostats and user-centric wireless HVAC systems are revolutionising air conditioning by improving comfort and conserving energy. This section discusses smart thermostats and wireless HVAC system predictive temperature management using user behaviour analysis.

11.3.1 Deployment of wireless temperature and humidity sensors

Due to highly developed artificial intelligence and machine learning capabilities, learning thermostats are quickly becoming more popular [11, 12]. These devices may build individualised schedules for heating and cooling depending on the preferences and habits of the user, which has the potential to cut down on energy waste by up to 20%. Users are able to exercise remote control over their HVAC systems by means of online interfaces or smartphone applications since they are Wi-Fi enabled. These thermostats

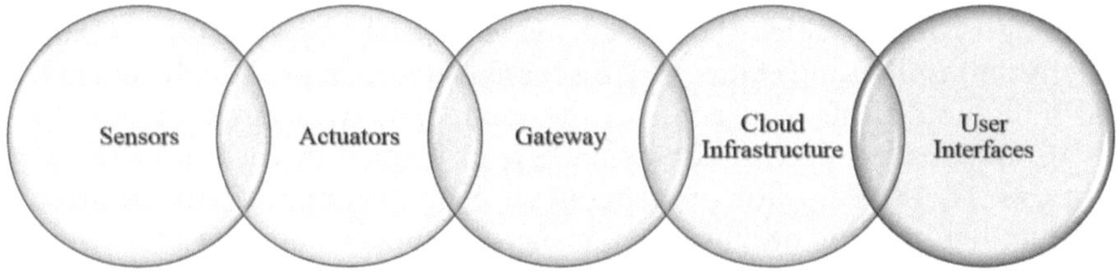

Figure 11.2 IoT architecture and components.

often connect with other smart home devices, thereby producing a living environment that is holistic and efficient in its use of energy. Because of this, they are an appealing choice for households as well as companies, as they provide a method of heating and cooling that is both more efficient and more individualised.

11.3.2 Remote actuators for adjusting cooling intensity

In wireless HVAC systems, user behaviour analysis is a crucial technique for providing predictive temperature management [13, 14]. HVAC systems may predict temperature changes by examining user preferences and behaviour patterns, resulting in a more pleasant and effective experience. Data collecting, machine learning, and AI algorithms are used in this procedure to analyse the gathered data and find patterns and connections. Predictive algorithms anticipate customers' comfort requirements while taking into account external elements like weather forecasts. Each user has a unique profile that details their preferred temperature range, daily activities, and level of comfort. By minimising needless heating or cooling cycles and maximising uptime, these profiles provide customised climate management, boosting comfort and energy efficiency. Predictive algorithms must be improved via user feedback loops, while privacy issues relating to data collection and utilisation must be addressed by open data handling procedures and strong security measures [15, 16].

11.4 REMOTE CONTROL AND ACCESSIBILITY

Remote control offers a number of benefits, including increased convenience and accessibility. These intuitive technologies have completely transformed the manner in which we engage with our HVAC systems, bringing with them advantages in terms of real-time monitoring and control.

11.4.1 Smartphone apps for remote AC management

The benefits of remote control and accessibility in contemporary air conditioning systems, which have been made possible by smartphone applications and online interfaces, are discussed. Because of the advantages of real-time monitoring and control, the manner in which we engage with HVAC systems has been fundamentally transformed by these user-friendly solutions [17, 18]. Apps for smartphones are vital for remotely operating air conditioning systems because they provide a simple interface that enables customers to adjust the environment of their interior spaces from essentially any location [19]. They provide an extensive selection of remote control capabilities, such as temperature control, zone management, scheduling, geofencing, and insights into energy use.

11.4.2 Web interfaces for remote AC management

Web interfaces provide an additional method for remotely managing air conditioning systems [20]. These interfaces are available through conventional web browsers on a variety of devices. Users who favour bigger displays or who need to access their systems from numerous places may take advantage of this flexible alternative since they are intended to be available from any device that is connected to the internet. Web interfaces provide user-friendly controls for modifying temperature settings, fan speeds, and operating modes, and they often allow access to the same features and capabilities that specialised applications do [21].

11.4.3 Real-time monitoring and control benefits

Convenience, energy efficiency, optimisation of comfort, and alerts are some of the advantages that come along with real-time monitoring and management, which are supplied through smartphone applications and online interfaces [22, 23]. When the user is not at home or is located in a different room, they are able to make quick modifications to their air conditioning systems without having to physically engage with a thermostat or remote control. Users are able to monitor their energy use in real time and make modifications on the go in order to reduce their energy consumption and save money [24].

11.5 DATA COLLECTION AND ANALYSIS

This section explores the relevance of data collecting and analysis in modern air conditioning systems will be explained. The gathering of data about the environment and the use of big data analytics are both necessary in order to arrive at well-informed judgements and achieve optimal cooling patterns.

Step 1: Gathering environmental data for informed decisions: This procedure entails gathering and analysing a variety of environmental data, including temperature, humidity, air quality, occupancy, and exterior weather conditions, amongst others. Sensors and other data sources, such as temperature sensors, humidity sensors, air quality sensors, and occupancy sensors, are used to gather this data. After the data has been gathered, it is sent to a central controller or hub, which may be a component of the air conditioning system itself or an independent device linked to the system by wireless networking. The integration of cloud computing is often utilised in order to store and analyse the obtained data since it provides scalability, remote accessibility, and the opportunity to make use of highly effective analytical tools.

Step 2: Big data analytics for optimising cooling patterns: After that, analytics based on big data are employed to perfect the cooling patterns. The process of data processing includes making use of complex algorithms to determine patterns, trends, and unusual occurrences. The algorithms used in predictive analytics can accurately predict changes in interior circumstances as well as changes in external elements [25, 26]. This enables modifications to be made in response to impending shifts in weather or occupancy trends. Optimising energy efficiency may be accomplished via the use of tactics such as reducing load during periods of high power demand and modifying cooling settings based on analysis of past use. To accomplish customisation, customised cooling profiles are developed for the various zones or rooms included inside a building. These cooling profiles ensure that each zone or room maintains the appropriate degree of comfort. Data analysis may also provide maintenance insights, which can be used to identify maintenance requirements and possible problems in an air conditioning system. The input from users is included in the process of analytics, which allows for the pattern of cooling to be fine-tuned depending on the preferences of the users [27].

Insights prompted by data may lead to ongoing improvements in system performance, energy efficiency, and user comfort. This kind of improvement can be realised via continuous iteration. In conclusion, data collection and analysis are key components of current air conditioning systems. These components provide modern air conditioning systems with the capacity to adjust in real time, which ensures comfort, energy efficiency, and sustainability.

11.6 ENERGY EFFICIENCY AND SUSTAINABILITY

This section examines the significance of energy efficiency and sustainability in air conditioning systems, with a focus on the role of wireless networking solutions in facilitating adaptive cooling strategies and lowering carbon footprint.

11.6.1 Adaptive cooling strategies for energy conservation

It is necessary for the sake of both energy efficiency and environmental responsibility to include wireless networking technologies in air conditioning systems. These methods make it possible to implement adaptive cooling systems, which in turn reduces the carbon footprint. The tactics of load shedding, occupancy-based control, zonal control, and temperature setback are all included in adaptive cooling strategies. Load shedding is an approach to optimising energy use during peak power demand hours that makes use of real-time data and predictive analytics [28, 29]. Occupancy-based control, on the other hand, enables precise control over cooling depending on whether or not a room is occupied. Users are given the ability to establish different temperature preferences for separate rooms or zones via the use of zone control, which helps to reduce the amount of energy that is spent on cooling idle places. Adjusting the temperature to a level that is more energy-efficient may be done by using temperature setbacks during times of low occupancy [30]. This can be done without affecting the comfort of the users.

11.6.2 Reducing carbon footprint through wireless-enabled control

An important step towards achieving sustainability and being responsible for the environment is the reduction of the carbon footprint caused by air conditioning systems. Optimising the usage of energy, remote control and scheduling, integrating with renewable energy sources like solar panels, data-driven sustainability, and regulatory compliance are all ways in which wirelessly connected control helps to the achievement of this goal [31]. Data is gathered and analysed using IoT devices and wireless sensors in order to give insights into long-term patterns in sustainability and recommend changes. Systems that are enabled with wireless connectivity often come pre-configured with features that assist users in complying with severe energy-saving rules and thereby avoiding fines and other penalties [32].

Thus the integration of wireless networking solutions empowers air conditioning systems to adopt energy-efficient strategies and reduce their carbon footprint. This benefits users by lowering energy bills and enhancing comfort, while also contributing to global sustainability efforts by curbing greenhouse gas emissions. As technology advances, wireless-enabled air conditioning systems are expected to play a pivotal role in achieving a more environmentally responsible approach to climate control.

11.7 SMART THERMOSTATS AND USER-CENTRIC APPROACHES

Smart thermostats are popular because they learn user behaviour and adjust temperature to save energy and money. They may be operated remotely through a smartphone app to keep a house at the correct temperature. Smart thermostats anticipate user preferences and adapt to assure comfort. Suppose you wake up at 6:00 AM and want a 72 °F temperature. A smart thermostat will start heating at 5:30 AM to reach and maintain that temperature. Customers seeking energy and cost savings choose smart thermostats [33, 34]. Smart thermostats learn human behaviour from occupancy trends, temperature preferences, and weather predictions. Using these patterns, they can automatically regulate the temperature for a pleasant household. Smart thermostats can also estimate climate control demands using machine learning for more accurate changes. They may discover that you warm your house before turning on the TV in the evening to watch TV.

11.8 CHALLENGES AND SECURITY CONSIDERATIONS

The use of wireless networking in HVAC systems has a number of potential advantages, but it also raises a number of potential difficulties and concerns about safety.

Potential interference and signal range issues: Problems with interference and signal range may interfere with communication between sensors, actuators, and the central control unit, which can result in the transmission of erroneous data or a delay in replies. It is possible for wireless signals to have a restricted range of coverage, particularly within huge buildings or in environments with many obstructions. It is vital for wireless networks to have scalability in order to meet rising demand; however, problems with scalability may lead to coverage gaps or performance deterioration.

Ensuring data security and privacy in wireless HVAC systems: In wireless HVAC systems, maintaining data privacy and security is of the utmost importance. Protecting sensitive data while it is being sent between devices and the central control unit requires the use of robust encryption techniques such as AES [35]. For the purpose of preventing unauthorised access, authentication measures, such as digital certificates or secure tokens, should be used. Access measures, including multi-factor authentication and role-based access control, should be in place to limit remote involvement.

As HVAC systems may gather data on the preferences and behaviours of occupants about their level of comfort, maintaining data privacy is vital. For the purposes of data collection and use, there should be clear privacy rules and user permission created. The administration of software patches and the implementation of regular software updates are both essential components of an effective software vulnerability mitigation strategy. Audits of the system's security as well as penetration tests have to be carried out on a regular basis in order to determine the system's vulnerabilities and flaws. To supplement the digital security that is in place, physical security measures need to be kept out of the hands of unauthorised persons.

Thus it is necessary to have a comprehensive incident response plan in place in order to provide a prompt and well-coordinated reaction in the event of a security breach or system breakdown [36]. It is essential, for the sake of ensuring the secure and dependable functioning of wireless HVAC systems, to find solutions to these issues and security concerns.

11.9 CASE STUDIES

This section offers real-world case studies that showcase effective implementations of wireless control systems for air conditioning. These examples illustrate the significant energy savings and increased user satisfaction that can be realised via the use of wireless networking technologies.

Case study 1: The effectiveness of energy conservation in commercial buildings

The subject of this case study is a major commercial office building that aimed to improve its energy efficiency while also ensuring that its inhabitants continued to enjoy the highest possible level of comfort. The management of the building built a wireless HVAC control system that was connected to IoT sensors and smart thermostats. This made it possible to monitor and regulate the heating and cooling remotely in real time. The system optimised temperature settings based on actual use by analysing occupancy patterns and environmental data. This led to a considerable decrease in energy consumption, which resulted in yearly savings of 30% on heating and cooling expenditures.

Case study 2: Integration of smart home technology in residential properties

This case study examines a domestic environment in which the residents wanted to modernise and improve the energy efficiency of their conventional HVAC system. Because the homeowners installed a wireless HVAC management system that included smart thermostats, occupancy sensors, and connection with a smart home ecosystem, their annual energy usage dropped by 25% as a direct consequence of their decision to use the system.

Case study 3: Providing a comfortable and sustainable retail environment

This case study focuses on a retail shop chain that worked towards enhancing the comfort of its customers while simultaneously reducing its overall carbon impact. As a result of the implementation of the system's wireless HVAC control system, which included wireless temperature sensors and capabilities for remote administration, overall energy consumption across all stores was reduced by 15%.

Case study 4: The precision of hospital climate control

This case study investigates the necessity of important locations inside a hospital, such as operating rooms and patient wards, having accurate temperature control. As a result of implementing a wireless HVAC management system with wireless sensors, real-time monitoring, and redundancy features, the hospital was able to reduce their energy consumption, improve the happiness of their users, and increase their level of comfort.

11.10 KEY INSIGHTS AND EMERGING OPPORTUNITIES: DISCUSSIONS AND FUTURE DIRECTIONS

Ultimately the goal of HVAC systems is to provide human comfort [37, 38]. If the indoor environment is good, the output of humans is also good, and they are more efficient in performing tasks [39–41]. This section explores the possibilities and emerging trends in wireless HVAC technology, as well as the integration of these systems with smart home ecosystems and building automation.

11.10.1 Emerging trends in wireless HVAC technology

The technology behind wireless HVAC is progressing at a fast pace, and a number of developing patterns indicate intriguing future prospects, as shown in Figure 11.3. Among these is improved sensor technology, which is able to assess things like air quality, occupancy, and individual preferences; this enables temperature management that is both more exact and more individualised. In order to optimise HVAC systems, artificial intelligence and machine learning will play a crucial role by analysing past data and user behaviour to forecast temperature preferences [42, 43]. Edge computing is becoming more popular in the HVAC industry and is being utilised to analyse data locally, which helps to reduce latency and improve reaction times [44]. The development of more energy-efficient HVAC technologies will continue, with the end goal of optimising energy savings while maintaining the same level of comfort as current systems. The elimination of old technology however produces e-waste [45]. The deployment of 5G networks will result in a connection that is both quicker and more dependable. This will result in improved real-time monitoring and control capabilities, particularly in big commercial and industrial environments. Figure 11.3 illustrates emerging trends in wireless HVAC technology.

11.10.2 Integration with smart home ecosystems and building automation

Users will be able to operate heating and cooling alongside other smart devices thanks to wireless HVAC systems, which integrate effortlessly with

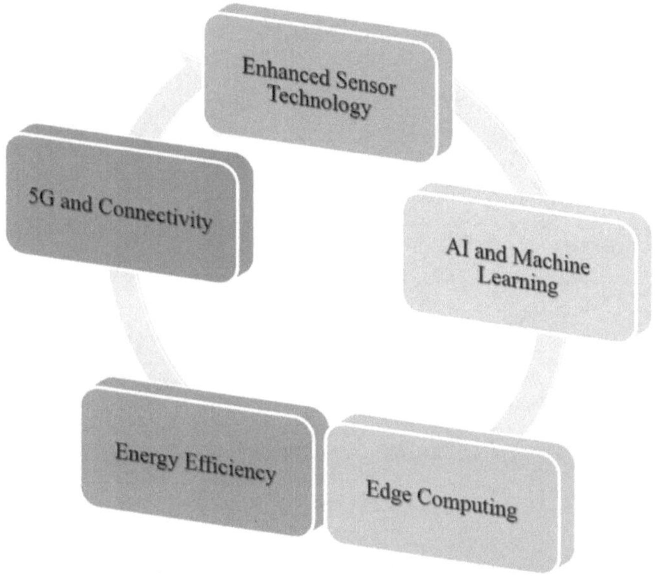

Figure 11.3 Emerging trends in wireless HVAC technology.

smart home ecosystems and are destined to revolutionise the heating, ventilation, and air conditioning market, as shown in Figure 11.4.

Speech-activated assistants such as Amazon Alexa and Google Assistant will become increasingly integrated with HVAC systems in the future [46, 47]. This will enable users to modify temperature settings using speech commands to the voice-activated assistant. Wireless HVAC systems will interface with building automation systems in commercial and industrial environments, allowing for centralised management and monitoring of all building systems. Recent works further explore machine learning's impact across healthcare, agriculture, and market intelligence [48–52]. In HVAC field also there are allied advances in context to machine learning, IoT, AI etc. Building automation systems will contain capabilities for predictive maintenance, which will use data from wireless sensors to detect problems with HVAC systems before they lead to failures. This will result in reduced downtime and expenses associated with maintenance. In addition, integration with building automation will make it possible to implement advanced energy management methods, therefore optimising HVAC operations in accordance with occupancy levels, weather predictions, and the cost of electricity. Through this integration, intelligent and responsive built environments will be created, which will improve quality of life while simultaneously reducing overall energy use [53–58]. Further optimisations of these systems also may improve system performances and efficiencies as in various engineering systems [59–62].

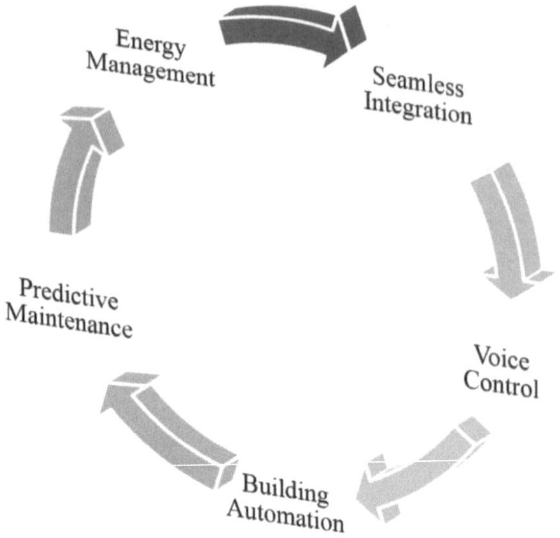

Figure 11.4 Integration patterns of smart home ecosystems and building automation.

11.11 CONCLUSION

This article provides an examination of the ways in which current air conditioning systems might benefit from the use of wireless networking technologies. The development of wireless connectivity and the IoT has profoundly transformed how we manage temperature and improve user comfort inside our built surroundings. This has been made possible by the IoT. Users are able to remotely operate their air conditioning systems via the use of smartphone apps and online interfaces as conventional control methods are rendered obsolete by the elimination of the physical limits imposed by wireless connection. Wireless sensors and actuators make it possible to monitor and regulate HVAC systems in real time, which contributes to the efficient saving of energy.

The collection of environmental data and the use of big data analytics, together with optimising cooling patterns and decreasing energy waste, make it possible to make decisions based on the collected data. Control techniques that make use of wireless connectivity may cut down on energy usage and carbon footprints. These tactics, which can include adaptive cooling and algorithms that are efficient with energy, are in line with global efforts to combat climate change. Personalised comfort settings may be provided through smart thermostats and user behaviour analysis, which both contribute to increased user happiness.

Building management that is both smooth and intelligent is something that might potentially be achieved in the future via the integration of wireless HVAC systems with smart home ecosystems and building automation.

Our interactions with air conditioning systems will continue to evolve as a result of developing trends in wireless HVAC technology. These developments include more sophisticated sensor capabilities, artificial intelligence, machine learning, and improved connection.

In conclusion, the convergence of wireless networking technologies with air conditioning systems represents a paradigm shift in the way we perceive and control the settings we find ourselves in indoors. This transition will ultimately result in increased comfort, less energy use, and more environmental responsibility. As we become more open to these shifts, we get closer to a future in which our houses and structures may actively respond to our requirements and inclinations, eventually elevating the standard of living we enjoy.

REFERENCES

1. Nazarian, N., Krayenhoff, E. S., Bechtel, B., Hondula, D. M., Paolini, R., Vanos, J., & Santamouris, M. (2022). Integrated assessment of urban overheating impacts on human life. *Earth's Future*, *10*(8), e2022EF002682.
2. Sarkar, S., Sengupta, A., Das, A., De, D., & Ahmed, N. (2023). Dew computing enabled consumer electronics for sustainable internet of agricultural things. In *Dew Computing: The Sustainable IoT Perspectives* (pp. 317–345). Singapore: Springer Nature Singapore.
3. Mazhar, T., Malik, M. A., Haq, I., Rozeela, I., Ullah, I., Khan, M. A., & Hamam, H. (2022). The role of ML, AI and 5G technology in smart energy and smart building management. *Electronics*, *11*(23), 3960.
4. Meena, C. S., Kumar, A., Jain, S., Rehman, A. U., Mishra, S., Sharma, N. K., & Eldin, E. T. (2022). Innovation in green building sector for sustainable future. *Energies*, *15*(18), 6631.
5. Wang, B. (2021). Interior decoration design and ecological reconstruction based on ecological energy saving technology. *Energy Reports*, *7*, 49–61.
6. Sodiq, A., Abdullatif, Y., Aissa, B., Ostovar, A., Nassar, N., El-Naas, M., & Amhamed, A. (2022). A review on progress made in direct air capture of CO2. *Environmental Technology & Innovation*, 102991. https://doi.org/10.1016/j.eti.2022.102991
7. Maddikunta, P. K. R., Hakak, S., Alazab, M., Bhattacharya, S., Gadekallu, T. R., Khan, W. Z., & Pham, Q. V. (2021). Unmanned aerial vehicles in smart agriculture: Applications, requirements, and challenges. *IEEE Sensors Journal*, *21*(16), 17608–17619.
8. Gaitan, N. C. (2021). A long-distance communication architecture for medical devices based on LoRaWAN protocol. *Electronics*, *10*(8), 940.
9. Kashyap, P. K., Kumar, S., Jaiswal, A., Prasad, M., & Gandomi, A. H. (2021). Towards precision agriculture: IoT-enabled intelligent irrigation systems using deep learning neural network. *IEEE Sensors Journal*, *21*(16), 17479–17491.
10. Aheleroff, S., Xu, X., Lu, Y., Aristizabal, M., Velásquez, J. P., Joa, B., & Valencia, Y. (2020). IoT-enabled smart appliances under industry 4.0: A case study. *Advanced Engineering Informatics*, *43*, 101043.

11. Antonopoulos, I., Robu, V., Couraud, B., Kirli, D., Norbu, S., Kiprakis, A., & Wattam, S. (2020). Artificial intelligence and machine learning approaches to energy demand-side response: A systematic review. *Renewable and Sustainable Energy Reviews*, *130*, 109899.
12. Campbell, R. W. (2020). Artificial intelligence in the courtroom: The delivery of justice in the age of machine learning. *Colorado Technology Law Journal*, *18*, 323.
13. Mariano-Hernández, D., Hernández-Callejo, L., Zorita-Lamadrid, A., Duque-Pérez, O., & García, F. S. (2021). A review of strategies for building energy management system: Model predictive control, demand side management, optimization, and fault detect & diagnosis. *Journal of Building Engineering*, *33*, 101692.
14. Bae, Y., Bhattacharya, S., Cui, B., Lee, S., Li, Y., Zhang, L., & Kuruganti, T. (2021). Sensor impacts on building and HVAC controls: A critical review for building energy performance. *Advances in Applied Energy*, *4*, 100068.
15. Shin, D., & Park, Y. J. (2019). Role of fairness, accountability, and transparency in algorithmic affordance. *Computers in Human Behavior*, *98*, 277–284.
16. Ngiam, K. Y., & Khor, W. (2019). Big data and machine learning algorithms for health-care delivery. *The Lancet Oncology*, *20*(5), e262–e273.
17. Saleem, M. U., Shakir, M., Usman, M. R., Bajwa, M. H. T., Shabbir, N., Shams Ghahfarokhi, P., & Daniel, K. (2023). Integrating smart energy management system with internet of things and cloud computing for efficient demand side management in smart grids. *Energies*, *16*(12), 4835.
18. Huseien, G. F., & Shah, K. W. (2022). A review on 5G technology for smart energy management and smart buildings in Singapore. *Energy and AI*, *7*, 100116.
19. Srivastava, H., & Akhai, S. (2022, November). The smart tapping identification model without installing a control program in modern wireless communication. In *2022 International Interdisciplinary Humanitarian Conference for Sustainability (IIHC)* (pp. 159–164). IEEE.
20. Al-Ali, A. R., Zualkernan, I. A., Rashid, M., Gupta, R., & AliKarar, M. (2017). A smart home energy management system using IoT and big data analytics approach. *IEEE Transactions on Consumer Electronics*, *63*(4), 426–434.
21. Ko, H. S., & Eshraghi, S. (2023). Automation in Home Appliances. In *Springer Handbook of Automation* (pp. 1311–1330). Cham: Springer International Publishing.
22. Jabbar, W. A., Kian, T. K., Ramli, R. M., Zubir, S. N., Zamrizaman, N. S., Balfaqih, M., & Alharbi, S. (2019). Design and fabrication of smart home with internet of things enabled automation system. *IEEE access*, *7*, 144059–144074.
23. Pan, J., Jain, R., & Paul, S. (2014). A survey of energy efficiency in buildings and microgrids using networking technologies. *IEEE Communications Surveys & Tutorials*, *16*(3), 1709–1731.
24. Mohsenian-Rad, A. H., & Leon-Garcia, A. (2010). Optimal residential load control with price prediction in real-time electricity pricing environments. *IEEE Transactions on Smart Grid*, *1*(2), 120–133.
25. Howarth, A., Ermanis, K., & Goodman, J. M. (2020). DP4-AI automated NMR data analysis: straight from spectrometer to structure. *Chemical Science*, *11*(17), 4351–4359.

26. Tian, H., Qin, P., Li, K., & Zhao, Z. (2020). A review of the state of health for lithium-ion batteries: Research status and suggestions. *Journal of Cleaner Production*, *261*, 120813.
27. Touvron, H., Martin, L., Stone, K., Albert, P., Almahairi, A., Babaei, Y., & Scialom, T. (2023). Llama 2: Open foundation and fine-tuned chat models. *arXiv preprint arXiv:2307.09288*.
28. Fu, Y., O'Neill, Z., Wen, J., Pertzborn, A., & Bushby, S. T. (2022). Utilizing commercial heating, ventilating, and air conditioning systems to provide grid services: A review. *Applied Energy*, *307*, 118133.
29. Aghniaey, S., Lawrence, T. M., Sharpton, T. N., Douglass, S. P., Oliver, T., & Sutter, M. (2019). Thermal comfort evaluation in campus classrooms during room temperature adjustment corresponding to demand response. *Building and Environment*, *148*, 488–497.
30. Goyal, S., Ingley, H. A., & Barooah, P. (2013). Occupancy-based zone-climate control for energy-efficient buildings: Complexity vs. performance. *Applied Energy*, *106*, 209–221.
31. Bibri, S. E. (2020). The eco-city and its core environmental dimension of sustainability: Green energy technologies and their integration with data-driven smart solutions. *Energy Informatics*, *3*(1), 1–26.
32. Mainetti, L., Palano, L., Patrono, L., Stefanizzi, M. L., & Vergallo, R. (2014, September). Integration of RFID and WSN technologies in a Smart Parking System. In *2014 22nd international conference on software, telecommunications and computer networks (SoftCOM)* (pp. 104–110). IEEE.
33. Tu, G., Faure, C., Schleich, J., & Guetlein, M. C. (2021). The heat is off! The role of technology attributes and individual attitudes in the diffusion of Smart thermostats–findings from a multi-country survey. *Technological Forecasting and Social Change*, *163*, 120508.
34. Ponce, P., Peffer, T., & Molina, A. (2017). Framework for communicating with consumers using an expectation interface in smart thermostats. *Energy and Buildings*, *145*, 44–56.
35. Rahman, Z., Yi, X., Billah, M., Sumi, M., & Anwar, A. (2022). Enhancing AES using chaos and logistic map-based key generation technique for securing IoT-based smart home. *Electronics*, *11*(7), 1083.
36. Akhai, S., & Balu, V. (2022, December). Code injection assault & mitigation model to prevent attacks. In *2022 11th International Conference on System Modeling & Advancement in Research Trends (SMART)* (pp. 1324–1328). IEEE.
37. Akhai, S., Thareja, P., & Singh, V. P. (2017). Assessment of indoor environment health sustenance in air conditioned class rooms. *Advanced Research in Civil and Environmental Engineering*, 4(1&2), 1–9.
38. Akhai, S., & Balu, V. (2022, December). Code injection assault & mitigation model to prevent attacks. In *2022 11th International Conference on System Modeling & Advancement in Research Trends (SMART)* (pp. 1324–1328). IEEE.
39. Akhai, S., Singh, V. P., & John, S. (2016). Investigating indoor air quality for the split-type air conditioners in an office environment and its effect on human performance. *Journal of Mechanical Civil Engineering*, *13*(6), 113–118.
40. Akhai, S., Singh, V. P., & John, S. (2016). Human performance in industrial design centers with small unit air conditioning systems. *Journal of Advanced Research in Production Industrial Engineering*, *3*(2), 5–11.

41. Akhai, S., Mala, S., & Jerin, A. A. (2021). Understanding whether air filtration from air conditioners reduces the probability of virus transmission in the environment. *Journal of Advanced Research in Medical Science & Technology (ISSN: 2394-6539)*, *8*(1), 36–41.
42. Akhai, S., Mala, S., & Jerin, A. A. (2020). Apprehending air conditioning systems in context to COVID-19 and human health: A brief communication. *International Journal of Healthcare Education & Medical Informatics (ISSN: 2455-9199)*, 7(1&2), 28–30.
43. Akhai, S. (2023). From black boxes to transparent machines: The quest for explainable AI. *Social Science Research Network*. doi: 10.2139/ssrn.4390887.
44. Chalapathi, G. S. S., Chamola, V., Vaish, A., & Buyya, R. (2021). Industrial internet of things (IIOT) applications of edge and fog computing: A review and future directions. In *Fog/edge computing for security, privacy, and applications* (pp. 293–325). doi: 10.1007/978-3-030-57328-7_12
45. Aggarwal, P., Rana, M., & Akhai, S. (2022). Briefings on e-waste hazard until COVID era in India. *Materials Today: Proceedings*, *71*, 389–393.
46. López, A. V. (2019). *Alexa & Google Assistant: The domestication and privacy implications of smart speakers and virtual personal assistants within the household* (Doctoral dissertation, Vrije Universiteit Brussel).
47. Akhai, S. (2023). Healthcare record management for healthcare 4.0 via blockchain: A review of current applications, opportunities, challenges, and future potential. *Blockchain for Healthcare 4.0*, *1*(1), 211–223.
48. Jain, V., & Chatterjee, J. M. (2020). Machine learning with health care perspective. *Cham: Springer*, 1–415.
49. Jain, V., Chatterjee, J. M., Hedayati, H., Krit, S., & Deperlioglu, O. (Eds.). (2021). *Deep Learning for Personalized Healthcare Services* (Vol. 7). Walter de Gruyter GmbH & Co KG.
50. Jain, V., Kumar, A., & Rathore, P. (2021). *Internet of Things and Machine Learning in Agriculture* (J. M. Chatterjee, Ed.). Nova Science Publishers, Incorporated.
51. Saroha, K., & Jain, V. (2024). Ethics in AI and Virtual Reality: Ensuring Responsible Development. In *Human-Centered Approaches in Industry 5.0: Human-Machine Interaction, Virtual Reality Training, and Customer Sentiment Analysis* (pp. 176–209). IGI Global.
52. Saroha, K., Sehrawat, M., & Jain, V. (2024). Machine learning and sentiment analysis for analyzing customer feedback: A review. *Big Data Analytics Techniques for Market Intelligence*, 411–440.
53. Tanwar, N., & Akhai, S. (2017). Survey Analysis for Quality Control Comfort Management in Air Conditioned Classroom. *Journal of Advanced Research in Civil and Environmental Engineering*, 4(1&2), 20–23.
54. Khang, A., & Akhai, S. (2024). Green Intelligent and Sustainable Manufacturing: Key Advancements, Benefits, Challenges, and Applications for Transforming Industry. *Machine Vision and Industrial Robotics in Manufacturing*, 405-417.
55. Akhai, S., & Khang, A. (2024). Energy Efficiency and Human Comfort: AI and IoT Integration in Hospital HVAC Systems. *Medical Robotics and AI-Assisted Diagnostics for a High-Tech Healthcare Industry*, 93–108.
56. Akhai, S. (2024). Towards Trustworthy and Reliable AI The Next Frontier. *Explainable Artificial Intelligence (XAI) in Healthcare*, *1*, 119–129.

57. Akhai, S., & Khang, A. (2024). Efficient Hospital Waste Treatment and Management Through IoT and Bioelectronics. *Revolutionizing Automated Waste Treatment Systems: IoT and Bioelectronics*, 126–140.
58. Akhai, S., & Khang, A. Innovations in Medical Manufacturing: A Review of 3D Printing, Robotics, and Internet of Things (IoT). *The Quantum Evolution*, 226–241.
59. Akhai, S. Innovating Healthcare Delivery: Harnessing Drones for Medical Services. In *Modelling of Virtual Worlds Using the Internet of Things* (pp. 232–250). CRC Press.
60. Akhai, S., & Kumar, V. (2024). Quantum Resilience and Distributed Trust: The Promise of Blockchain and Quantum Computing in Defense. In *Sustainable Security Practices Using Blockchain, Quantum and Post-Quantum Technologies for Real Time Applications* (pp. 125–153). Singapore: Springer Nature Singapore.
61. Kumar, H., Wadhwa, A. S., Akhai, S., & Kaushik, A. (2024). Parametric optimization of the machining performance of Al-SiCp composite using combination of response surface methodology and desirability function. *Engineering Research Express*, 6(2), 025505.
62. Kumar, H., Wadhwa, A. S., Akhai, S., & Kaushik, A. (2024). Parametric analysis, modeling and optimization of the process parameters in electric discharge machining of aluminium metal matrix composite. *Engineering Research Express*, 6(2), 025542.

Chapter 12

Malicious node detection in IoT networks using artificial neural networks

A machine learning approach

Kazi Kutubuddin Sayyad Liyakat

12.1 INTRODUCTION

Healthcare, public safety, monitoring, smart home, and environmental safety are some of the possible uses for IoT, and its uses are predicted to increase.[1] In the IoT, several devices are connected and exchange data. The sensory information is transmitted quickly by numerous devices before it reaches the aggregation node in a multi-hop network, where it can be utilised to inform the operator's decision-making. Several IoT protocols, like Thread, Z-Wave, and Zig-Bee/IEEE802.15,[2] now use the mesh architecture. Through multi-hop routing, devices with these characteristics can connect to other nearby devices[3] and even the base station.

Although IoT is emerging at a rapid pace, there are several security vulnerabilities related to it. Many IoT devices could be at risk of attack because of their simplicity. An attacker conducting a passive attack merely listens in on node-to-node communications using a wireless control. Subsequently, these can be difficult to recognise, as a passive attack works by collecting data without changing how a protocol typically operates.[4]

Businesses must test, run, debug, and secure these connected devices in real time as the number of IoT devices increases across the world. However, this job is difficult because:

I. First, there is little control over the devices because they are remotely deployed;
II. Second, the diverse environment makes the network susceptible to errors, flaws, and malicious threats.

To build adaptive and novel anomaly detection algorithms and malicious node detection to deal with the second issue, the anomalous activity of IoT devices must be identified. To address these issues, device fingerprinting methods including offline log analysis, statistical irregularity discovery, and rule-based identification have become popular. Security and privacy are important considerations. Behavioural traits can be monitored by machine learning (ML) based on habits, accuracy, confidentiality, and other useful

 DOI: 10.1201/9781003541363-12

parameters.[5] The real-time device classification solutions offered by ANN-based technologies that feature extraction-driven methods, however, are not available using the methodologies discussed above. To our knowledge, there is no satisfactory approach for investigating the literature concerning the long-term behaviour of devices; as a result, extracting statistical characteristics from the behavioural snapshot of raw IoT traffic data is necessary.[6]

Security attacks, both active and passive, put the security of the system at risk. Security attacks refer to unauthorised attempts to steal, alter, or destroy sensitive info or data coming from the system. During an active attack, an attacker tries to alter the message content. A passive attack entails the attacker viewing the messages, replicating them, and potentially utilising them maliciously. An attacker utilises the material within the message, copies it, and saves it for later use. In passive attacks, the attacker doesn't change the content of the initial message. While inactive attacks have no impact on the system, they may be damaging to message confidentiality. The victim is unaware of the attack, which is another crucial aspect of a passive attack. A passive attack doesn't alter the message, making it challenging to identify these attacks. By employing encryption methods, we might be able to stop passive attacks. An active security attack occurs when the attacker accesses the network's connection and attempts to change the message's information. Therefore, active attacks pose a serious threat to the message's accessibility and integrity. System assets may be altered by active attacks. Active attacks frequently involve deception, denial of service, content changes to messages, rejection, replication, etc. This harms the framework and its assets as a result. Keep in mind that in cases of ongoing attacks, the victim is made aware of the attack. The usage of one-time passwords (OTPs), the creation of random session keys, and other techniques may be used to stop attacks.

An active attack[3] (for instance, Blackhole attack,[4] Sybil attack) is where an attacker node actively violates the security protocols and poses a hazard to the availability and integrity of the network by swiftly introducing incorrect information, discarding and deploying data packets, and negatively affecting network protocol. Depending on where attacks originate from, they are categorised as either internal or external.

i. Internal attacks come through nodes already connected to a network and thus use its resources illegally. When a person or a group in an organisation tries to sabotage operations or take advantage of organisational resources, it is called an internal attack. In many instances, the attacker makes extensive use of resources, tools, and expertise to launch a complex computer attack and possibly even remove any traces of that attack. Internal threats might include, for example, data breaches occurring when staff improperly handle sensitive data or access private information without authorisation.

ii. External attacks originate from a node outside the network.[7] External threats are dangers that come from outside the company. Cyberattacks, natural disasters, economic fluctuations, and even regulatory developments are instances of these. While external threats are often out of managers' direct control, it is vital to proactively identify them and provide strategies to mitigate their impact.

Due to the significant harm malicious nodes can do to the structure and network's functioning, it is imperative that they are swiftly recognised. Several studies on the identification and recognition of unsafe behaviour and stream of traffic have been published in the literature. In this research, we formally address this issue by highlighting that a characteristic of regression analysis becomes appropriate for characterising nodes before establishing the trustworthiness of the network's nodes and routes.[8] The node's reliability is then determined using a support vector machine (SVM)[9] and linear regression (LR) methods.[10] Both approaches are adept at changing the recognition model by providing statistics, qualifying the detection of nodes having the best reputations.[11, 12]

Organisations are required to handle their affairs, evaluation, debugging, and safety in real time because of the widespread use of IoT devices.[13] Nonetheless, this task is difficult due to three important factors:

I. First, devices become challenging to manage because they have been installed remotely, and people find it difficult to effectively communicate with others due to different locations.[14, 15]
II. Second, itself is difficult.
III. Third, and possibly more importantly, is the issue of safeguarding structures and information from dangers, attacks, and other issues.

This third difficulty requires the development of adaptive and novel abnormality detection methods along with malicious node detection.[16–18] To do this, one must first recognise the specific behaviour of IoT devices. Some common fingerprinting of device techniques, such as in-person log analytics,[19] mathematical abnormality identification,[20] and rule-based detection,[21] have been applied to tackle these issues. They judge ML through the traditional CIA model[22] to collect behavioural characteristics that could be used in decisions where confidentiality and safety are crucial.[5, 23]

The aforementioned techniques cannot provide a real-time device classification solution, as opposed to ANN-based extraction procedures.[24] However, the analysis of long-term device behaviour requires statistical feature extraction using IoT traffic information to capture behaviour in a single image, which we were unable to find in the literature.[25] Our research attempted to identify features from headers of IoT traffic information to classify risky nodes. Figure 12.1 shows various IoT threat types, as well as

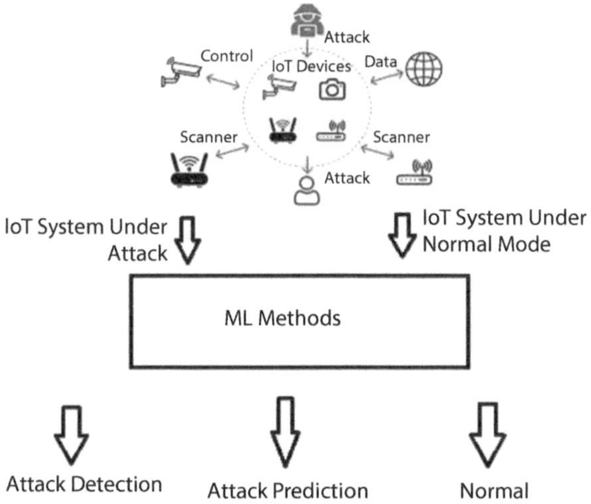

Figure 12.1 ML clarification process for detection of IoT attacks.

the related ML countermeasures.[26–28] Our technique is based on the idea that attributes can be merged with those of devices that have undergone long-term behavioural research, and traits can be derived from the packet headers of random nodes. This strategy enables us to develop a method that can identify malicious devices in IoT networks from nodes acting normally. To establish a malicious node detection approach based on ANN and build a threat model to predict future attacks, this chapter analyses data, including frames, from IoT devices. Additional security applications for this technique include device isolation, comprehensive log analysis, and packet header inspection. The author suggests another approach for IoT security, known as the KK approach which is suggested by Dr Kazi K.

12.2 PROBLEM STATEMENT

With the increasing use of wireless communication, experts predict that there will be a rise in undiscovered attacks as well as vulnerabilities.[29] Wi-Fi networks, mainly public networks, and connected smart machines in airports, hostels, restaurants, and other locations are therefore likely to face security issues. Nevertheless, it does not completely rule out the potential of attacks being conducted against private networks and household equipment. Any interface that allows users to connect instantly without requiring a password or that has a weak password might make it possible for an attacker to pretend to be a device and try to destroy the system. Smart device attacks are not difficult. To launch IoT-based cyberattacks, the attackers can employ home equipment with network access. Numerous fraudulent emails have recently been sent via devices including home routers and switches, TVs, refrigerators, PlayStations, and so forth.[30, 31] Let's say a network has several linked devices. A hacker may access the system and start exchanging data with it as if it were a typical device. There is no simple way to find such an abnormality, especially if there is a lot of equipment. This issue prompts us to propose a simple strategy capable of spotting security vulnerabilities in the environment of smart appliances on the network itself.

12.3 CONTRIBUTIONS

Here, we present a technique for identifying IoT device nodes that are malicious. In doing so, we created the dataset needed to put our technique into practice and then used an ANN classifier to analyse performance ratings based on the attributes that were extracted from the traffic information of each smart network. The following are the work's contributions:

I. Our technique exclusively uses NS2 networks with particular traffic statistics for categorisation. This could be a novel attempt to exploit network traffic data by employing an NS2 experimentation network for ANN classifiers in IoT contexts to detect rogue nodes.
II. Using intelligent IoT devices, we created the targeted system and demonstrated the performance of the proposed classifier. It reflects usage in the actual world.
III. The recommended approach is imitated as we utilise the NS2 programme. We assess the proposed technique and show that a variety of metrics, including throughput, latency, and packet loss, show good performance for our system.

12.4 THREAT MODEL

In IoT systems, the concept of threat is a helpful tool for understanding the system and avoiding risks. Because IoT settings are so complex, choosing an application might be helpful.[32–34] Public Wi-Fi networks[35] were the main area of attention while creating the threat model. Originally, these networks were meant to be local area networks (LANs) with mesh architecture with every device linked to a fixed access point (AP).[36, 37] For instance, restaurant information stored in the machine is sensitive and secure, and consumer smartphone[38] data is essential. IoT devices may be misused and used for illegal activities in such situations. When intelligent devices exchange data, an attacker may exploit them and utilise them for a variety of attacks.[39] In this research, we take into account IoT safety threats. These are depicted in Figure 12.2.

i. Spoofing: To gain illicit access into the IoT system, a spoofing node uses information such as MAC address and IP address to imitate a legitimate IoT device.[40]
ii. Sniffing: A popular network security attack known as sniffing involves a device removing sensitive data over the network.[13, 41] In wireless communications and settlements that are used for broadcasts, this attack occurs frequently. Attackers will attempt to sniff to gain crucial information such as passwords (computer, email,[42] FTP, databases) when they join a network via spoofing or imitating an IoT device.
iii. Packet sniffing: When an attacker attempts to see each packet that moves over the network, this is known as packet sniffing. The attacker may steal the data and analyse it if the data packets do not have encryption with robust network security. These IoT devices are susceptible to sniffer attacks, as they don't employ any encryption on the network.
iv. Malware: An attack involving malware is a malicious programme that an attacker can employ to commit a crime. The process starts with making GET calls for the server to retrieve port-mapping parameters by NAT punch holes and disclose the gathered local devices. Many different types of malware have recently been created to target IoT

Attack type	Attack Name
Spoofing	IP Spoofing MAC Spoofing
Sniffing	Packet Sniffing
Impersonation	Impersonation
DoS	Ping of Death Land Attack

Figure 12.2 IoT attack types.

devices. This kind of attack combines several harmful programmes, including backdoors, information thieves, spammers, and worms. Trojan horses and viruses represent a few of the serious kinds of malware employed by adversaries to abuse users, and backdoor attacks are capable of sending destructive data to any system (computer, router, database, etc.) at any moment.

v. Impersonation attack: Attackers seek to pass for either a physical node or component of an implicit node. These attacks may seriously affect system performance and are especially simple to initiate in IoT contexts.[17] Attacks that pretend to be someone else are occasionally the first stages of a full-throttle attack.

vi. DoS attack: A denial of service (DoS) attack aims to disable a service such that clients are unable to access the system or network. The goal of a DoS attack is to cause the targeted system to crash or restart by blocking massive amounts of memory on it. An attacker sends malicious data to the router by pretending to be a legitimate user and attacks the system in this way. Let's say the attacker delivers a data packet, and the firewall releases it by mistaking it as a data packet from a legitimate user. The payload of the packet can have enough data to allow an attack against the router.

   a. Ping of death: A hacker attempts to collapse the targeted system by sending an excessively large ping packet. The system is unable to manage such huge ping packets (Figure 12.3).
   b. Land attack: Another sort of DoS attack, a land attack send an unusually harmful data packet to a system or network in an effort to jam it.

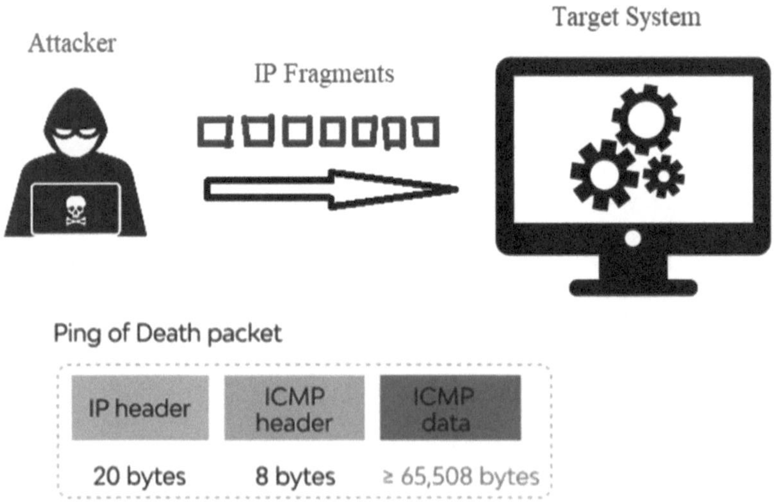

Figure 12.3 Ping of death.

12.5 PROPOSED METHOD

We established a procedure for recognising hypothetically harmful IoT systems based on ANN.[34] The employment of some supervised approaches, comprising ANN classifiers, and assessment of the consequences were prepared. We estimated the usefulness of the designated algorithms using our dataset. Algorithm 1 shown in Figure 12.4 is a pseudo-code which demonstrates the strategy.

Phase 1: First, we've got an illustration of the home (source) node and destination nodes of the network. RREQ messages swamp a network as the outcome of the home (source) node. The originating node defines routing time of routes (RRT), request (RREQ), and reply message (RREP). Once the RREQ flood starts, the timer at home (source) is started, and is stopped after all of the nodes get the RREP in order

ALGORITHM 1. DETECTION OF MALICIOUS NODES USING ARTIFICIAL NEURAL NETWORK(ANN) **Input:** *Classification training attributes are stored in a dataset, together with the 'learning rate' parameter L and the multi-layer neural classifier (ANN).*

Result: Detection of Malicious nodes using ANN Techniques

Begin

1. Implement WSN with limited no. of nodes

2. Locate the network nodes that will serve as both destinations and origins.

3. Establish a Path

3.1. If the path exits amid source and destination

3.1.1. Start data transmission

Else

3.1.2. Source flood RREQ packets in the network

3.1.3. Source start timer for noticing RRT

3.1.4. Source maintains list of every node with the time after receiving route reply message at source

3.1.5. Source selects the finest path amid source &target on the basis of hop count and sequence number

3.1.6. The path having least hop count and extreme sequence number is chosen as best path amid source & target

4. Malicious node Detection

4.1. The source starts to transmit data across the chosen path

4.2. The source notices data PAT at every hop

4.3. if node has higher time than RRT

4.3.1 Node marked as malicious

4.4. Else

4.4.1 source transmits data across the chosen path

4.4.2. Repeat step 4 till data transmission

Completed

End

Figure 12.4 Algorithm for proposed system.

to calculate the RRT. The source route records the entire round trip time(s) (RREQ, RREP) for respective nodes. The focus of the home (source) is on finding an ideal travel path by maintaining two factors. The best route of two locations is indicated by the fewest hops and uppermost sequence integer. The source node also computes the distance that exists between the source and the destination, taking into account the overall number of hops finished.

Phase 2: The process takes place to find any malevolent behaviour in the network. The information will then be transmitted via the route chosen to link the two locations, according to the source (home) node. The period that is required for a data packet to travel from its source to each intermediate node and then to its destination is its transit time. We look at the transit time between nodes and the round trip time (RTT) of the RREP messages. A small number of network nodes, particularly individuals with high temporal criticality for delivering data packets, are identified as malicious. The network is kept separate from the malicious swarm using a multipath routing technique. The network architecture will disregard a different route taken by those nefarious flies as a result

12.6 FEATURE EXTRACTION

In every ML design procedure, choosing the most important features is very important. Prior to feature extraction, we construct foundational profiles from the raw data packets of the devices for each data packet that comes in. We take the set of 18 attributes from the baseline profile. This feature extraction approach is the foundation of the malicious node detection technique. We identified the six characteristics from our baseline profile that are most crucial. The unprocessed data packets comprise details such as source/destination addresses (IP address), source/destination port, and protocols that are not enough as input to the learning method and result in noise and overfitting during the learning process. We developed a feature extraction technique as a result, and it is useful for IoT device learning algorithms. Our chosen feature includes the following:

i. *Period time*: This is the period of time in question. A packet that includes the interlude of time is crucial.
ii. *Previous captured frame (PCF)*: The time associated with the prior frame that was captured.
iii. *Previous displayed frame (PDF):* The period of time before the previous frame that was displayed.
iv. *Time since reference (TSR):* The time until reference serves as the starting point for all computations of the following packet's time.

v. *UDP payload length*: This refers to how much data is carried inside a UDP packet as a payload. In other words, this is the amount of information that a particular device sends as a whole.
vi. The total length of the packet is 250 packets with 1000 Byte bodes.

12.7 DEVELOP SYNTHETIC NEURAL NETWORK

For each NS2 network, we employed an artificial neural network (ANN) classifier as our primary method of detecting malicious nodes. This classifier replicates a biologically influenced computational technique that makes extensive use of neurons. Neurons are linked together and work simultaneously. An ANN is made up of several intricately interconnected process parts that transform a set of inputs into a set of desired outputs. Every neuron is connected to every additional neuron through robust communication links, representing the neural structure and giving it a total weight. Weights are what the network uses to resolve a problem as it evolves, although often, layers fall under the neuron category. A self-organising feature map was created using a neural network design that uses a single layer of neurons to encode input from a chosen domain inside the structure of a geometrically prearranged feature map.[29] The first step in categorising devices is the training phase. The ANN classifier is initially trained using the labelled dataset utilised in supervised approaches for device categorisation. The training dataset has restrictions on every feature that is needed to classify malicious nodes. In order to methodically organise the training data, backpropagation networks typically try to strengthen the relationships between neurons during the training process. Particularly close patterns are associated with the backpropagation's hidden layers as a consequence of the training. We note that there might be multiple hidden layers within the network.

12.8 FINDING MALICIOUS NODES

In this section, ANN conducts dot product summation with a bias for the test dataset using a weight set. Both hidden layers and output layers make use of the activation function. We used the sigmoid activation function to determine the various interaction weights.

$$\frac{1}{1+\exp(-x)} \tag{12.1}$$

Whether harmful nodes are found or not, the activation function provides the result. To assess how accurately the neural network was trained, we employed the fourth hidden layer. Our suggested technique is malicious node detection using a Python-based organisational codebase.

12.9 EXPERIMENTATION

Here, we examine the performance of our suggested ANN-based attack discovery method in terms of certain important metrics, like packet loss, accuracy, and delay. To assess how well our suggested system would work, we used the NS2 simulator. A "packet loss" denotes data packets that fail to reach their destination in a network. The additional time required for the data packet to reach its destination is the delay. The quantity of packets received that equals the entire quantity of data packets is referred to as network throughput. With 250 packets transmitted every session and a 1000-byte packet size, a test is done across a 500 x 500-metre network area. The prearranged total bandwidth (BW) ranges from 10 to 20 Megahertz.

Only BWs of 2 MHz, 4 MHz, or 6 MHz are manageable for IoT nodes.[75] The shared control channel is allotted a bandwidth of 2 Mbps. The information from ten distinct simulation runs was averaged to arrive at the final results. Figure 12.5 demonstrates methods for creating a path between the initial points to the finishing position. The latency is augmented by unfriendly nodes in the system. An improved ANN-based ML recognition methodology is engaged to detect these malicious behaviours when the system latency is too severe.

Figure 12.6 illustrates how the proposed ANN-based hybrid verification technique detects malicious nodes with less packet loss than the present system. The comparison shows that our suggested approach is more effective in packet loss. The grey line represents our system.

Figure 12.7 illustrates how the novel approach and the old way differ in terms of how time-consuming it is to complete a task. The suggested technique demonstrates an insignificant delay in comparison to the well-known scheme (displayed in blue) for recognising malicious node in a network, as designated by the grey line in the research results.

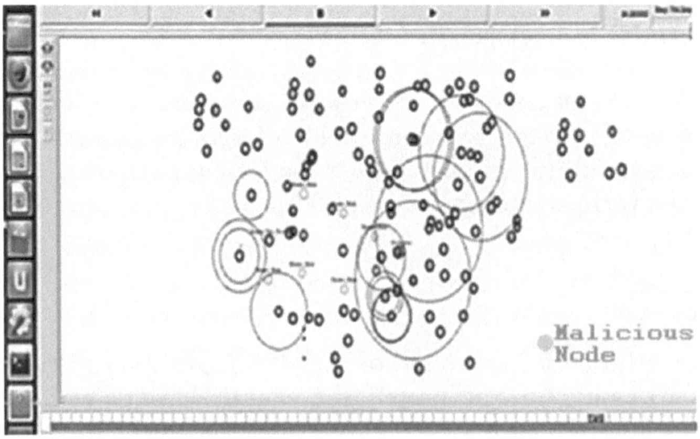

Figure 12.5 Node setup design using NS2.

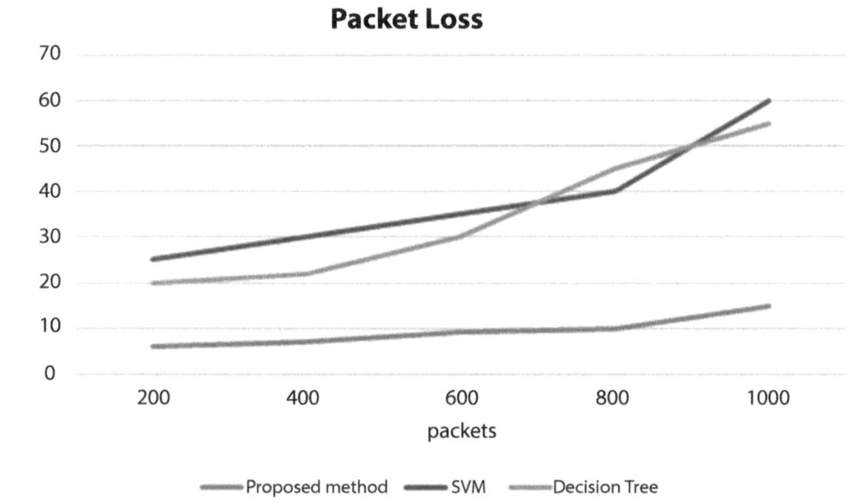

Figure 12.6 Packet loss comparison between different methods incorporating the suggested technique.

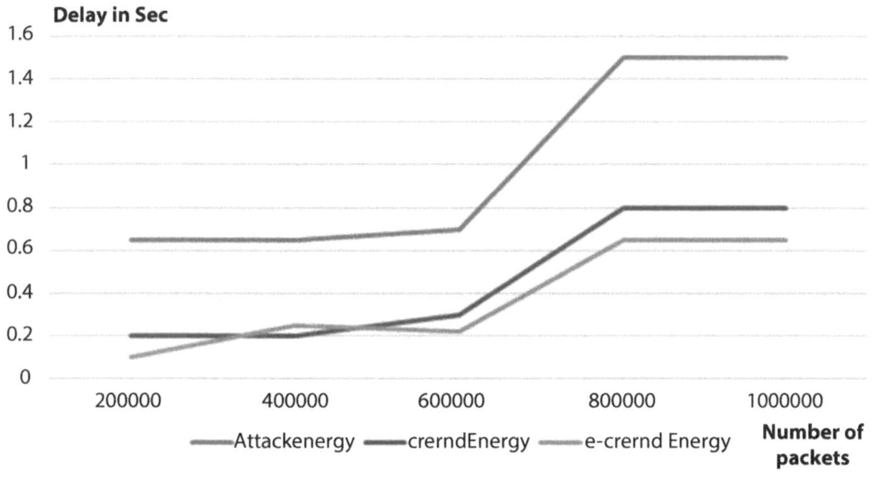

Figure 12.7 Delay comparison between different methods incorporating the suggested technique.

Figure 12.8 displays the outcomes of contrasting the suggested strategy with others. It proves that ANN's accuracy is superior to, or at least comparative with, prior investigations. The comparison shows that our suggested approach is the most appropriate for all uses. In the future, we intend to compare the effectiveness of the suggested system with established systems.

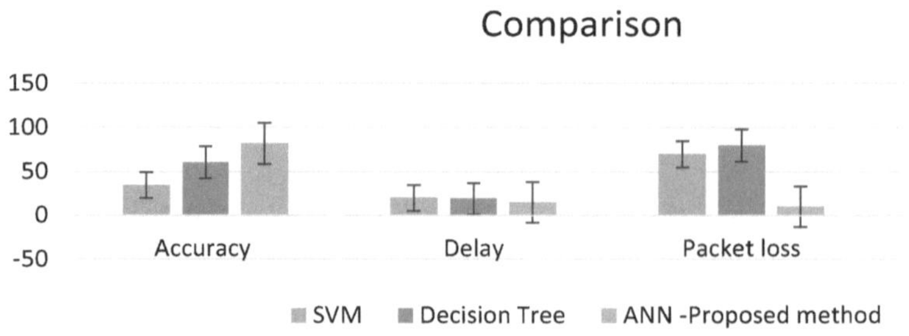

Figure 12.8 Comparative assessment of different methods incorporating the suggested technique.

The author also suggests another approach for IoT security, known as the KK approach.

12.10 CONCLUSION

The goal of this chapter was to explore a cutting-edge methodology that can help in spotting malicious activity in a network. This proposed approach is built on an ANN structure that employs ML to recognise dangerous nodes. The proposed strategy is contrasted with the existing trust-based system. In terms of latency and packet loss, the recommended approach performs better than the existing trust-based techniques. This suggested technique delivers circumstances for both anti-attack and trust-based networks. In conclusion, this chapter has shown how ANN may be used to explore network traffic sequentially for identifying rogue nodes in an IoT context. We likewise demonstrated using classification accuracy that this method was trustworthy. Once testing and training took place on the dataset, accuracy was improved. The identification of malicious IoT nodes was 81.51% accurate thanks to the ANN classifiers. In addition, we investigated and created a threat model to identify attempted adversarial attacks; however, these were more theoretical than actual. In the future, we would like to create an embedded system that queries blockchain network nodes for updates to the software and then analyses the response to determine if the requested firmware is there or not. If it is not already present, the most recent firmware is downloaded to the embedded device via the peer-to-peer node network. Although the latest version will be the most recent, the firmware's validity or correctness is still confirmed.

REFERENCES

1. Tseng, F.-H., Chiang, H.-P., and Chao, H.-C. Black hole along with other attacks in MANETs: a survey. *Journal of Information Processing Systems* 2018; 14:56–78.

2. Jan, M. A., Nanda, P., and Liu, R.P. A sybil attack detection scheme for a forest wildfire monitoring application. *Future Generation Computer Systems* 2018; 80:613–26.
3. Papernot, N., McDaniel, P. Sinha, A., and Wellman, M. P. "Sok: Security and privacy in machine learning," in 2018 IEEE European Symposium on Security and Privacy (EuroS P), April 2021, pp. 399–414.
4. Meidan, Y., Bohadana, M., . Mathov, Y., Mirsky, Y., Shabtai, A., Breitenbacher, D., and Elovici, Y., N-baiot—network-based detection of iot botnet attacks using deep autoencoders. *IEEE Pervasive Computing* 2019; 17(3):12–22.
5. Mendhurwar, S., and Mishra, R., Integration of social and IoT technologies: architectural framework for digital transformation cyber security challenges. *Enterprise Information Systems* 2021; 15(4):565–584.
6. Ashraf, S., Shawon, M. H., Khalid, H. M., and Muyeen, S. Denial-of-service attack on IEC 61850-based substation automation system: a crucial cyber threat towards smart substation pathways. *Sensors* 2021; 21:6415.
7. Khalid, H. M., and Peng, J. C. H. Immunity toward data-injection attacks using multisensor track fusion-based model prediction. *IEEE Trans. Smart Grid* 2017; 8:697–707.
8. Alsharif, M., and Rawat, D. B. Study of machine learning for cloud assisted IoT security as a service. *Sensors* 2021; 21:1034.
9. Choi, C., and Choi, J. Ontology-based security context reasoning for power IoT-cloud security service. *IEEE Access* 2019; 7:110510–110517.
10. Ge, C., Susilo, W., Liu, Z., Xia, J., Szalachowski, P., and Fang, L. Secure keyword search and data sharing mechanism for cloud computing. *IEEE Trans. Dependable Secur. Comput.* 2021; 18:2787–2800.
11. Luo, T., and Nagarajan, S. G. Distributed anomaly detection using autoencoder neural networks in WSN for iot. In: 2018 IEEE International Conference on Communications (ICC); 2018, pp. 1–6.
12. Liu, X., Abdelhakim, M., Krishnamurthy, P., and Tipper, D. Identifying malicious nodes in Multi-hop IoT networks using diversity and unsupervised learning. In: IEEE International Conference on Communications; 2018, pp. 1–6. https://doi.org/10.1109/icc.2018.8422484.
13. Liyakat, K. K. S. Machine Learning Approach Using Artificial Neural Networks to Detect Malicious Nodes in IoT Networks. In: Udgata, S. K., Sethi, S., and Gao, X. Z. (eds) *Intelligent Systems. ICMIB 2023. Lecture Notes in Networks and Systems*, vol. 728. Springer, Singapore. https://doi.org/10.1007/978-981-99-3932-9_12.
14. Halli, U. M. Nanotechnology in IoT security. *J Nanosci Nanoeng Appl* 2022; 12(3):11–16.
15. Wale Anjali, D., et al. Smart agriculture system using IoT. *Int J Innov Res Technol* 2019; 5(10): 493–497.
16. Kazi, K. S. Detection of malicious nodes in IoT networks based on throughput and ML. *J Electr Power Sys Eng* 2023; 9(1):22–29.
17. Kazi, K. Model for agricultural information system to improve crop yield using IoT. *J Open Source Dev* 2022; 9(2):16–24.
18. Akansha, K. et al. Email security. *J Image Proc Intel Remote Sens* 2022; 2(6). https://doi.org/10.55529/jipirs.26.23.31.
19. Kazi, K. S. L. Nanotechnology application in neural growth support system. *J Nanote Its Appl* 2022; 24(2):47–55.

20. Kazi, K. S. L. A novel design of IoT based 'love representation and remembrance' system to loved one's. *Grad Rev J* 2022; 8(12):377–383.
21. Kazi, K. Multiple object Detection and Classification using sparsity regularized Pruning on Low quality Image/ video with Kalman Filter Methodology (Literature review). 2022
22. Kazi, K. Smart grid energy saving technique using machine learning. *J Instrument Technol Innov* 2022; 12(3): 1–10.
23. Pradeepa, M. et al. Student Health Detection using a Machine Learning Approach and IoT. 2022 IEEE 2nd Mysore sub section International Conference (MysuruCon), 2022.
24. Kazi, K. S. L. Business mode and product life cycle to improve marketing in healthcare units. *E-Com Fut Trend* 2022; 9(3): 1–9.
25. Kazi, K. S. IoT based healthcare system for home quarantine people. *J Instrument Innov Sci* 2023; 8(1): 1–8.
26. Kazi, K. S. L. IoT-based weather Prototype using WeMos. *J Cont Instrum Eng* 2023; 9(1):10–22.
27. Kazi, K. Detection of malicious nodes in IoT networks based on packet loss using ML. *J Mobile Comp, Comm Mob Net* 2022; 9(3):9–16.
28. Kazi, K. S. IoT-based healthcare monitoring for COVID-19 home quarantined patients. *Recent Trends Sens Res Tech* 2022; 9(3):26–32.
29. Kazi, K. Blockchain-enabled iot environment to embedded system a self-secure firmware model. *J Telecommun Stud* 2023; 8(1).
30. Liyakat, K. K. S. "Detecting Malicious Nodes in IoT Networks Using Machine Learning and Artificial Neural Networks," 2023 International Conference on Emerging Smart Computing and Informatics (ESCI), Pune, India, 2023, pp. 1–5, doi: 10.1109/ESCI56872.2023.10099544.
31. Kasat, K., Shaikh, N., Rayabharapu, V. K., and Nayak, M. "Implementation and Recognition of Waste Management System with Mobility Solution in Smart Cities using Internet of Things," 2023 Second International Conference on Augmented Intelligence and Sustainable Systems (ICAISS), Trichy, India, 2023, pp. 1661–1665.
32. Liyakat, K. K. S. IoT based Smart Health Care Monitoring. In: Rhituraj Saikia (eds), *Liberation of Creativity: Navigating New Frontiers in Multidisciplinary Research*, Vol. 2, July 2023, pp. 456–477, ISBN: 979-8852143600
33. Liyakat, K. K. S. IoT based Substation Health Monitoring. In: Rhituraj Saikia (eds), *Magnification of Research: Advanced Research in Social Sciences and Humanities*, Volume 2, October 2023, pp. 160–171, ISBN: 979-8864297803
34. Nerkar, P. A., et al. Monitoring fresh fruit and food using Iot and machine learning to improve food safety and quality. *Tuijin Jishu/J Propul Technol* 2023; 44(3):2927–2931. https://propulsiontechjournal.com/index.php/journal/article/view/914
35. Liyakat, K. K. S. Integrating IoT and mechanical systems in mechanical engineering applications. *J Mech Robot* 2023; 8(3):1–6.
36. Liyakat, K. K. S. IoT in the electric power industry. *J Control Convert* 2023; 8(3):1–7.
37. Liyakat, K. K. S. ML in the electronics manufacturing industry. *J Switch Hub* 2023; 8(3):9–13.
38. Liyakat, K. K. S. IoT in electrical vehicle: A study. *J Contr Instrum Eng* 2023; 9(3):15–21.

39. Kazi, K. AI-Driven IoT (AIIoT) in Healthcare Monitoring. In Nguyen, T. and Vo, N. (eds), *Using Traditional Design Methods to Enhance AI-Driven Decision Making*. IGI Global, 2024, pp. 77–101. https://doi.org/10.4018/979-8-3693-0639-0.ch003 available at: https://www.igi-global.com/chapter/ai-driven-iot-aiiot-in-healthcare-monitoring/336693
40. Magadum, P. K. Machine learning for predicting wind turbine output power in wind energy conversion systems. *Grenze Int J Eng Technol* 2024; 10(1):2074–2080. Grenze ID: 01.GIJET.10.1.4_1
41. Kazi, K. Modelling and Simulation of Electric Vehicle for Performance Analysis: BEV and HEV Electrical Vehicle Implementation Using Simulink for E-Mobility Ecosystems. In Nagpal, L. D. N., Kassarwani, N., Varthanan, V., and Siano, P. (eds), *E-Mobility in Electrical Energy Systems for Sustainability*. IGI Global, 2024, pp. 295–320. https://doi.org/10.4018/979-8-3693-2611-4.ch014 Available at: https://www.igi-global.com/gateway/chapter/full-text-pdf/341172
42. Kazi, K. S. Computer-Aided Diagnosis in Ophthalmology: A Technical Review of Deep Learning Applications. In Garcia, M., and de Almeida, R. (eds.), *Transformative Approaches to Patient Literacy and Healthcare Innovation*. IGI Global, 2024, pp. 112–135. https://doi.org/10.4018/979-8-3693-3661-8.ch006 Available at: https://www.igi-global.com/chapter/computer-aided-diagnosis-in-ophthalmology/342823
43. Kazi, K. S. (2024b). IoT Driven by Machine Learning (MLIoT) for the Retail Apparel Sector. In T. Tarnanidis, E. Papachristou, M. Karypidis, & V. Ismyrlis (Eds.), Driving Green Marketing in Fashion and Retail (pp. 63-81). IGI Global. Available at: https://doi.org/10.4018/979-8-3693-3049-4.ch004
44. Kazi, K. S. (2024). Artificial Intelligence (AI)-Driven IoT (AIIoT)-Based Agriculture Automation. In S. Satapathy & K. Muduli (Eds.), *Advanced Computational Methods for Agri-Business Sustainability* (pp. 72–94). IGI Global. Available at: https://doi.org/10.4018/979-8-3693-3583-3.ch005
45. K. Rajendra Prasad, Santoshachandra Rao Karanam (2024). AI in public-private partnership for IT infrastructure development, *Journal of High Technology Management Research*, Volume 35, Issue 1, May 2024, 100496. Available at: https://doi.org/10.1016/j.hitech.2024.100496

Chapter 13

Blockchain technology-based framework to build cyber-secured smart cities

Nashra Javed, Halima Sadia, Tasneem Ahmed, and Mohammad Faisal

13.1 INTRODUCTION

Urbanization is a global trend, and when it comes to developing cities with smart tools we have a plethora of ICT-based applications that can help in the transformation. The term "smart city" has lots of meanings associated with it. When the word "smart" is connected to devices, this means that the machines can work efficiently and intelligently with human monitoring. With the utilization of smart innovations, general increases have been found in the populace, economy, and productivity of city life. It was predicted by the United Nation's study on global urbanization that in 2050, 68% of the world's population will reside in metropolitan areas [1]. When the population of an urban area surpasses 1.5 million, it is considered a megacity [2]. The smart city's intelligence is defined by its ability to seamlessly integrate all its resources, efficiently and flawlessly achieving objectives and goals aimed at constructing safer smart cities [3]. But with all the advancements, the risks associated with breaches of confidential information are one of the major concerns for the government and individuals. The requirement for an infrastructure that can support cyber security is the topmost priority for all smart cities [4]. Considering the needs and challenges for urbanization, this study includes the major components that are primarily required for the transformation of any place into a smart city. A multi-tier architecture is proposed for the implementation of advanced technologies necessary for urbanization. A rigorous literature review is conducted to find out the sub-components of smart city multi-tier architecture.

13.2 RELATED STUDY

A literature review is required to analyze the different architectures proposed by researchers for the implementation of smart cities and how data integrity is ensured through blockchain technology. Many strategies are proposed by various authors for confronting the urbanization challenges

DOI: 10.1201/9781003541363-13

and exploring the opportunities for building safer smart cities. Table 13.1 provides a literature review with research focuses and limitations.

13.3 SMART CITY MAJOR COMPONENTS

There is a wide variety of components that help in providing an intelligent and efficient solution for services required by society [10]. The combination of these components allows us to effectively control and monitor different products and services for users. Figure 13.1 demonstrates how components are merged and linked with the importance of connectivity and data for creating a smart city [3]. The major elements of a smart city comprise the services and contemporary developments that might contribute to urbanization [11]. These components will be key in the transformation of cities into smart cities:

- *Smart living:* This includes all the innovative technological advancements that help to improve the standard of living in urban areas.
- *Smart healthcare:* This includes remote patient monitoring, telemedicine, wearable devices for health monitoring, ingestible sensors, robotic surgery, etc., for enhancing healthcare systems.
- *Smart infrastructure:* To facilitate the expansion of smart infrastructure, including water and energy networks, roadways, and buildings, among others, this integrates sensors and smart grid technology [3].

Table 13.1 An overview of the literature review conducted

References	*Year*	*Research focus*	*Limitations*
[5]	2023	The study shows the present and future prospects of smart cities along with a detailed review of challenges and opportunities.	No strategy is proposed for dealing with the challenges faced during smart city implementation.
[6]	2023	Aims to define and implement smart public policies for urbanization and identification of low-cost solutions.	The proposed model has limited indications for defining large urban areas.
[7]	2021	Provides a blockchain-based solution for data security when transforming cities into smart cities.	Inclusion of more complex business networks having multiple sensors in the vehicles.
[8]	2020	Integration of blockchain with IoT to provide a decentralized architecture for secured communication.	No implementation and testing of architecture was performed with general analysis.
[9]	2019	Focus on e-governance in concurrence with smart contracts.	For data security needs of blockchain protocols, implementation is required.

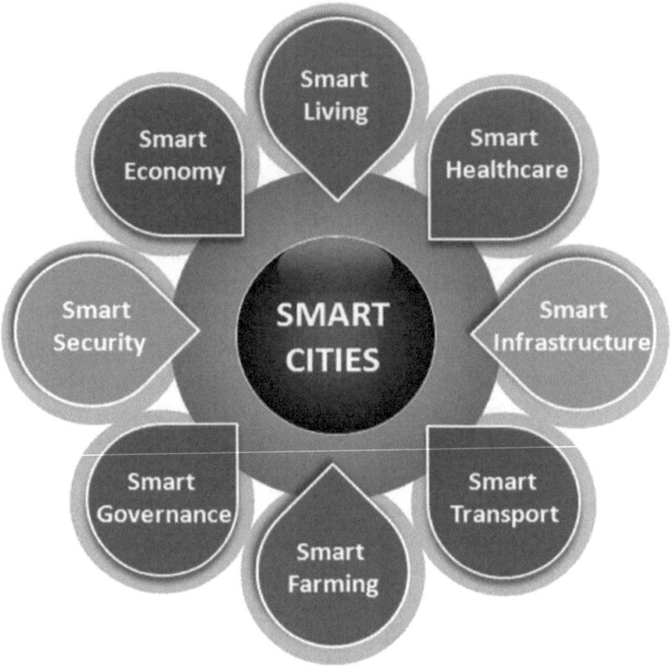

Figure 13.1 Important components required for the implementation of a smart city.

- *Smart transport:* This consists of smart parking, real-time monitoring and control systems, and smart traffic management systems for city surveillance.
- *Smart farming:* This includes sensors for soil, water, light, humidity, temperature management, IoT-based software, telematics, positioning technologies, and so on for improving farming methods.
- *Smart governance:* To effectively govern smart cities for public services, this includes e-governance, a system that utilizes technology and innovation to aid and streamline improved decision-making and planning processes [12].
- *Smart security:* This includes security features for e-banking, e-commerce, home security, city surveillance, and so on.
- *Smart economy:* This comprises technological advancements and innovation that boost economic development, job creation, and urban expansion [3].

All these parts work together to give smart services to city residents. In addition, smart governance supports the implementation of smart city missions for urbanization [4]. Proper infrastructure is required to execute these components for smooth functioning.

13.4 SMART CITY ARCHITECTURE

With recent advancements in emerging Internet-based technologies like smartphones, data analytics, cloud technologies, cyber security, and the Internet of Things, the emphasis is now primarily turning toward sustainability for smarter and better user interfaces. In the literature on intelligent smart cities, the concept of smart cities has been assessed from the perspective of technology and elements that display certain distinctive characteristics [3]. It centers on the latest advancements in agent technologies, mobile computing, and their growing integration with urban physical environments. Intelligent cities, which encompass knowledge-intensive activities, collaborative institutions, and web-based collective intelligence applications to establish territorial innovation systems, differ from smart cities in their focus on intelligent embedded devices [13, 14]. The foremost challenge is to address urbanization issues and prioritize improvements in a globally innovation-focused world [3, 15]. The initial step toward developing smart cities involves establishing a robust broadband network environment capable of supporting digital applications. This environment should encompass: (a) establishing a robust broadband infrastructure that utilizes cables, wireless networks, and optical fibers for delivering high-speed connectivity to both residents and businesses; (b) provision of real-time data management, such as alerts, and information processing by integrating embedded systems, sensors, actuators, and smart devices into the physical environments and infrastructures of cities [3]; and (c) creating software that facilitates data collection and processing, fosters web-based collaboration, and promotes the collective intelligence of citizens [3, 16]. The second goal involves initiating extensive participatory innovation initiatives aimed at developing software that will function to improve every sector, urban center, and infrastructure. All economic activities and utilities within the city, essentially functioning as innovation ecosystems, entail the involvement of both citizens and organizations in the generation, delivery, and utilization of goods and services [17]. Different countries have different plans for smart city implementation, according to their population, resources, and economy. A solution for developing smart cities can be provided by a layered architecture that is information and communication technology (ICT) based, for enhancing services and support for citizens [14]. Smart city architecture can be understood through the multi-tier communication system that has varied technologies across the layers that work in an integrated manner to deliver smart city services [15, 18]. Figure 13.2 shows the multi-tier architecture of the smart city, which consists of the various components that help in the transformation of smart cities. The four layers of smart city multi-tier architecture comprise a plethora of components that build a smart city in its true sense.

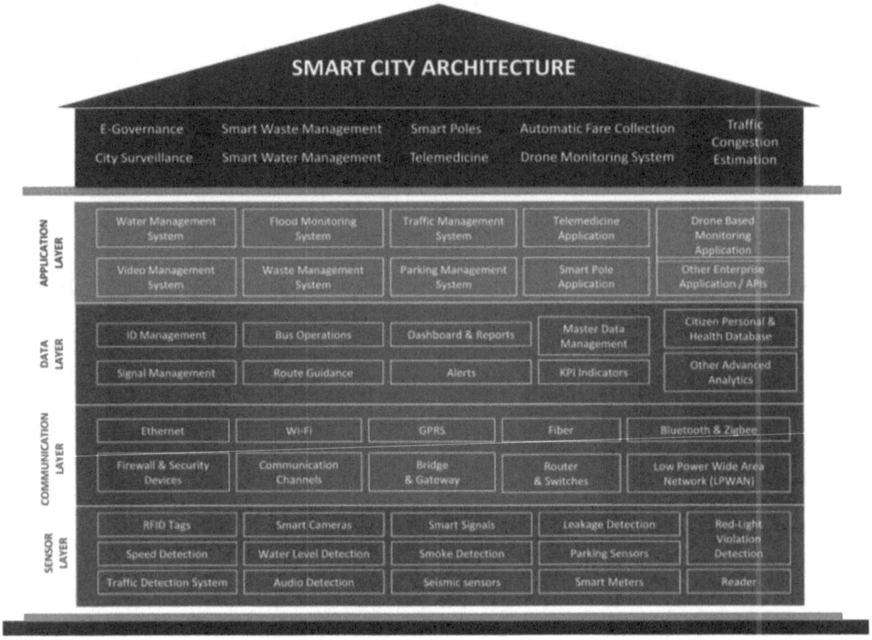

Figure 13.2 A four-layer multi-tier architecture for building smart cities.

13.4.1 Application layer

This layer enables smart city administration and provides smart services to the citizens that are included in the application layer for efficient operation [14]. The components that work at the application layer to support the government for urbanization are as follows. (a) Water management system: smart cities can use a circular strategy to minimize their dependency on freshwater by recycling water multiple times [19]. (b) Flood monitoring system: ultrasonic depth sensors assist in flood and water monitoring systems to measure water levels. (c) Traffic management system: this system can help to program traffic lights in real-time traffic conditions through advanced traffic control technology based on artificial intelligence. (d) Telemedicine applications: these enable asynchronous doctor's appointments and the store-and-forward functionality that helps to transfer diagnostic images or other health-related data to an expert/doctor. Through online video conferencing, patient counseling, disease diagnosis, and therapy management can be done with ease [20]. (e) Drone-based monitoring application: the use of drones in smart cities will significantly improve the quality of urban living. Several applications like traffic monitoring, security, transportation, and monitoring of infrastructure development will assist in the creation and maintenance of smart cities in an efficient way. (f) Video management system: video management software allows several activities that are important in today's big and small cities to be carried out, such as maintaining the safety of residents and improving the quality of city services. (g) Waste management system: the present state of smart waste management technology involves the use of sensors that measure the fill level of trash bins. (h) Parking management system: in order to maximize the use of available parking spots in big parking lots, this system will deliver cutting-edge sensing and monitoring technology. (i) Smart pole application: delivering wireless connectivity such as 4G LTE, 5G, Wi-Fi, LoRa, and Sigfox by making use of the infrastructure of street lights. A larger range of connections for Internet of Things-based applications is made possible by the wireless networking capabilities of smart poles [21]. (j) Other enterprise applications/APIs: in order to give flexibility and access to real-time information for the design and delivery of effective services through multiple APIs, city system integration is crucial [22].

13.4.2 Data layer

Data is essential to smart cities and may be created or utilized. It can also be kept centrally or distributed. The application layer for managing smart cities is supported by a variety of data-based tools and capabilities in the data layer. The components that work at the data layer for managing data flow are as follows. (a) ID management: devices require their own digital identities and digital identity management systems to securely identify each other, share relevant data, and form relationships between devices and users to interact. (b) Bus operations: by minimizing traffic congestion on roadways and strategically assigning and routing buses to locations with more travelers, you can assist individuals to their destinations much faster. (c) Dashboard and reports: these are important for monitoring and management of smart city-related services. They are primarily used to track events and monitor critical cases across the city. (d) Master data management: this is utilized to orchestrate operational and analytical big data applications efficiently in urban cities. (e) Citizen personal and health database: information about a living individual who may be recognized utilizing such information is referred to as personal data [23]. Cities must have reliable metrics and long-term monitoring tools to respond quickly to potential risks to maintain community health and safety. Cities may use open data to safeguard their population from crime, violence, drug epidemics, and other emergencies. (f) Signal management: advanced traffic management technology, such as adaptive traffic control and traffic analytics, can improve safety and reduce traffic congestion and greenhouse gas (GHG) emissions. Signal management is done to control traffic and avoid congestion. (h) Route guidance: due to constantly changing road and traffic circumstances, driving in modern cities is becoming increasingly difficult, necessitating the use of navigation systems for successful route planning. In evaluating road and traffic conditions, navigation systems are becoming increasingly sophisticated and precise. Through this, not only will travel times be minimized but also the efficiency of roads in the city network and traffic performance at the route level of analysis will be enhanced by using routing advice. (i) Alerts: the intelligent traffic alert system employs visual and audible cues to alert drivers to potential dangers on the road. The transmitter units on the roadside send crucial information to motorists in real time over the commercial FM radio frequency range. It can provide information like speed limit, the state of the road, and unexpected traffic circumstances to the drivers [24]. (j) KPI indicators: key performance indicators (KPIs) help to develop criteria for evaluating ICT's contribution to making cities smarter and more sustainable. Also, this equips cities with self-assessment tools. Each indicator is part of a three-dimensional picture of a city's performance: economy, environment, society, and culture. (k) Other advanced analytics: more advanced analytics can assist in building smart cities. Artificial intelligence-based applications and analysis can easily optimize the data layer services.

13.4.3 Communication layer

Multiple access and aggregation networks ranging from public to private and cable to wireless are used to connect people, objects, and sensors in smart cities. Effective communication infrastructure is the major requirement for cities to transform themselves into smart ones [14]. Many of the smart city advances – such as in energy, transportation, and the effective delivery of public services – rely on the availability of ubiquitous communication services. It is difficult for a city to become more efficient, sustainable, and secure for its residents without the proper communication networks [25]. (a) Ethernet: this is a communication technology primarily employed for creating local area networks, utilizing wired connections to transmit and receive data. (b) Wi-Fi: this is the wireless technology that links computers, tablets, cell phones, and various other devices to the internet. GPRS: the general packet radio system, also known as GSM-IP (global-system mobile communications internet protocol), keeps all users connected, allowing them to make phone conversations and browse the internet while on the go. (c) Fiber: this involves the transmission of data using optical fibers that transport light from one end to the other. (d) Bluetooth and Zigbee: devices can connect over Bluetooth without the need for cables or wires. Any device can interact using the short-range radio frequency technology known as Bluetooth as long as it is close enough to the other device. Zigbee is commonly employed in low-data-rate applications that require extended battery life. Smart living: this includes all the innovative technological advancements that help to improve the standard of living in urban areas. (e) Firewall and security devices: these filter incoming and outgoing network traffic as per the organization's security guidelines; a network security tool known as a firewall monitor is used [26]. (f) Communication channels: a network path that links the sender and recipient for the transfer of data is called a communication channel, often referred to as a communication medium or transmission media. (g) Bridge and gateway: a gateway is used to improve communication interoperability between two networks that utilize different protocols. A bridge always works with frames, but a gateway works with packets. (h) Routers and switches: a router functions at the network layer, connecting multiple networks and facilitating the transfer of data packets between them. (i) Low power wide area network (LPWAN): these systems need high receiver sensitivities of up to -130 dBm to provide long-range capabilities, as opposed to -90 to -110 dBm in many typical wireless technologies. As a result, the energy per bit is higher, which causes the modulation rate to be slower [27].

13.4.4 Sensor layer

The major elements of these layers are sensors or actuators that can monitor or control a physical item [25]. They are utilized to keep track of data on the status of a process or the environment, including temperature, chemical composition, humidity, fluid levels in tanks, fluid flow in pipes, and the speed of an assembly line [28]. Through an actuator, corrective action can be taken in real time by, for instance, changing the motions of an industrial robot, whenever a sensor detects a condition or event that requires a nearly instantaneous response [14]. (a) RFID tags: both passive and active RFID tags can store data, and when an RFID reader is close to any passive tags, it produces an electromagnetic field. It gives the devices power (longer ranges for active tags), enabling them to access information from the memory of the tag. (b) Smart cameras: Artificial Intelligence-based security cameras will become a major component for applications that are meant for intelligent traffic management systems and proactive danger identification. (c) Smart signals: intelligent transportation systems blend with smart signaling in urban traffic systems, enabling consumers to control the city's transportation system algorithmically and according to context via a single user interface. (d) Leakage detection: fog technology, IoT-based smart water distribution, and monitoring systems will act as key components for the detection of water leakages. Their scope will range from customer water usage, water quality management, fault prediction, water theft detection, and problem localization. (e) Red-light violation: to enhance the traffic management system the red-light violation detection (RLVD) system will be able to deliver safe driving of vehicles on the road, and it will also improve adherence to traffic regulations and smooth vehicle movement along the stop line by lowering collision and fatality rates at junctions. The overhead camera records the violation scenario, including the license plate of the violating vehicle captured by an automatic number-plate recognition (ANPR) camera and read by optical character recognition (OCR) to identify the violator. (f) Speed detection: smart microsensors are used for detecting the speed of vehicles to automatically detect the speed of moving vehicles on the road. (g) Water level detection: ultrasonic depth sensors can be used in flood and water monitoring systems to measure water levels. Early warning systems notify when the water level is rising so that smart cities can be prepared beforehand for any flood risks [29]. (h) Smoke detection: in smart cities, smoke and fire detection is a key element of early disaster-related event detection. A multi-sensor-based scheme with an open-source edge computing framework and automated next-generation emergency call functionality can be used for a high-end smoke detection system. (i) Parking sensors: with the use of guided parking applications, smart parking sensors, and parking data analysis, smart parking sensors provide an ingenious option for affordably cost enforcement. (j) Traffic detection system: a traffic management system can be effectively deployed using artificial intelligence-powered software to instantly evaluate data from traffic cameras and alert transport management systems [30]. (h) Audio detection: these work by detecting environmental sound events, such as the sound of breaking glasses, to safeguard commercial streets or public buildings from theft and looting, or monitoring traffic status by measuring noise levels at various points across. (k) Seismic sensors: these sensors operate as either velocity sensors or accelerometers, detecting earth vibrations. They play a crucial role in earthquake detection and mitigating risks to both property and human safety. (l) Smart meters: a common IoT gadget among utility companies is the smart meter. These devices, which are mounted to buildings and connected to smart energy infrastructure, allow utility providers to control the flow of energy. (m) Reader: this is utilized to collect data from diverse sensors and provide it to monitoring and control management systems for taking actions or analysis.

It is clear that various layers in smart city services perform on-the-ground evaluations of vulnerabilities that are common and pose a threat to social, economic, health, or reputational aspects [31]. There is a major issue in handling the presence of inherent challenges that come with the implementation of smart city services. There should be proper guidelines and regulations for the proper working of smart cities.

13.5 CYBERSECURITY CHALLENGES IN SMART CITIES

To recapitulate, smart cities should be able to handle major global crises connected to climate change, urbanization, resource scarcity, and rapid population growth [32]. However, several security issues must be addressed [3]. There is an absence of technology that is capable of handling enormous amounts of data. The Internet of Things will necessitate a huge number of services, applications, and linked nodes. These elements can highlight the variations in their operation, which might expose security weaknesses. The lack of well-defined standards is a matter of concern that is already on the minds of ICT professionals. Presently, there is no all-encompassing security framework available as a point of reference for the gathering, handling, processing, and distribution of data [33]. To improve the user experience and enable more innovative applications, the future generation of e-government systems requires integration with services like geospatial data, regulatory publications, and public discussion data [34]. The primary issues that require resolution include the following. (a) Numerous connected devices: the vast number of interconnected devices and the physical spread of infrastructure can create a large attack surface. (b) Physical dispersion: given the widespread nature of a smart city and the accessibility of its fundamental infrastructure to various stakeholders, there may be instances where devices need to be installed in locations with limited physical space [16, 34]. The privacy and security concerns in e-government within smart cities can potentially contribute to attackers' entry points, thereby jeopardizing the security of the entire system even when some segments of the infrastructure are effectively managed and regulated. Some crucial security challenges in smart cities comprise providing hardware protection, protecting smart cars, securing Wi-Fi networks, optimizing bandwidth usage, and mitigating risks linked to various smart city applications [35]. Hardware security is critical as it provides a foundational layer of protection that complements software-based measures, helping to ensure the reliability, integrity, and safety of smart city systems [16]. Concerns regarding the possible vulnerability of a vehicle's navigation and control systems are raised by the connection of self-driving smart vehicles, which is another crucial factor to consider. Certain Ethernet protocols used by self-driving cars need network aggregation (broadcast), which can make the network more vulnerable and complicated. To evaluate the efficacy of the security and privacy measures implemented for mitigating various threats in a smart city, it is imperative to identify the attack surface within the environment. In this regard, the existence of Wi-Fi devices with default passwords or unencrypted connections adds to the threat landscape. Additionally, concerns arise from compromised Wi-Fi encryption methods like WEP and vulnerabilities such as the WPA2 KRACK weakness, which was recently uncovered [34]. Smart cities are at risk from software products with security defects [34]. Different firms' applications for accessing company data could be vulnerable to security breaches. The created security metric should be future-proof as well as focus on the present situation of an e-government system [31]. Integrating global initiatives, adhering to regulations, embracing ecosystem frameworks, and fostering collaboration and capacity building are essential for developing successful smart cities. These elements help ensure that smart city technologies are effective, sustainable, and aligned with the needs of communities. Cybersecurity and privacy legislation have been enacted to guarantee that security is prioritized. Existing rules have been revised regularly to include the security standpoint of smart cities [16]. In the context of smart cities, some countries have developed cybersecurity frameworks and set security and privacy rules. A favorable atmosphere has also been created to encourage the development of cyber start-up centers. Smart cities across the world have started a slew of cyber security efforts based on three pillars: effective legislation, ecosystem structure and standards, and collaboration and capacity building.

13.6 BLOCKCHAIN IN SMART CITIES FOR SECURITY ENHANCEMENT

As discussed, the smart city paradigm not only focuses on the sustainable development of the cities around us along with its digitization to harness stakeholder service delivery but also requires that these integrated digital technologies be secure enough to protect stakeholders against system vulnerabilities. The current scenario necessitates the construction of a cyber-secured smart city framework utilizing blockchain technology to minimize potential risks and susceptibilities in city infrastructures and digitized smart services [7]. Blockchain has proved to be a popular technology to mitigate risks related to cyber threats, in addition to its implementation as the technology behind Bitcoin. It can provide cyber protection, as it uses a secure mechanism to prevent and stop fraudulent activities through the use of consensus. The core working features of blockchain are [9]: (a) distributed ledger technology (DLT), (b) decentralized network, (c) encrypted blocks consisting of digital assets that are chained together, (d) forms a chronological truth for data blocks, (e) ledgers are transparent, ensuring integrity and trust as each change that will be made in the block is well documented, (f) blockchain technology is widely used in the areas of the supply chain, healthcare, cybersecurity, and digital payments, and (g) provides fault tolerance, improved reliability, efficient working, and enhanced scalability. These features will be used to build smart cities and provide individuals with the necessary cyber security to feel safe and secure.

13.6.1 Blockchain use cases for implementing cyber-secure and safe smart cities

Although not infallible, blockchain has a reputation of being a reliable technology that results in an unaltered digital asset maintaining the integrity and transparency of the system by using an ordered cryptographic hashing technique. This property of blockchain makes it the most popular technology for use against vulnerabilities of several systems that require security against theft of data. It is a practical and feasible solution that has found success in many areas of use. These features make it a popular choice when considering building a cyber-secure and safe smart city [8].

Smart city multi-tier architecture is shown in Figure 4.1. Is exposed to several vulnerabilities and inherent challenges at different levels of implementation. It is powered by different technologies, such as operational technology (OT) systems and devices, Internet of Things (IoT) sensors, and traditional information technology (IT), that are functional at the four main layers: application layer, data layer, communication layer, and sensor layer. There are several vulnerabilities that the layers of the proposed smart city framework have. These are detailed in Table 13.2, as well as the possible solutions that blockchain technology can offer to obtain a transparent, smart, and secure digital network in and around the smart city.

Privacy and security breach issues have far exceeded the apparent benefits of information exchange in smart city architecture. Analyzing the vulnerabilities related to various smart city services may be categorized into three groups according to the effect level and how easily the vulnerabilities can be exploited. Table 13.3 summarizes various smart city services, and their impact level on each service is placed into three category bands: (a) catastrophic band, (b) critical band, and (c) marginal band [8]. CCTV city surveillance, e-governance, and telemedicine are the most critical services for any smart city.

Through blockchain technology, the integrity and confidentiality of data can be ensured. A collaborative approach is essential, emphasizing shared responsibility in upholding data trustworthiness. Establishing reliable processes for data gathering, processing, and sharing is critical, requiring constant vigilance and proactive measures to ensure data integrity.

13.7 CONCLUSION

Smart cities will only be feasible if security, authentication, and authorization of data can be assured. To achieve security goals, governments, businesses, application/software providers, device creators, energy/electricity providers, and network service providers must play their part in developing integrated solutions. The success of smart cities solely depends on access to actionable, real-time, and trustworthy data. Interactions with sensitive systems must be recorded and linked to an individual. These logs must maintain reliable integrity protection and must be hard to counterfeit. To achieve security standards, the ecosystem must be connected through strong authentication and ID management systems that ensure that the data is exchanged only with those who are authorized. When integrated at various levels with the segments of smart cities, blockchain technology offers the increased prospect of extra revenue with safe contracts and secure interactions between equipment and services provided. Future efforts to improve data reliability in smart cities require a multifaceted strategy. Stakeholders should work together to create safe and flourishing urbanization by emphasizing improved security,

Table 13.2 Blockchain-based smart city framework

Layer	*Smart city service*	*Smart city layer functionality*	*Blockchain security at different layers of the smart city architecture*
Application Layer	Water management system Flood monitoring system Traffic management system Telemedicine application Drone-based monitoring Video management system Waste management system Parking management Smart pole application Enterprise application	The application layer comprises various applications that facilitate smart city management and provide smart services to people for seamless functioning.	This layer encompasses a collection of intelligent programs that collaborate to make informed decisions. For example, an application of a smartphone can provide location information to a smart home system, enabling it to activate the air conditioner five minutes before someone arrives home. Nevertheless, these programs must be closely integrated, as vulnerabilities in one application could potentially grant attackers access to other interconnected operations.
Data Layer	ID management Bus operation Dashboard and reports Master data management Citizen personal and health database Signal management Route guidance Alerts KPI indicators Other advanced analytics	The data layer includes various data-based tools and features that support the application layer for smart city management.	A distributed ledger is a type of decentralized database utilized in blockchain technology, where records are stored in a sequential manner. Each entry in the ledger is accompanied by a timestamp and a unique cryptographic signature. The authorized users can access and verify the entire transaction history record in the ledger. In empirical applications, distributed ledgers can be either permissionless or permissioned. The permissionless ledger offers considerable benefits, including transparency and resistance to censorship. However, compared to private ledgers, public ledgers require complex shared record maintenance and take more time to achieve consensus. Additionally, public ledgers may be susceptible to attacks by anonymous adversaries. In real-time applications like traffic management in smart cities, the utilization of private ledgers is recommended for ensuring scalability, performance, and security of data.

(*Continued*)

Table 13.2 (Continued) Blockchain-based smart city ramework

Layer	*Smart city service*	*Smart city layer functionality*	*Blockchain security at different layers of the smart city architecture*
Communication Layer	Ethernet Wi-Fi GPRS Fiber Bluetooth and Zigbee Firewall and security devices Communication channels Bridge and gateway Router and switches Low power wide area network (LPWAN)	Multiple access and aggregation networks ranging from public to private and cable to wireless are used to connect people, objects, and sensors in smart cities. Effective communication infrastructure is the major requirement for cities to transform themselves into smart ones.	To transmit information across different devices like Wi-Fi, LPWAN, Bluetooth, 3G, 4G, and Ethernet, the smart city network uses a range of communication protocols at this layer. Ensuring the confidentiality and privacy of transmitted data requires the integration of blockchain protocols with this layer. For example, using technologies like Telehash, transaction records can be transformed into blocks and broadcast across the network. Peer-to-peer communication is made possible through protocols such as Bit Torrent, while smart contract functionality is enabled through platforms like Ethereum. The specific requirements can vary among different applications, making it challenging to seamlessly integrate existing communication protocols with blockchain technology. One potential solution is to implement multiple blockchains with the assistance of a blockchain access layer to enable application-specific functionalities. This approach allows for greater flexibility and customization within the smart city network.
Sensor Layer	RFID tags Smart cameras Smart signals Leakage detection Red-light violation Speed detection Water level detection Smoke detection Parking sensors Traffic detection system Audio detection Seismic sensors Smart meters Reader	To record information about the status or state of a process, sensors are utilized. They can give data like temperature in degrees, humidity values, chemical composition, tank fluid levels and movement in a pipe, or the assembly line's exact speed. Sensors or actuators in this layer monitor or control a physical item, respectively.	For urbanization, the entire system is composed of sensors and actuators responsible for collecting and transmitting data to upper-layer protocols. Certain devices within this system, such as the Nest thermostat and the Acer Fitbit, are susceptible to security attacks due to inadequate encryption and access control measures. Additionally, there is currently no universal standard for smart devices, hindering the seamless integration and exchange of data for cross-functionality. To address these challenges associated with smart devices, vendors must establish a set of communication and implementation standards.

Table 13.3 Smart city services and impact levels

Smart city service	*Impact level*	*Category description*
E-governance	Catastrophic zone	The services under this category, if compromised, will affect residents' privacy, safety, and security, as well as their health and confidence in the government.
City surveillance	Catastrophic zone	
Telemedicine	Catastrophic zone	
Smart waste management	Critical zone	Critical services will only challenge delivery when they are compromised. The damage may, however, be increased, and the catastrophic zone could be reached by further utilizing these services.
Smart poles	Critical zone	
Smart water management	Critical zone	
Enterprise GIS application	Critical zone	
Automatic fare collection system	Marginal zone	Citizens will experience significant inconveniences if the zone's services are affected.
Intelligent traffic management system	Marginal zone	

interoperability, privacy preservation, decentralized identity, ongoing blockchain integration, cooperation, citizen engagement, regulatory frameworks, and resilience against emerging threats.

REFERENCES

1. "United Nations | Peace, Dignity and Equality on a Healthy Planet." 2023. Accessed December 4. https://www.un.org/en/.
2. Angelidou, Margarita. 2014. "Smart City Policies: A Spatial Approach." *Cities* 41 (September). Elsevier Ltd: S3–11. doi:10.1016/j.cities.2014.06.007.
3. Das, Amardeep, Sumanta Chandra Mishra Sharma, and Bikram Kesari Ratha. 2019. "The New Era of Smart Cities, From the Perspective of the Internet of Things." *Smart Cities Cybersecurity and Privacy*, January. Elsevier, 1–9. doi:10.1016/B978-0-12-815032-0.00001-9.
4. "IoT Solutions for 2023 (Connectivity & Security)." 2023. Accessed December 4. https://www.thalesgroup.com/en/markets/digital-identity-and-security/iot.
5. Gracias, Jose Sanchez, Gregory S. Parnell, Eric Specking, Edward A. Pohl, and Randy Buchanan. 2023. "Smart Cities—A Structured Literature Review." *Smart Cities* 6 (4): 1719–43. doi:10.3390/smartcities6040080.
6. Ionescu, Romeo Victor, Monica Laura Zlati, and Valentin Marian Antohi. 2023. "Smart Cities from Low Cost to Expensive Solutions under an Optimal Analysis." *Financial Innovation* 9 (1). Springer Berlin Heidelberg. doi:10.1186/s40854-023-00448-8.
7. Abbas, Khizar, Lo'Ai A. Tawalbeh, Ahsan Rafiq, Ammar Muthanna, Ibrahim A. Elgendy, and Ahmed A. Abd El-Latif. 2021. "Convergence of Blockchain and IoT for Secure Transportation Systems in Smart Cities." *Security and Communication Networks* 2021. doi:10.1155/2021/5597679.

8. Bekkali, Abla El, Mohammed Boulmalf, Mohammad Essaaidi, and Driss El Majdoubi. 2020. "Towards Blockchain-Based Architecture for Smart Cities Cyber-Security." 2020 International Conference on Electrical and Information Technologies, ICEIT 2020, March. Institute of Electrical and Electronics Engineers Inc. doi:10.1109/ICEIT48248.2020.9113223.
9. Theodorou, Sophocles, and Nicolas Sklavos. 2019. "Blockchain-Based Security and Privacy in Smart Cities." In *Smart Cities Cybersecurity and Privacy*, edited by Danda B Rawat and Kayhan Zrar Ghafoor, 21–37. Elsevier. doi:10.1016/B978-0-12-815032-0.00003-2.
10. Huang, Chao, and Shah Nazir. 2021. "Analyzing and Evaluating Smart Cities for IoT Based on Use Cases Using the Analytic Network Process." *Mobile Information Systems* 2021. doi:10.1155/2021/6674479.
11. Khan, Uzair, and Vikash Yadav. 2021. "A Study Based on Structural Needs and Smart Tools Using IoT for Developing a Smart City." *International Journal of Advanced Intelligence Paradigms* 20 (1–2). Inderscience Publishers: 210–20. doi:10.1504/IJAIP.2021.117617.
12. İkizer, İhsan. 1AD. "Do Smart City Solutions Contribute to the Achievement of the Sustainable Development Goals?: Case of Istanbul." *IGI Global*, 22–44. doi:10.4018/978-1-7998-7785-1.CH002.
13. Holubava, Volha. 2019. "Smart Cities." In *Modeling Economic Growth in Contemporary Belarus*, edited by Bruno S Sergi, 223–33. Emerald Publishing Limited. doi:10.1108/978-1-83867-695-720191015.
14. Sivarama Krishnan, R., R. Aggarwal, A. Viquar, V. Sood, S. Bhunia, A. Verma, and M. Bhattacharya. 2018. *Creating Cyber Secure Smart Cities*. PricewaterhouseCoopers. Accessed October 27, 2023. https://www.pwc.in/assets/pdfs/publications/2018/creating-cyber-secure-smart-cities.pdf.
15. Skouby, Knud Erik, Per Lynggaard, Iwona Windekilde, and Anders Henten. 2014. "How IoT, AAI Can Contribute to Smart Home and Smart Cities Services: The Role of Innovation." International Telecommunications Society Regional Conference, 13.
16. Varfolomeev, Alexander A., Liwa H. Alfarhani, and Zahraa Ch Oleiwi. 2021. "Overview of Five Techniques Used for Security and Privacy Insurance in Smart Cities." *Journal of Physics: Conference Series* 1897 (1). doi:10.1088/1742-6596/1897/1/012028.
17. Gopalaiyengar, Krishnarajanagar Srinivasa, Abhishek Kumar, Nabeel Siddiqui, and B. J. Sowmya. 2015. "Smart System Design in Cyber-Physical Systems." *Cyber-Physical Systems: A Computational Perspective*, January. CRC Press, 313–30. doi:10.1201/B19206-20/SMART-SYSTEM-DESIGN-CYBER-PHYSICAL-SYSTEMS.
18. *Smart Cities and Academia towards Action & Research (SAAR) | IBEF*. India Brand Equity Foundation. Accessed October 27, 2023. https://www.ibef.org/blogs/smart-cities-and-academia-towards-action-research-saar.
19. "Sustainable Water Management for Smart Cities. - Top Construction & Infrastructure Magazine." 2023. Accessed December 4. https://construction-times.co.in/sustainable-water-management-for-smart-cities/.
20. Faisal, M., H. Sadia, T. Ahmed, and N. Javed. 2022. "Blockchain Technology for Healthcare Record Management." *EAI/Springer Innovations in Communication and Computing*. doi:10.1007/978-3-030-77746-3_17.

21. "Connectivity Management - IoT Connectivity Management | Aeris." 2023. Accessed December 4. https://www.aeris.com/aeris-intelligent-iot-network/connectivity-management/.
22. Yin, Chuan Tao, Zhang Xiong, Hui Chen, Jing Yuan Wang, Daven Cooper, and Bertrand David. 2015. "A Literature Survey on Smart Cities." *Science China Information Sciences* 58 (10): 1–18. doi:10.1007/s11432-015-5397-4.
23. "Low-Carbon Transport, Individual Wellbeing & Planetary Health in the Era of Smart Cities & New Mobilities | Department of Geography and Resource Management (GRM), The Chinese University of Hong Kong." 2023. Accessed December 4. https://www.grm.cuhk.edu.hk/en/research-collaboration/low-carbon-transport-individual-wellbeing-planetary-health-in-the-era-of-smart-cities-new-mobilities/.
24. Ksiksi, Assil, Saeed Al Shehhi, and Rashad Ramzan. 2015. "Intelligent Traffic Alert System for Smart Cities." Proceedings - 2015 IEEE International Conference on Smart City, SmartCity 2015, Held Jointly with 8th IEEE International Conference on Social Computing and Networking, SocialCom 2015, 5th IEEE International Conference on Sustainable Computing and Communic, no. December 2015: 165–69. doi:10.1109/SmartCity.2015.65.
25. Kaluarachchi, Yamuna. 2022. "Implementing Data-Driven Smart City Applications for Future Cities." *Smart Cities* 5 (2): 455–474. doi:10.3390/smartcities5020025.
26. "Firewall Monitoring & Management Support Services | Firewall Security Company India." 2023. Accessed December 4. https://firewall.firm.in/firewall-monitoring-management-support-services/.
27. Pal, Arpan, Hemant Kumar Rath, Samar Shailendra, and Abhijan Bhattacharyya. 2018. "IoT Standardization: The Road Ahead." In edited by Jaydip Sen, Ch. 3. Rijeka: IntechOpen. doi:10.5772/intechopen.75137.
28. "Singapore: The World's Smartest City | Thales Group." 2023. Accessed December 4. https://www.thalesgroup.com/en/worldwide-digital-identity-and-security/iot/magazine/singapore-worlds-smartest-city.
29. Bodendorf, Frank, and Jörg Franke. 2020. "Risk Assessment of Vehicle Sensor Data as a Vending Object or Service." In *Internet of Things - ICIOT* 2020, edited by Wei Song, Kisung Lee, Zhisheng Yan, Liang-Jie Zhang, and Huan Chen, 141–151. Cham: Springer International Publishing.
30. Doku, Ronald, and Danda B Rawat. 2019. "Big Data in Cybersecurity for Smart City Applications." In *Smart Cities Cybersecurity and Privacy*, edited by Danda B Rawat and Kayhan Zrar Ghafoor, 103–12. Elsevier. doi:10.1016/B978-0-12-815032-0.00008-1.
31. Thilakarathne, Navod Neranjan, and W D Madhuka Priyashan. 2022. "An Overview of Security and Privacy in Smart Cities." In *IoT and IoE Driven Smart Cities*, edited by Samarendra Nath Sur, Valentina Emilia Balas, Akash Kumar Bhoi, and Anand Nayyar, 21–44. Cham: Springer International Publishing. doi:10.1007/978-3-030-82715-1_2.
32. Ghosal, Amrita, and Subir Halder. 2018. "Building Intelligent Systems for Smart Cities: Issues, Challenges and Approaches," no. September: 107–25. doi:10.1007/978-3-319-76669-0_5.
33. Yang, Longzhi, Noe Elisa, and Neil Eliot. 2019. "Privacy and Security Aspects of E-Government in Smart Cities." In *Smart Cities Cybersecurity*

and Privacy, edited by Danda B Rawat and Kayhan Zrar Ghafoor, 89–102. Elsevier. doi:10.1016/B978-0-12-815032-0.00007-X.
34. Piscini, E, and L Kehoe. 2018. "Blockchain & Cyber Security. Let ' s Discuss For More Information Please Contac.:" *Deloitte*, 14.
35. Din, Zarina, Dian Indrayani Jambari, Maryati Mohd Yusof, and Jamaiah Yahaya. 2019. "Challenges in Managing Information Systems Security for Internet of Things-Enabled Smart Cities." In 2019 6th International Conference on Research and Innovation in Information Systems (ICRIIS), 1–6. doi:10.1109/ICRIIS48246.2019.9073661.

Chapter 14

Greening the wireless world

Energy-efficient and sustainable communication

M. Anandaraj, P. Ganeshkumar, S. Naganandhini, and K. Selvaraj

14.1 INTRODUCTION

Wireless communication is the foundation of our modern interconnected world, transforming the way we interact, exchange information, and perform business. At its core, wireless communication enables the transmission of data, voice, and multimedia content without the constraints of physical wires and cables [1]. This method of communication relies on the use of electromagnetic waves, such as radio waves and microwaves, to establish connections between devices, ranging from small handheld gadgets to large-scale network infrastructures. Cellular networks, demonstrated by generations like 3G, 4G, and the emerging 5G, enable mobile phones to communicate with base stations and other devices over vast geographical areas. Wi-Fi, another ubiquitous technology, empowers devices to connect to local networks and the internet within specific areas of coverage [2]. Bluetooth facilitates short-range communication between devices, enabling wireless headphones, keyboards, and other peripherals to connect to computers and smartphones [3]. Internet of Things (IoT) devices, ranging from smart thermostats to wearable fitness trackers, depend on wireless communication to transmit data and receive commands. In industrial settings, wireless sensors enable real-time monitoring of equipment, helping prevent failures and optimise processes. As technology evolves, wireless communication continues to shape our world in new ways. Wireless communication serves as the backbone of modern connectivity, enabling us to communicate, collaborate, and innovate in ways that were once unimaginable [4]. This chapter serves as a comprehensive guide to navigating the intricate landscape of green wireless communications. It provides a roadmap for industry stakeholders, policymakers, and researchers, offering insights into the modern pathways that lead to energy-efficient, environmentally conscious, and sustainable wireless communication systems. By integrating technical advancements with ecological responsibility, the chapter illustrates the transformative power of green wireless communications in shaping a more conscientious and resilient digital future [5]. Figure 14.1 shows the alternate power source for the variety of wireless communication towers.

DOI: 10.1201/9781003541363-14

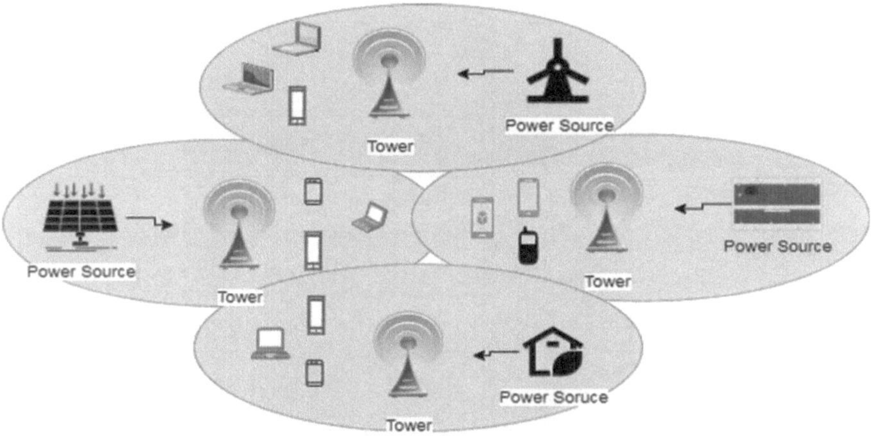

Figure 14.1 Green power source for towers.

The main contribution of the proposed chapter lies in its comprehensive examination and implementation of strategies and technologies within the area of green wireless communications. It elaborates on innovative solutions, including dynamic spectrum management, sleep mode techniques, energy-efficient protocols, and advanced power management strategies. Notably, the research emphasises the pivotal role of renewable energy sources, such as solar and wind power, in establishing sustainable and environmentally friendly communication networks. It also explores the potential of cutting-edge technologies such as network virtualisation and software-defined networking (SDN) to facilitate dynamic energy management, showcasing a forward-looking technique for addressing challenges in energy. The proposed chapter is also substantiated through case studies and real-world examples, illustrating its concrete impact on both operational costs and environmental preservation.

14.2 ENERGY CHALLENGES POSED BY TRADITIONAL WIRELESS NETWORKS

Traditional wireless networks, while integral to modern connectivity, cause significant energy challenges that deserve attention. One notable challenge is the high power consumption associated with these networks [6]. Microcell base stations, essential for coverage over vast areas, demand considerable energy to maintain connectivity, leading to elevated electricity usage and operational costs. The centralisation of data processing in data centres, prompted by data-intensive applications, introduces further energy challenges. These centres require energy-intensive cooling mechanisms to

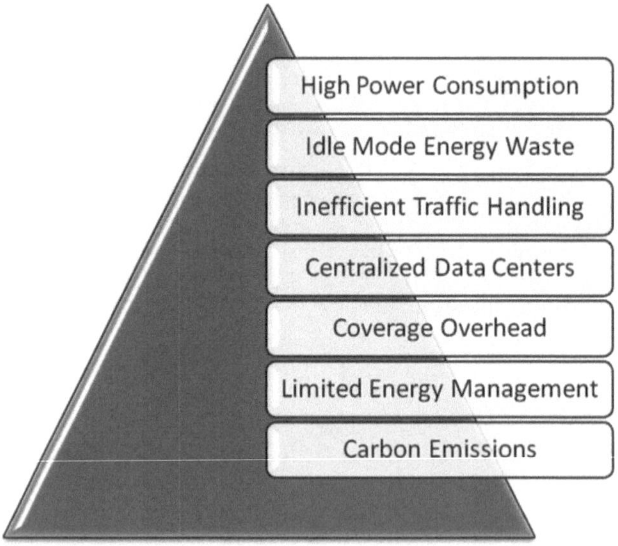

Figure 14.2 Issues in wireless communication.

maintain optimal conditions for operation. The spatial coverage of conventional networks is another energy-related consideration. High-density deployment of base stations ensures uninterrupted connectivity although introduces coverage overhead due to potential interference among neighbouring stations [7]. These configurations lead to energy inefficiencies and overlap in network coverage. The reliance on non-renewable energy sources, such as fossil fuels, contributes to carbon emissions and environmental degradation. Addressing these energy challenges requires a multi-pronged approach. Innovations in energy-efficient hardware, dynamic power management techniques, and the incorporation of renewable energy sources hold promise for mitigating energy consumption. The evolution towards next-generation technologies like 5G also presents an opportunity to restructure networks for enhanced energy efficiency, guiding in a more sustainable era of wireless communication [8]. Figure 14.2 represents the various important issues that need to be addressed in wireless communication.

14.3 GREEN WIRELESS COMMUNICATIONS

Green wireless communications represent an idea shift within the field of telecommunications, where the essentials of energy efficiency and environmental responsibility take centre stage. The convergence of technological innovation and environmental awareness has prompted a re-evaluation of how wireless communication systems are designed, operated, and also sustained. The objective of green wireless communications is to reduce energy consumption, minimise carbon emissions, and mitigate the ecological impact of wireless networks at the same time as maintaining the high standards of connectivity and performance that users expect. By exploiting solar, wind, and other renewable resources, communication infrastructure such as cellular towers, base stations, and data centres operate on clean energy, substantially lowering their carbon footprint. This transition to renewable not only contributes to environmental sustainability but also improves the resilience of communication networks by reducing dependency on conventional energy grids [9]. Network architecture also plays a crucial role in energy conservation. The deployment of small cells, which are low-power access points that provide localised coverage, reduces the need for high-power transmission across enormous areas. Ultimately, the quest for green wireless communications goes beyond technological considerations. By prioritising energy efficiency, integrating renewable resources, and embracing smart management practices, the wireless communication industry can contribute considerably to global sustainability goals. The journey towards greener wireless communications is marked by innovation, collaboration, and also a shared commitment to minimising the ecological footprint of our digital connectivity [10].

14.3.1 Key aspects of green communication

Green communication, also known as sustainable communication, refers to the practice of designing, implementing, and using communication technologies and networks in an environmentally responsible and energy-efficient way. One of the major focuses of green communication is energy efficiency. This involves optimising the energy consumption of ICT infrastructure, such as data centres, network equipment, and user devices, through the use of energy-efficient hardware, software, and cooling systems. Another essential aspect is the use of renewable energy sources such as hydroelectric power, solar, and wind to supply electricity to ICT facilities. Finally, raising awareness about the environmental impact of ICT and educating users and organisations about green communication practices play an essential role in driving positive change [11]. Figure 14.3 shows the key aspects of green communication.

14.4 ENERGY CONSUMPTION AND ENVIRONMENTAL SUSTAINABILITY

Energy consumption and environmental sustainability meet at the crossroads of wireless communication, where technological development converges with the imperative of responsible resource management. Manufacturers are designing energy-efficient hardware components to extend battery life,

Figure 14.3 Key aspects of green communication.

while network operators are optimising networks with energy-efficient technologies such as small cells and smart antennas. Implementing sleep mode techniques conserves energy during periods of inactivity, enhancing overall efficiency. Yet, the nexus of energy and wireless communication extends beyond power considerations to encompass broader environmental sustainability concerns. The carbon footprint generated by energy production and utilisation contributes to climate change and air pollution, demanding environmentally conscious solutions. In response, the drive for sustainability in wireless communication encompasses several crucial facets. The inclusion of renewable energy sources, such as wind turbines and solar panels, into wireless infrastructure is essential to decrease reliance on fossil fuels and alleviate greenhouse emissions of gas. Governments and regulatory bodies are instrumental in setting energy efficiency standards and incentivising the adoption of environmentally friendly technologies. The dynamic relations between energy consumption and environmental sustainability within wireless communication require a holistic and thoughtful approach. An appropriate balance between technological advancement and ecological responsibility is essential to guarantee that wireless communication continues to evolve while minimising its environmental footprint. The convergence of these objectives holds the assurance of a more resilient, efficient, and sustainable wireless communication ecological unit.

14.4.1 Energy efficiency (EE)

For the outstanding wireless network technology of today, 5G networks, there are ambitious visions of achieving remarkable performance improvements. These improvements include a range of capabilities including supporting a thousandfold increase in throughput compared to existing networks, facilitating connectivity for roughly 7.6 billion mobile subscribers along with at least 100 billion devices on a global scale, delivering individual user broadband speeds of up to 10 Gbps, enabling Internet of Everything (IoE) communications, empowering tangible internet applications, and achieving network latencies less than 1 ms. These performance goals have prompted the emergence of various novel technologies to meet these demands. Energy efficiency (EE), a metric expressed in bits per Joule (b/J), quantifies the amount of information (bits) that can be transmitted or processed per unit of energy (Joule) consumed [12]. It is written as

$$\mathrm{EE} = \frac{FR \times SS \times BW \times \log_2(1 + SINR(D))}{P_c + P_T}[b/j] \tag{14.1}$$

Energy efficiency (EE) in Equation (1) considers several parameters to assess the energy efficiency of a wireless communication system, particularly in cellular networks. The parameters in the equation have specific meanings and implications as given below.

a) SS (number of spatial streams): Represents the spatial multiplexing factor.
b) BW (bandwidth): Refers to the signal's bandwidth in Hertz (Hz).
c) FR (frequency reuse factor): Signifies the ratio of reused available frequency bands.
d) D (distance): Denotes the distance between the communicating devices.
e) PC (circuit power consumption): Represents the static power consumption.
f) PT (transmit power consumption): Refers to the dynamic power consumption.

The EE in Equation (1) highlights how the energy efficiency of cellular networks can be enhanced through various strategies. Increasing the BW, SS, and FR can contribute to higher data transmission rates without a proportionate increase in energy consumption. Additionally, reducing the PC and PT can further enhance energy efficiency. This understanding has led to several strategies and advancements being pursued to enhance energy

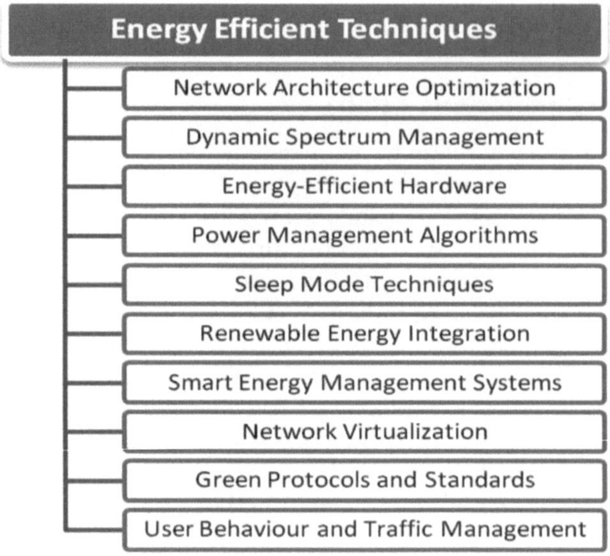

Figure 14.4 Strategies and advancements.

efficiency in wireless communication, especially in paradigms for 5G networks, as depicted in Figure 14.4. The challenge lies in developing strategies that facilitate the desired performance enhancements while simultaneously mitigating the potential increase in energy consumption [13, 14]. The trade-off between these two critical aspects underscores the complexity of the 5G landscape.

14.5 INTEGRATION OF RENEWABLE ENERGY SOURCES FOR WIRELESS COMMUNICATION INFRASTRUCTURE

The integration of renewable energy sources into wireless communication infrastructure marks an important stride towards sustainability and environmental responsibility. Solar power, wind energy, hydropower, and other renewables are transforming the way wireless networks are powered, reducing their reliance on conventional energy grids and fossil fuels. One of the primary advantages of renewable energy in this context is its suitability for off-grid and remote locations. Solar panels can be deployed atop cell towers, harnessing the sun's energy to keep these critical communication nodes operational. Similarly, wind turbines can provide a consistent source of electricity in regions with dependable wind patterns. The adoption of renewable energy in wireless communication is not just an environmentally conscious choice; it's also a cost-effective one. Over time, the decreasing

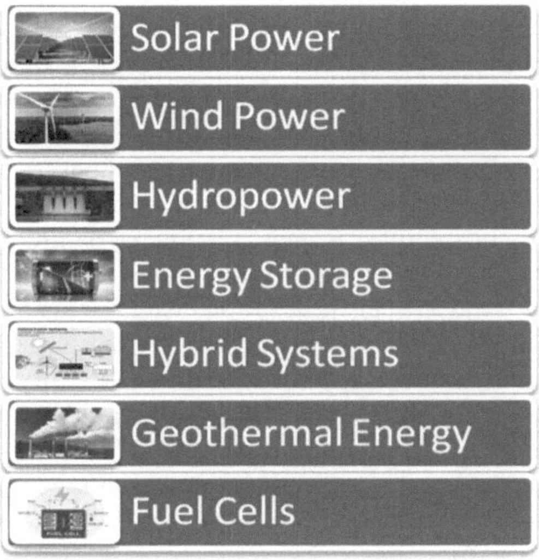

Figure 14.5 Energy sources.

costs of renewable technologies, coupled with energy savings and reduced carbon emissions, make renewable-powered networks financially advantageous. As the world shifts towards a more sustainable future, the integration of renewable energy into wireless communication infrastructure exemplifies the commitment of industry to reduce its carbon footprint. Figure 14.5 shows the alternate power source for the green communication system.

14.5.1 Advantages of the green integration

Integrating renewable energy sources into wireless communication infrastructure provides a multitude of advantages, ranging from environmental benefits and cost savings to enhanced reliability and alignment with sustainability objectives. The advantages of integrating renewable energy sources into wireless communication infrastructure are significant and wide-ranging, as given below.

1. Environmental sustainability: Renewable energy sources, such as solar and wind power, generate electricity without emitting greenhouse gases and other pollutants.
2. Cost savings: Solar and wind power systems require minimal maintenance and incur no fuel costs. This cost predictability and reduction in energy expenses can positively impact the bottom line for wireless network operators.
3. Energy independence: By generating their own electricity, wireless networks become less reliant on conventional power grids. This is

especially beneficial in remote or off-grid areas where access to reliable electricity may be limited.

4. Reliability and resilience: Renewable energy systems, when paired with energy storage solutions like batteries, enhance network reliability. During grid outages or adverse weather conditions, renewable-powered networks can continue to operate, providing vital communication services during emergencies.
5. Long-term sustainability: Renewable energy sources are sustainable in the long term. Unlike finite fossil fuels, renewable resources like sunlight and wind are naturally replenished. This ensures a continuous and sustainable energy supply for wireless networks, reducing concerns about resource depletion or price volatility.
6. Reduced operating costs: Once installed, solar and wind power systems require relatively low maintenance.
7. Public image: Embracing renewable energy enhances the public image of wireless network operators, positioning them as environmentally responsible organisations.
8. Technological advancements: The adoption of renewable energy encourages innovation in energy storage technologies, solar panels, wind turbines, and related systems.
9. Scalability: Renewable energy solutions are highly scalable, allowing operators to tailor the energy generation capacity to meet specific network requirements.
10. Emergency preparedness: Renewable energy systems equipped with energy storage provide critical power during emergencies and natural disasters.
11. Global sustainability goals: Wireless communication operators that reduce their carbon footprint and embrace renewable energy contribute to global sustainability goals.

14.5.2 Implementation challenges

Achieving energy efficiency is at the forefront of green wireless communication confronts. In order to minimise the environmental impact, wireless networks and devices must be designed with energy conservation in mind [15, 16]. The goal is to extend the battery life of mobile devices while simultaneously reducing the overall energy consumption of network infrastructure. Optimising its use can result in significant energy savings. The implementation of technologies such as dynamic spectrum sharing, cognitive radio, and intelligent spectrum allocation mechanisms is essential. These innovations enable better spectrum utilisation, reducing energy consumption. The energy consumption of network infrastructure, including base stations and data centres, presents a substantial obstacle to achieving green wireless communication. These facilities are accountable for a major portion of the energy consumed by wireless networks. To address this challenge, the industry must

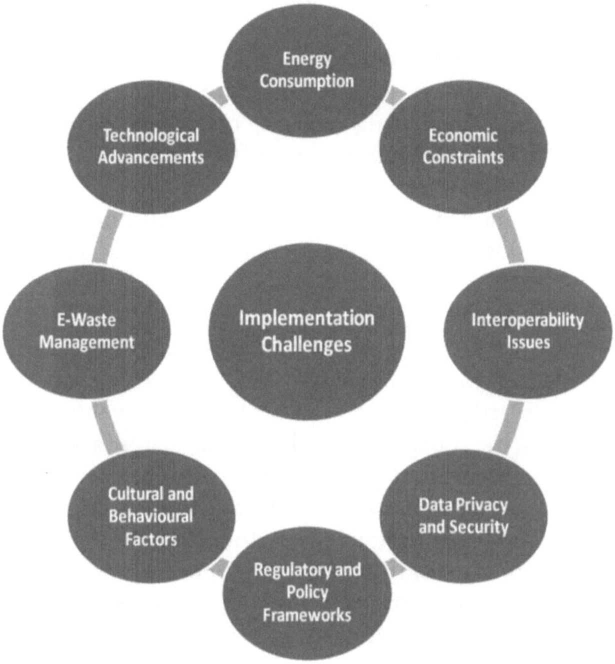

Figure 14.6 Implementation challenges.

invest in green technologies such as renewable energy sources, energy-efficient cooling systems, and eco-friendly construction materials. The increasing heterogeneity of wireless networks introduces complexity in their design and management. Integrating diverse technologies like Wi-Fi, 5G, and IoT into an organised, energy-efficient network poses an important technical challenge. Coordinating the operation of these diverse technologies while maintaining energy efficiency requires sophisticated network management, and optimisation strategies are needed. These standards have an impact on the design and operation of wireless networks. Staying informed about and adapting to changing regulatory requirements is crucial for sustainability. Implementing green wireless communication technologies may initially come with higher upfront costs [17]. Figure 14.6 shows the important issues to be considered while implementing green communication.

14.6 OPTIMISATION OF ENERGY-EFFICIENT DESIGN OF NETWORK

Optimising energy-efficient network design for wireless communication is a multifaceted process that combines various strategies to minimise energy consumption while maintaining or improving network performance [18]. Resource allocation and management strategies are essential. Maximising spectrum efficiency through spectrum sharing and dynamic allocation reduces energy consumption. Load balancing ensures the even distribution of network traffic, preventing resource overuse and minimising energy consumption in overloaded components. Effective quality of service (QoS) management allocates resources based on service priorities, reducing energy consumption during non-priority tasks while maintaining service quality for critical applications [19]. Energy-efficient protocols and standards, like LPWAN options for IoT and Wi-Fi 6 with features like Target Wake Time, enhance energy efficiency in various network scenarios. Monitoring and optimisation tools, including real-time energy monitoring and predictive analytics, provide insights into energy usage patterns and help network operators make informed decisions. Integrating renewable energy sources, such as solar and wind power, contributes to sustainability and reduces reliance on non-renewable resources. Optimising energy-efficient network design for wireless communication involves a multifaceted approach that encompasses hardware selection, network topology design, dynamic power management, resource allocation, and adherence to energy-efficient protocols. This comprehensive strategy aims to reduce energy consumption, lower operational costs, and promote environmental sustainability while ensuring reliable wireless communication services.

14.7 DYNAMIC ENERGY MANAGEMENT

Network virtualisation and SDN are two innovative technologies that are reshaping the telecommunications industry. Network virtualisation abstracts physical network resources into virtual entities, allowing for the formation of several virtual networks on a shared physical infrastructure. SDN separates the network's control plane from the data plane. This, coupled with resource optimisation, results in energy-efficient networks that can adapt to varying workloads. Improved network management, policy-based management, network slicing, and compatibility with emerging technologies such as IoT and 5G further harden the importance of network virtualisation and SDN in the telecommunications landscape. As networks continue to change, these technologies will remain at the forefront, driving innovation and efficiency in network infrastructure [20]. Network virtualisation and SDN hold great promise in revolutionising energy management within the telecommunications industry. These technologies provide the means to dynamically control and allocate network resources in ways that were previously unattainable. SDN controllers can also be programmed with algorithms that prioritise energy-efficient routing, considering the consumption of energy profiles of network elements. Predictive analytics, powered by the wealth of data generated within these systems, enables network operators to anticipate traffic patterns and proactively optimise energy consumption. Dynamic energy management systems can intelligently allocate network tasks to coincide with peak renewable energy generation periods, thereby reducing dependence on non-renewable energy sources.

14.8 ENERGY-EFFICIENT PROTOCOLS

Energy-efficient protocols are a foundation of green wireless communication, designed to decrease energy consumption and promote sustainability without compromising performance [21]. These protocols aim to address the pressing need to minimise the environmental impact of wireless networks. They employ techniques such as low-power sleep modes, short data packets, low-duty cycle operations, and efficient communication mechanisms to achieve energy efficiency. Depending on the specific application and use case, selecting the right protocol significantly contributes to sustainability and environmental conservation in wireless communication systems. Some of the energy-efficient protocols are discussed here.

14.8.1 IEEE 802.11 Power Save Mode (PSM)

The IEEE 802.11 PSM is often referred to as Wi-Fi power save mode. It is a protocol strategically designed to minimise energy consumption in wireless networks. PSM operates by allowing Wi-Fi-enabled devices to enter a low-power sleep state during periods of inactivity [22]. This means that when a device is not actively transmitting or receiving data, it periodically goes into sleep mode and wakes up at predefined intervals to save any pending data or network activity. The major advantage of IEEE 802.11 PSM is its dual role in the conservation of battery life and optimisation of the network. It significantly prolongs the battery life of devices such as laptops, smartphones, and IoT sensors by reducing the time the radio is active and consuming power. Additionally, it reduces overall congestion of network and interference, as devices aren't continually vying for access to the wireless medium, ultimately making it a basic component of green wireless communication.

14.8.2 Bluetooth Low Energy (BLE)

BLE is a wireless communication protocol designed with a strong focus on power efficiency, particularly in the context of IoT and wearable technology [23]. It achieves its energy efficiency through some key mechanisms. Firstly, it utilises short data packets, which inherently need less energy to transmit than longer packets, making it perfect for battery-operated devices. Secondly, BLE devices normally function on a low-duty cycle, meaning they spend important periods in a low-power sleep state and only wake up periodically to exchange data. This approach minimises the overall consumption of power. Finally, BLE ensures fast connection establishment, reducing the time devices spend in active, power-hungry states. BLE operates by allowing devices to advertise their presence and capabilities periodically. Other devices, acting as scanners, can discover and interact with these advertising devices. It uses a client-server architecture based on the GATT (Generic Attribute Profile) protocol, allowing for efficient data exchange. Figure 14.7 shows the BLE mode of transferring the data between the devices and central nodes.

14.8.3 Zigbee

Zigbee is a wireless communication protocol designed for low-power, short-range communication, mainly for IoT and home automation applications [24]. Zigbee's energy-efficient operation revolves around a low-duty cycle approach. Zigbee devices spend the majority of their operational time in a low-power sleep state and periodically wake up to check for data or network activities. This sleep and wake cycle reduces power consumption. Zigbee also uses efficient communication techniques to guarantee that when devices are active, they make the most of that time, reducing unnecessary usage of energy. Moreover, Zigbee supports mesh networking, allowing devices to relay data through the network, reducing the need for direct communication between distant devices and further conserving energy. Zigbee's design makes it well-suited for energy-efficient applications like

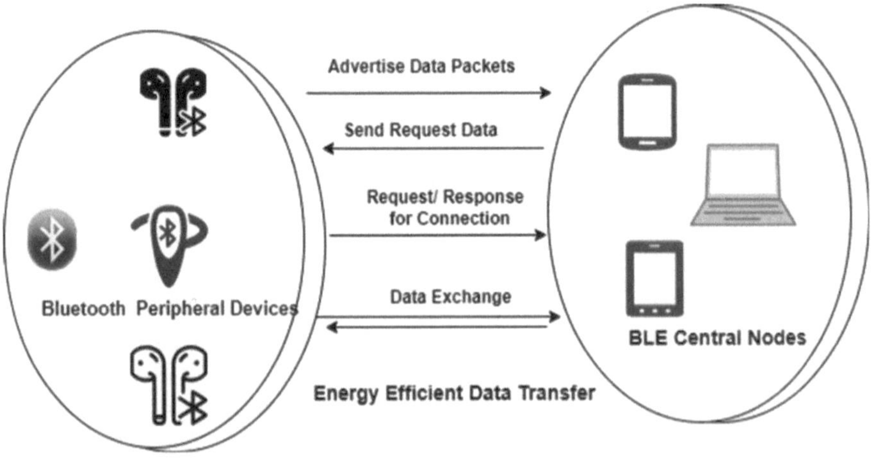

Figure 14.7 Bluetooth Low Energy (BLE).

smart lighting, thermostat control, and also remote monitoring, underscoring its importance in green wireless communication [25].

14.8.4 6LoWPAN

6LoWPAN is a protocol designed for energy-efficient communication in IoT devices, particularly those with limited power and computational resources [26, 27]. It achieves energy efficiency through several strategies. One key aspect is reducing the overhead associated with IPv6 packets, making them more efficient for small, low-power devices. By minimising this overhead, 6LoWPAN ensures that the energy used for communication is as efficient as possible. Additionally, it focuses on efficient packet delivery, ensuring that data packets are delivered effectively and without unnecessary redundancy, further conserving energy. Lastly, 6LoWPAN devices can operate in a low-power mode when they are not actively communicating, minimising energy consumption during periods of inactivity. This makes 6LoWPAN a suitable choice for IoT applications where energy-efficient communication is critical, thereby contributing to greener and more sustainable wireless communication systems [28].

14.9 IMPACT OF GREEN WIRELESS COMMUNICATION

One of the most significant impacts of green wireless communication is the substantial reduction in the carbon footprint. By using energy-efficient technologies and practices, such as optimising network protocols and employing low-power hardware components, the telecommunications sector can considerably reduce its emissions of carbon [29, 30]. This is particularly crucial as the world grapples with the challenges of climate change, and industries strive to reduce their contributions to global warming. Energy efficiency is a cornerstone of green wireless communication. It encompasses a range of strategies to reduce energy consumption across the entire wireless ecosystem. From the operation of cellular towers to the functioning of end-user devices, green wireless initiatives aim to minimise energy use. This has direct implications for reducing operational costs for network operators and promoting energy-conscious consumer behaviour [31]. Furthermore, green wireless communication underscores the importance of resource conservation. In wireless networks, efficient allocation of resources, such as spectrum and bandwidth, is pivotal. Another critical impact of green wireless communication is its promotion of renewable energy integration. The telecommunications industry is a significant consumer of energy, and the transition to cleaner, more sustainable power sources is paramount. Many green wireless initiatives advocate for the integration of renewable energy, such as solar and wind power, into the operation of wireless infrastructure.

This not only reduces the industry's reliance on fossil fuels but also contributes to a greener and more sustainable energy mix. Green wireless technologies also have a tangible impact on battery life. This, in turn, reduces the frequency of battery replacements, leading to a decrease in electronic waste generation, a critical environmental concern. In addition to environmental benefits, green wireless communication positively influences network reliability. Energy-efficient technologies can improve network performance and reduce energy-related failures, making the infrastructure more dependable [32]. This not only benefits consumers by providing uninterrupted connectivity but also enhances the overall sustainability of the network.

In addition, as environmental regulations become more stringent, green wireless communication aids network operators and manufacturers in complying with these standards [33]. In an era where consumers and investors increasingly value sustainability efforts, organisations that prioritise green wireless technologies can enhance their reputation and competitiveness in the market [34, 35]. Moreover, green wireless communication plays a pivotal role in supporting emerging technologies like the Internet of Things (IoT) and the development of smart cities. Energy-efficient wireless networks are essential for connecting and managing the vast number of IoT devices and applications, making them a cornerstone of future urban development and connectivity. By reducing greenhouse gas emissions, promoting sustainability, and supporting technological advancements, green wireless communication is a critical component of building a more sustainable and environmentally friendly future for the telecommunications industry and the world as a whole. Based on the prediction made in [36], the projected mobile communication's carbon footprint is shown in Figure 14.8.

14.10 CONCLUSION

This chapter aims to shed light on a transformative shift within the domain of wireless technology. The main theme revolves around the essentials of striking a comfortable balance between technological innovation and environmental responsibility. As traditional wireless networks struggle with challenges in energy consumption and ecological impacts, the paradigm of green wireless communications emerges as a practical and feasible solution. The journey through this chapter has revealed the multifaceted dimensions of green wireless communications systems. From energy-efficient hardware design to the integration of renewable energy sources, the spectrum of strategies is open. The importance of network optimisation, smart energy management, and the incorporation of advanced technologies underscores the comprehensive nature of the approach. The potential of green wireless communications is not only confined to energy conservation. It also extends to fostering a culture of sustainability within the industry, influencing policy formulation, and championing awareness among stakeholders.

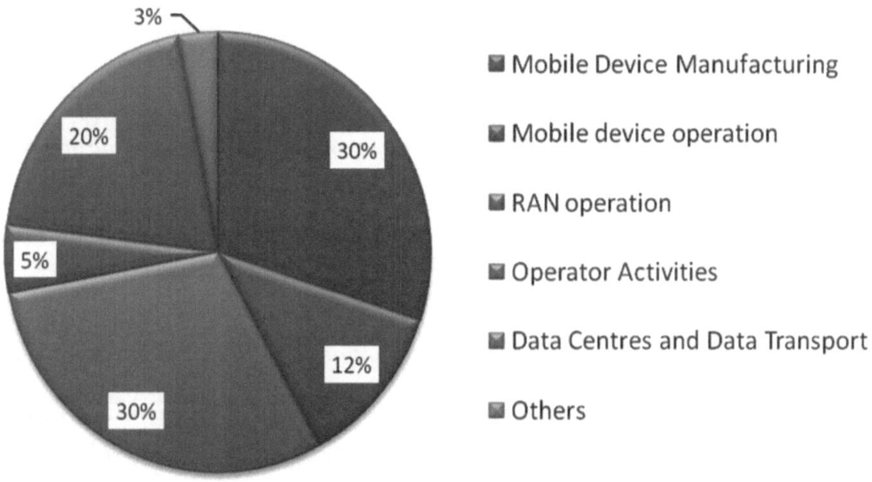

Figure 14.8 Mobile communication's carbon footprint (expected in 2025).

The integration of renewable resources, alongside innovative energy management, serves as evidence of the power of collaborative efforts in reducing carbon emissions and reducing environmental impact. Looking ahead, the chapter leaves us with a vision of a more energy-efficient and economically responsible wireless communication landscape. The opportunities for reducing energy consumption, improving operational efficiency, and contributing to a greener future are abundant. As this transformative journey towards green wireless communications begins, there is the potential for benefits, beyond simple connectivity, that align technology with sustainability, leading to an era where wireless communication is not only a driving force of progress but also a guardian of our planet's well-being.

REFERENCES

1. M. Khasawneh, O. Kaiwartya, L. M. Abualigah, and J. Lloret, "Green computing in underwater wireless sensor networks pressure centric energy modeling," *IEEE Systems Journal*, vol. 14, no. 4, pp. 4735–4745, 2020.
2. M. J. Usman, A. S. Ismail, G. Abdul-Salaam et al., "Energy-efficient nature-inspired techniques in cloud computing data centers," *Telecommunication Systems*, vol. 71, no. 2, pp. 275–302, 2019.
3. S. Kumar, O. Kaiwartya, M. Prasad, N. Kumar, and H. Song, "Green computing in IoT: time slotted simultaneous wireless information and power transfer," *Computer Communications*, vol. 168, pp. 155–169, 2021.
4. S. Kumar, O. Kaiwartya, and A. H. Abdullah, "Green computing for wireless sensor networks: optimization and Huffman coding approach," *Peer-to-Peer Networking and Applications*, vol. 10, no. 3, pp. 592–609, 2017.

5. S. Kumar, O. Kaiwartya, M. Rathee, N. Kumar, and J. Lloret, "Toward energy-oriented optimization for green communication in sensor enabled IoT environments," *IEEE Systems Journal*, vol. 14, no. 4, pp. 4663–4673, 2020.
6. F. Engmann, F. A. Katsriku, J. D. Abdulai, K. S. Adu-Manu, and F. K. Banaseka, "Prolonging the lifetime of wireless sensor networks: a review of current techniques," *Wireless Communications and Mobile Computing*, vol. 2018, Article ID 8035065, 23 pages, 2018.
7. L. Huang, S. Bi, L. P. Qian, and Z. Xia, "Adaptive scheduling in energy harvesting sensor networks for green cities," *IEEE Transactions on Industrial Informatics*, vol. 14, no. 4, pp. 1575–1584, 2018.
8. A. Khatri, S. Kumar, O. Kaiwartya, N. Aslam, N. Meena, and A. H. Abdullah, "Towards green computing in wireless sensor networks: controlled mobility–aided balanced tree approach," *International Journal of Communication Systems*, vol. 31, no. 7, article e3463, 2018.
9. S. Chakrabarti and S. Das. "Energy harvesting enabled adaptive mode selection for cognitive device-to-device communication in a hybrid wireless network: a stochastic geometry perspective," *Wireless Personal Communications: An International Journal*, vol. 129, no. 3, 2023, pp. 1693–1716. Online publication date: 1-Apr-2023.
10. M. Qin, W. Wu, Q. Yang, R. Zhang, N. Cheng, H. Zhou, R. R. Rao, and X. Shen. "Green-oriented dynamic resource-on-demand strategy for multi-RAT wireless networks powered by heterogeneous energy sources," *IEEE Transactions on Wireless Communications*. doi: 10.1109/TWC.2020.2994367.
11. M. N. O. Sadiku, A. E. Shadare, and S. M. Musa, "Green communication," *International Journal of Trend in Research and Development*, vol. 5, no. 4, pp. 211–212, July–August 2018.
12. Q. Wu, G. Y. Li, W. Chen, D. W. K. Ng, and R. Schober, "An overview of sustainable green 5G networks,' *IEEE Wireless Communications*, vol. 24, pp. 72–80, 2017.
13. P. K. Kashyap, S. Kumar, A. Jaiswal, O. Kaiwartya, M. Kumar, U. Dohare, A. H. Gandomi. "DECENT: Deep learning enabled green computation for edge centric 6G networks," *IEEE Transactions on Network and Service Management*, vol. 22, pp. 1–15, 2022.
14. S. Naganandhini and D. Shanthi, "Optimizing Replication of Data for Distributed Cloud Computing Environments: Techniques, Challenges, and Research Gap", 2023, 2nd International Conference on Edge Computing and Applications (ICECAA), Namakkal, India, 2023, pp. 35–41, doi: 10.1109/ICECAA58104.2023.10212287.
15. M. Anandaraj, K. Selvaraj, P. Ganeshkumar, K. Rajkumar and S. Sriram. "Genetic algorithm-based resource minimization in network code-based P2P network,' *Journal of Circuits, Systems, and Computers*, vol. 30, no. 8, 2021.
16. I. K. Okakwu, O. E. Olabode, A. S. Alayande, O. O. Ade-Ikuesan, and A. M. Sulaiman, "A comparative analysis of techno-economic viability of hybrid renewable systems as sustainable alternative for energizing selected base transceiver station in Ogun State, Nigeria," *Mindanao Journal of Science and Technology*, vol. 18, p. 1, 2020.
17. A. Imoize, O. Adedeji, N. Tandiya, and S. Shetty, '6G enabled smart infrastructure for sustainable society: opportunities, challenges, and research roadmap," *Sensors*, vol. 21, p. 1709, 2021.

18. J. O. Ogbebor, A. L. Imoize, and A.A.-A. Atayero, "Energy efficient design techniques in next-generation wireless communication networks: emerging trends and future directions," *Wireless Communications and Mobile Computing*, vol. 2020, pp. 1–19, 2020.
19. O. Alamu, A. Gbenga-Ilori, M. Adelabu, A. Imoize, and O. Ladipo, "Energy efficiency techniques in ultra-dense wireless heterogeneous networks: an overview and outlook," *Engineering Science and Technology, an International Journal*, vol. 23, pp. 1308–1326, 2020.
20. A. E. Ibhaze, P. E. Orukpe, and F. O. Edeko, "High capacity data rate system: review of visible light communications technology," *Journal of Electronic Science and Technology*, vol. 18, p. 100055, 2020.
21. A. Boukerche, Q. Wu, and P. Sun, "Efficient green protocols for sustainable wireless sensor networks", *IEEE Transactions on Sustainable Computing*, vol. 5, no. 1, pp. 61–80, 2019.
22. M. Wang and Y. Hao, "Key Technologies of Green Communication for 5G Mobile Network", 2021 IEEE International Conference on Electronic Technology Communication and Information (ICETCI), pp. 101–104, Oct. 2021.
23. S. Kumari and D. Gupta, "Review on Energy Management for Green IoT in Smart World," 2023 8th International Conference on Communication and Electronics Systems (ICCES), Coimbatore, India, 2023, pp. 322–327, doi: 10.1109/ICCES57224.2023.10192811.
24. Vini Gupta, Sharda Tripathi and Swades De, "Green sensing and communication: A step towards sustainable IoT systems", *Journal of the Indian Institute of Science*, vol. 100, pp. 383–398, 2020.
25. Mahmoud A. Albreem, Abdul Manan Sheikh, Mohammed H. Alsharif, Muzammil Jusoh, and Mohd Najib Mohd Yasin, "Green Internet of Things (GIoT): applications practices awareness and challenges", *IEEE Access*, vol. 9, pp. 38833–38858, 2021.
26. M. Maksimovic, "Greening the future: Green Internet of Things (G-IoT) as a key technological enabler of sustainable development", Internet of Things and Big Data Analytics Toward Next-Generation Intelligence, pp. 283–313, 2018.
27. R. Salama, F. Al-Turjman, D. Bordoloi, and S. P. Yadav, "Wireless Sensor Networks and Green Networking for 6G communication- An Overview," 2023 International Conference on Computational Intelligence, Communication Technology and Networking (CICTN), Ghaziabad, India, 2023, pp. 830–834, doi: 10.1109/CICTN57981.2023.10141262.
28. X. Jiang, M. Sheng, Z. H. A. O. Nan, X. I. N. G. Chengwen, L. U. Weidang and W. A. N. G. Xianbin, "Green UAV communications for 6G: a survey", *Chinese Journal of Aeronautics*, vol. 35, no. 9, pp. 19–34, 2022.
29. S. Verma, S. Kaur, M. A. Khan, and P. S. Sehdev, "Toward green communication in 6G-enabled massive internet of things", *IEEE Internet of Things Journal*, vol. 8, no. 7, pp. 5408–5415, 2020.
30. B. Mao, F. Tang, Y. Kawamoto, and N. Kato, "Al models for green communications towards 6G", *IEEE Communications Surveys & Tutorials*, vol. 24, no. 1, pp. 210–247, 2021.
31. J. He, Y. Mao, Y. Zhou, T. Wang, and Y. Shi, "Reconfigurable Intelligent Surfaces Empowered Green Wireless Networks With User Admission

Control," in IEEE Transactions on Communications, vol. 71, no. 7, pp. 4062–4078, July 2023, doi: 10.1109/TCOMM.2023.3270898.
32. R. Olu-Ajayi, H. Alaka, I. Sulaimon, F. Sunmola and S. Ajayi, "Building energy consumption prediction for residential buildings using deep learning and other machine learning techniques", *Journal of Building Engineering*, vol. 45, p. 103406, 2022.
33. M. Aljebreen, M. Obayya, H. Mahgoub, S. S. Alotaibi, A. Mohamed and M. A. Hamza, "Chaotic equilibrium optimizer-based green communication with deep learning enabled load prediction in internet of things environment," *IEEE Access*, 2023. doi: 10.1109/ACCESS.2023.3345803.
34. P. Segara, E. Hikmawati and K. Surendro, "Towards Sustainable Higher Education: A Framework for Implementing Green IT Strategies," 2023 International Conference on Electrical Engineering and Informatics (ICEEI), Bandung, Indonesia, pp. 1–6, 2023. doi: 10.1109/ICEEI59426.2023.10346921.
35. H. Li, K. Xiong, Y. Lu, B. Gao, P. Fan and K. B. Letaief, "Energy-Efficient Coordinated Beamforming in Multi-Pair MISO Networks With CDI and Eavesdroppers," *IEEE Transactions on Mobile Computing*, vol. 23, no. 1, pp. 437–452, Jan. 2024, doi: 10.1109/TMC.2022.3217121.
36. P. Gandotra, R. K. Jha, and S. Jain, "Green communication in next generation cellular networks: a survey," *IEEE Access*, vol. 5, pp. 11727–11758, 2017.

Chapter 15

Communication networks and media convergence in the age of intelligent systems

Archan Mitra and Sayani Das

15.1 INTRODUCTION

Most of the recent media shifts are due to the rapid proliferation of intelligent technologies. The paradigm shift caused by these technologies in communication networks and media convergence requires further study. The researcher's main interest is how intelligent systems are affecting media convergence and global communication networks.

Conceptualize media convergence mechanisms in the age of intelligent systems. Media convergence is not new, but artificial intelligence (AI) and machine learning have expanded it (Jenkins, 2006). Jenkins suggested media convergence affects technology, industry, culture, and society. These transitions are being redefined and expanded by intelligent systems (Couldry & Hepp, 2017).

The environmental communication argument shows how intelligent systems have merged traditional and digital media, reshaping media creation, dissemination, and consumption (Anderson, 2017). These innovations have produced new landscapes that media and communication researchers and practitioners, especially environmental communicators, must study.

Communication systems have also benefited from intelligent systems. Intelligent systems provide dynamic, autonomous, and adaptable systems with unprecedented communication network robustness and sophistication (Castells, 2009). Castells said that hybrid intelligent systems have greatly enhanced flexibility, flexibility, and scalability – key qualities of digital networks.

The study will imagine how networking and press conferences might change in a future where intelligent systems are prevalent. Increased cross-functional contact blurs the borders between producer and consumer in the current era of hyper-connectivity and media flow, transforming conventional forms of communication.

DOI: 10.1201/9781003541363-15

15.2 LITERATURE REVIEW

Media convergence and smart systems research has only hinted at the virtual era's impact on discussion networks. Jenkins (2023) examines the moving components of media convergence in the internet age, focusing on how sensible arrangements improve media accessibility and person involvement. Liu and Silverman (2023) examined how AI affects media production and transmission, focusing on AI's ability to personalize content based on individual preferences. Kapoor et al. (2023) analyze how smart systems are changing communication networks. Their findings suggest that modern networks are becoming more flexible and green. This helps people manage digital media's growing complexity.

AI and intelligent systems are transforming media in many ways. AI-powered systems can organize news feeds, generate music and video, and make targeted suggestions (Li et al. 2023). Intelligent systems are also boosting CDN performance, improving service quality and user experience across platforms and devices (Aghdaei et al., 2021). AI systems analyze user preferences and behaviors to personalize advertising (Huang et al., 2022). Smart technologies optimize media consumption depending on user choice, enabling tailored content (Knijnenburg et al., 2020).

15.3 CHALLENGES AND OPPORTUNITIES

Although sensible structures have the capability to significantly improve communique networks and media convergence, there are numerous problems that should be resolved. Ethical concerns are apparent in the context of AI algorithms because of their ability to perpetuate bias and discrimination. Therefore, it is essential to carefully examine the ethical outcomes associated with their use (Noble, 2018). With regard to issues related to the protection of personal information, the extensive utilization of AI for tailored content distribution gives rise to apprehensions regarding user privacy and data security (Ohm, 2017). The difference between people or groups who have access to and are proficient with digital technologies – like computers and the internet – and those who do not is known as the "digital divide." Inequalities in digital technologies and high-speed internet connections have the potential to exacerbate already-existing inequalities in digital access (World Bank, 2022).

15.4 THEORETICAL FOUNDATION

15.4.1 The concept of media convergence

A process called "convergence" occurs when diverse pieces move toward a common point, integrating previously distinct entities (Meikle & Young,

2011, page 2). In media and communications, "media convergence" has remained popular and is used to describe improvements and changes in technological, industrial, cultural, social, spatial, and political sectors. Jenkins (2006), Latzer (2013), and Miller (2011) have devoted significant time and effort to this subject. Media convergence has been recognized as a critical transition process and the result of digitization from its inception. This is especially true with the rise of the internet and other online technologies. This has made it prominent in academic, policy, and business literature. Digitization has eased data capture, storage, and transmission and made it possible to separate technology from media services for the first time. A universal digital standard allows us to do this. Digital media, including speech, sound, text, and film, was no longer bound to one device. This was true across media formats. According to Figure 15.1, digital media is rising and conventional media is declining. Most media devices could simultaneously display many media types and services. It was true whether it was speech, sound, text, or film. "This is the core of technological convergence: all media being increasingly stored and transferred on the same format and becoming completely interchangeable" (Miller, 2011). Many consider media convergence a comprehensive transformational process and a major consequence of digitization, according to Peli and Sparviero (2017).

Figure 15.1 illustrates the trends in media convergence spanning a period of two decades from 2000 to 2020. It analyzes the relative utilization rates of traditional media and digital media throughout the years.

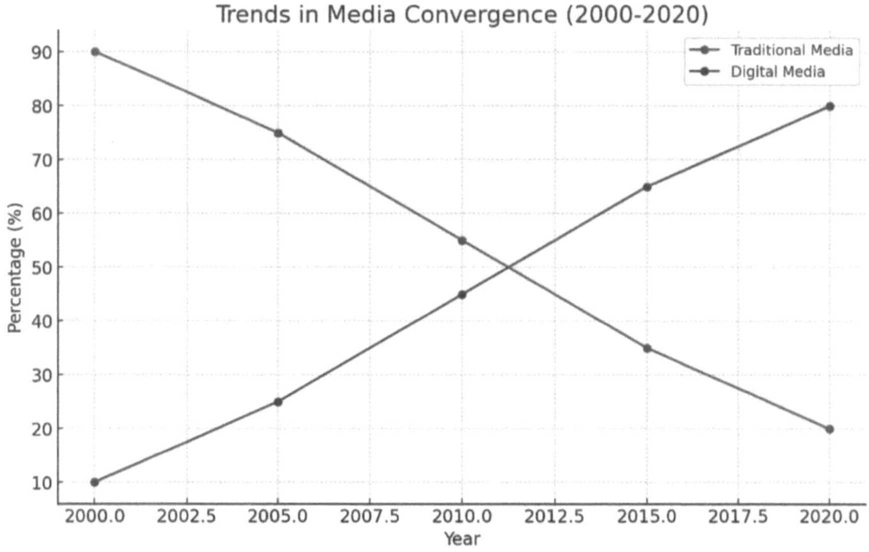

Figure 15.1 Trends in media convergence (2000–2020).

The x-axis denotes the temporal dimension, specifically the years from 2000 to 2020. The y-axis indicates the percentage scale, ranging from 0% to 90%. The lines intersect at the halfway point of the graph, indicating that around the year 2010, the usage of traditional and digital media was roughly equivalent.

This figure depicts a notable change in the way people consume media, showing that digital media has become the predominant medium over the course of 20 years, while the use of traditional media has significantly decreased. The shift can be ascribed to technical improvements, alterations in consumer behavior, and the development of media delivery platforms.

15.4.2 Communication networks in the digital age

Communication technology help people engage and share information across time zones. This applies even when people are apart. Ramanujam (2008) and Daramola (2012) investigated this topic. Communication equipment includes computers, phones, mobile phones, email, the internet, social media, and digital technology. Hancock, Thom-Santelli, and Ritchie (2004) state that modern communication technology allows people geographically apart to communicate. One of the most significant affordances and powers of recent communication technology is its capacity to enhance human interactions even without physical presence. Thus, technology has become an important part of people's lives, enabling relationships that were unthinkable a decade ago. Communication tools allow people across time zones to engage and share information quickly. Compacting space and time gives users more control over information transmission (Onyeator & Okpara, 2019).

The internet is needed for unique mediated interactions in many formats. These include Facebook, Instagram, FaceTime, Skype, and Twitter videos. Technoself academics say mobile phones and telephony have made the internet more accessible, merging mobile culture into daily life and maybe physiology (Harper, Rodden, Rogers, & Sellen, 2008). Technology has shaped human communicative identity and culture, according to Watkins (2012).

15.5 THE DYNAMICS OF MEDIA CONVERGENCE

15.5.1 Media synergy and cross-promotional strategies

The concept of media synergy is akin to the will-o'-the-wisp. Everyone agrees that it is good, but few can actually prove it. There is no established way to gauge the value of media synergy, yet it must add something. Everyone who buys media for advertising purposes hopes to achieve that, but no one is quite sure how to do it. Professor Don Schultz of Northwestern University's Medill School examines and demonstrates why this "value-added" mystery-marketing communication notion has eluded marketers for so long and why a solution might be on the horizon.

The bar chart in Figure 15.2 presents the adoption rate of intelligent systems in various media industries as of 2020. It shows how sectors like social media, entertainment, and gaming have relatively high adoption rates, reflecting the significant impact of AI, machine learning, and data analytics tools in these areas. The news industry and online advertising also demonstrate considerable utilization of intelligent systems, albeit at slightly lower rates compared to entertainment and social media.

The concept of media synergy has been long-standing. The objective of what was formerly referred to as cross-media research has consistently been to determine the additional impact of diverse media formats or messages on a specific brand being observed or adopted by a particular consumer.

According to Schultz, Block, and Raman (2012), and a concept of media synergy that is now widely recognized,

> Media synergy arises when the combined effect or impact of a number of media activities is different from the sum of their individual effects on individual consumers. Thus, synergy is a phenomenon in which the whole is not always exactly equal to the sum of the parts, but could be greater than, or lesser than that sum.

In its simplest form, synergy refers to working together to accomplish a goal that cannot be accomplished alone. Synergy can benefit from cross-media

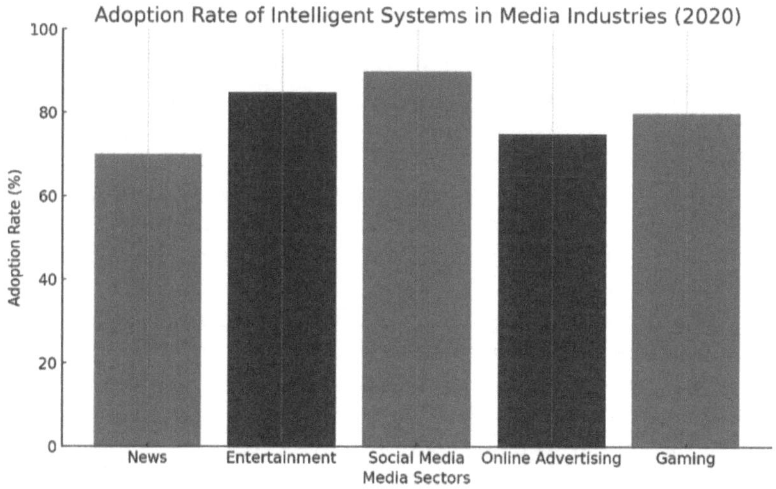

Figure 15.2 Adoption rate of intelligent systems in media industries.

convergence if businesses are smart enough to capitalize on the connections they have created. Disney exemplifies a synergistic company structure, with its various divisions, such as the film studio, kids' TV channel, and Disney Store, working together to promote and reinforce each other's products. This includes the cross-promotion of films on the TV channel and the sale of related merchandise in the Disney Store, both in physical stores and online. As a result, children are exposed to Disney products through multiple channels, leading to requests for the toys and DVDs/CDs they have seen on TV, the internet, or in theaters.

15.5.1.1 Case study

The LEGO video game series exemplifies a prosperous instance of cross-promotional branding. The games are founded around well-known licensed intellectual properties, including Star Wars, Harry Potter, and Batman, and they prominently showcase the distinctive aesthetic of LEGO building blocks. The amalgamation of familiarity and innovation has rendered LEGO video games highly sought-after among individuals of all age groups, youngsters and adults alike.

The LEGO video games transcend their role as mere adaptations of films and other media, offering a more complex and multifaceted experience. At the same time, the game maintains great gameplay where you are able to adapt your game characters and still have the experience of a challenge. These games also have hidden jokes and references; they always maintain a connection to fans of the original stories.

LEGO video games have be en a great success for both LEGO and its media partners. Consequently, this has allowed both organizations to make more money from introducing new audiences to other aspects of the enterprise.

There are several crucial elements that have shaped the development of the LEGO video game franchise.

- LEGO video games are derived from widely recognized and popular forms of entertainment, making them more appealing to a global audience This trend gives the games a natural audience of enthusiasts who, when first playing them, become familiar with the characters and the plot.
- LEGO video games are intentionally designed to be family-friendly, making them accessible to individuals of all ages. Easy-to-use controls and humorous elements contribute to immediate accessibility and playability, but challenging and progressively more challenging puzzles keep players engaged and engrossed.
- LEGO video games offer unique and original gameplay that stands out from other licensed video games. The use of LEGO building

bricks allows players to create and personalize their own levels, thus bringing an increased level of creativity and fun to the game.
- LEGO video games have various cross-promotional strategies for LEGO and its affiliated media agencies. For example, games can be a way to gain support and promote recently released movies, television shows, or merchandise.

The LEGO video game franchise is an example of the effectiveness of cross-promotional branding in establishing a synergistic partnership betwe en two companies. LEGO video games have successfully capitalized on the huge appeal of licensed titles, bringing in more customers and generating significant financial returns for LEGO and its affiliated media agencies.

15.5.2 The audience in the convergent media landscape

Media experts have extensively documented the changing media landscape in recent years, leading to a widely accepted consensus. The proliferation of mass media has created a dynamic digital media environment, resulting in audience fragmentation across existing and emerging platforms. The proliferation of media outlets has created several opportunities for brands to invest and engage. However, target segments have different preferences for specific media channels, making media planning for businesses more complicated.

Recent years have seen a convergence of traditional and digital media channels. Television, print, and radio are adopting interactivity, on-demand content, and algorithmic approaches. Digital channels are also converging.

Due to growing competition, digital platforms copy popular formats and features. People want to keep up with the latest trends, from the Story format and AR filters to the race to establish a Clubhouse competitor. Additionally, the copying format was adopted to facilitate creators, as explained in the next section. Undoubtedly, not all such efforts succeed. Instagram successfully integrated Stories, but Twitter recently announced it would terminate Fleets. However, these occurrences add to the growing uniformity of digital platform user experiences. It is anticipated to help brands create cross-platform creatives, but different platforms have varied interpretations of these formats and features, which companies must account for.

The growing usage of algorithmic feeds is another sign of digital platform convergence. This method can create individualized material streams for each user, but it can also create echo chambers and perpetuate bad habits. This method is cost-effective for extracting cultural aspects and categorizing the audience by interest group. TikTok's decision to monetize its highly praised algorithms suggests that more digital content platforms may exploit algorithmic consumption, despite the potential risks.

Businesses must understand the many characteristics that various media channels offer for different audience segments when considering conversion. Traditional media companies gave YouTuber Lilly Singh and Instagram viral phenomenon Ziwe their own TV shows. However, adding additional media outlets to their channels brings many challenges. In this setting, digital platforms may have an edge, but they also face valid concerns about becoming too big and alienating consumers.

15.6 INTELLIGENT SYSTEMS AND FUTURE COMMUNICATION NETWORKS

15.6.1 The advent of intelligent systems

Intelligent systems possess the capability to autonomously and optimally address intricate challenges within designated contexts. Collaborative efforts among individuals and technological progress – including but not restricted to Big Data, the Internet of Things (IoT), mobile networks (3G, 4G, and 5G), AI, robotics, video analytics, computer vision, and augmented reality – are utilized to establish these entities. According to Molina (2020),

> An intelligent system operates in an environment with other agents, possesses cognitive capabilities such as perception, action control, deliberative reasoning or language, follows principles of behavior based on rationality and social norms, and has the ability to adapt by learning.

The ability to independently select their path of action based on perceptions, facts, and experiences that have been pre-established by human agents is a characteristic that is possessed by intelligent systems. Understanding the ways in which intelligent systems contribute to companies by offering practical answers is of utmost importance.

Organizations utilize intelligent systems in various ways to improve their operations, goods, and services. Here, we present five applications of intelligent systems commonly employed by organizations.

Autonomous robots: Autonomous robots possess the ability to make autonomous decisions and then take appropriate actions, similar to humans. An authentically autonomous robot possesses the capability to detect its surroundings, make judgments based on its perception or programmed recognition, and subsequently initiate movement or manipulation within said surroundings. Regarding robot mobility, decision-based activities encompass essential duties such as initiating, ceasing, and evading obstacles along their trajectory.

Natural language processing: Natural language processing (NLP) is a significant discipline within the broader domain of AI. This technology has been successfully applied to personal assistants in various business sectors. The system works by analyzing what the user voluntarily provides, breaking it down into interconnected parts to ensure compliance, and processing it. In this order, the established systems are implemented. This method is now in high demand in the market due to its reputation and efficiency. The discipline of natural language processing is gaining in popularity, providing capabilities such as seamless integration of intelligent devices and interactive communication with humans. Important features of AI applications in NLP include knowledge extraction, rational reasoning, and constraint satisfaction.

Expert systems: An expert system is a computer program that applies artificial intelligence methods to resolve issues in a particular field where human expertise is typically needed. This technique utilizes collected knowledge and prior experience to develop and simulate problem-solving. These programs have been specifically intended to replicate and simulate the cognitive processes of the human brain.

Information is being gathered to resolve the most challenging aspects. Data interpretation is crucial for determining suitable solutions, considering unique project objectives such as patient and physician evaluation, treatment, and diagnosis.

Sentiment analysis: Sentiment analysis is a field that combines artificial intelligence and applied linguistics to study and analyze the interactions between machines and humans.

This technique is currently being used on multiple social networks to determine the underlying ideas and intents expressed by the message author. This system employs advanced artificial intelligence techniques, including natural language processing, text analysis, and data science, to identify, extract, and analyze subjective information.

As a result of this technological advancement, corporations have acquired the ability to discern client responses to a particular product or service through the categorization of textual data into positive, negative, or neutral sentiments. Intelligent systems play a crucial role in enabling organizations to optimize diverse processes, hence facilitating the development of goods that effectively address societal needs and enhance the overall quality of products and services (Algotive, 2022).

15.6.1.1 Case study: Amazon's Alexa

Alexa is the name of Amazon's voice-activated smart home system. Amazon developed the Alexa Voice Service (AVS) to mimic authentic conversations, but in actuality, it operates through straightforward voice instructions.

It is a technology that utilizes natural language processing and intelligent voice recognition to enable voice commands on any connected device with a microphone and speaker. Alexa has the ability to understand and respond to human interactions, continuously learning about human behavior and interests.

This tool has significantly transformed home technology through advancements in human-computer interaction. It is now capable of accurately identifying our interests and initiating meaningful interactions.

15.6.2 Future trajectories: predictive analyses and recommendations

Given the ongoing evolution of multimedia communication and communication networks, driven by breakthroughs in intelligent systems, it is crucial to anticipate possible future technologies within this domain. Based on a comprehensive review of contemporary scholarly works and authoritative evaluations, this section aims to provide a prognostic examination of emerging patterns that are likely to influence the overall environment in the future. Additionally, it offers guidance and suggestions for young actors within the creative field.

15.7 PREDICTIVE ANALYSES

15.7.1 Artificial intelligence and deep learning

Technological advances, especially AI and deep learning, have the potential to bring about profound changes in communication networks and reshape the modern landscape, according to Schäfer and van Es (2017). This is anticipated as AI systems gradually take over complex tasks, streamlining business data analysis. AI advances will lead to more sophisticated tools for content creation being developed, which will lead to dramatic changes in media behavior.

15.7.2 Decentralized communication networks

Defined communication can communicate thoughts and ideas in many ways and participate in decision-making. Such networks are ideal for organizations that want to promote creativity, collaboration, and diversity in communication and boost employee morale and engagement by giving them greater autonomy and influence. However, digital communication can also lead to inconsistent and unclear communication due to conflicting information from multiple sources. Tandoc Jr. et al. (2018) argue that the concept of decentralization is a viable option for a modern environment characterized by larger democratic principles. As opposed to focusing on this process, people are distributed, making it easier to build a wide range of ideas and information.

15.8 RECOMMENDATIONS

15.8.1 Embracing collaborative models

Given the ever-changing nature of today's landscape, it's really important for everyone involved to work together and think of ways to involve audiences in the process. By doing this, we can use the collective knowledge and perspective of everyone involved. In their study, Jenkins et al. (2018) emphasized the importance of collaboration in networking. They argue for an environment that facilitates cultural assimilation, aiming to make the modern landscape more inclusive.

15.9 DISCUSSION

This study's investigation of information networks and networks in the era of intelligent systems provides several important insights. Integration of intelligent systems with media networks dramatically increases efficiency and improves productivity, which is consistent with Jenkins' (2023) observation on media accessibility. Research also highlights the role of intelligent systems in communication, with an emphasis on network reorganization, which aligns with the findings of Kapoor et al. (2023).

Some aspects of environmental communication suggest that intelligent systems and multimedia communication play an important role in the dissemination of environmental information, where, as Smith and Gomez (2023) emphasize, the public domain will be involved. However, this also raises concerns about data privacy and the ethical use of AI in medicine, as highlighted by Liu and Silverman (2023).

Limitations of the current study include its focus on developed areas, which may overlook the nuances of media encounters in developing countries. Future research should address this gap and explore the global implications of intelligent systems in information and communication systems.

This study opens the door for further research on the ethical implications of AI in the media, the impact of media on global environmental communication, and achieving equity in media channels using information worldwide.

15.10 CONCLUSION

AI and deep learning advances have the potential to fundamentally change media and communication networking. This suggests a future middle ground that is likely to drive innovation and high user engagement and is based on the delivery of tailored content and microcontent. Furthermore, digital communication networks are increasingly leading to a fundamental shift in multimedia delivery and process management, which will advance a radical shift from traditional models. This decentralization may mean that a rich tapestry of different perspectives and narratives exists in a democratic media landscape where power is distributed rather than concentrated. This democratization of media discourse highlights the potential for transforming power dynamics, enabling a more diverse and inclusive media environment. To better understand the potential ethical issues and implications that arise as technology advances in AI-enabled information production, it is important to look to the future in this growing area of research. In addition, ways to better encourage the use of responsive AI in media, including data binding and user privacy, should be explored. Calls for appropriate policy adoption will establish defined networks to promote healthy ecosystems to mitigate risk. This area can provide fertile ground for ideas and real success in environmental negotiations and beyond. Scholars should focus on this dynamic intersection with a view to the future because it is not only technological progress but AI and the enhancement of digital networks that are formidable tools that can change the boundaries of media and communication.

REFERENCES

5G Americas. (2022). *5G Americas Market Data & Forecasts.* https://www.5gamericas.org/

Aghdaei, M. H., Abolfazli, S., & Shehabi, A. (2021). Intelligent content delivery networks: A survey. *IEEE Transactions on Network and Service Management, 18*(4), 2882–2900.

Algotive. (2022). Intelligent systems: What are they, how do they work and why are they so important. https://www.algotive.ai/blog/intelligent-systems-what-are-they-how-do-they-work-and-why-are-they-so-important

Anderson, A. A. (2017). Media convergence and the interconnectedness of digital platforms: A case study of environmental communication. *Journal of Media and Communication Studies, 9*(5), 69–78.

Castells, M. (2009). *Communication power.* Oxford University Press.

Couldry, N., & Hepp, A. (2017). *The mediated construction of reality.* Polity.

Council of Europe. (2018). Artificial Intelligence and its impact on freedom of expression. https://www.coe.int/en/web/freedom-expression/-/artificial-intelligence-and-its-impact-on-freedom-of-expression

Dal Zotto, C., & Lugmayr, A. (2016). Media convergence as evolutionary process. In A. Lugmayr and C. Dal Zotto (Eds.), *Media convergence handbook-Vol. 2: Firms and user perspectives* (pp. 3–16). Heidelberg: Springer.

Daramola, I. (2012). *Introduction to mass communication.* Rothan Press.

Diehl, S., & Karmasin, M. (Eds.). (2013). *Media and convergence management.* Springer.

Dwyer, T. (2010). *EBOOK: Media convergence.* McGraw-Hill Education (UK).

Fagerjord, A., & Storsul, T. (2007). *Questioning convergence.* In T. Storsul & D. Stuedahl (Eds.), *Ambivalence towards convergence: Digitalization and media change.* Nordicom.

Hancock, J. T., Thom-Santelli, J., & Ritchie, T. (2004, April). Deception and design: The impact of communication technology on lying behavior. In *Proceedings of the SIGCHI conference on human factors in computing systems* (pp. 129–134).

Harper, R., Rodden, T., Rogers, Y., & Sellen, A. (2008). *Being Human: HCI in the Year 2020.*

Hu, Y., Zhang, H., Li, G., Li, Z., & Wu, W. (2023). Artificial intelligence-enabled software-defined networking: A survey. *IEEE Transactions on Network and Service Management, 20*(1), 1–24.

Huang, S., Wang, F., & Xiao, B. (2022). A survey of artificial intelligence for targeted advertising. *ACM Transactions on Intelligent Systems and Technology (TIST), 13*(6), 1–40.

Jenkins, H. (2008). *Convergence culture: Where old and new media collide.* NYU Press.

Jenkins, H. (2023). *Media convergence in the digital age: Exploring the role of intelligent systems.* Academic Press.

Jenkins, H., Ito, M., & boyd, d. (2018). *Participatory culture in a networked era.* Polity.

Jensen, K. B. (2022). *Media convergence: The three degrees of network, mass, and interpersonal communication.* Routledge.

Jin, D. Y. (2013). *De-convergence of global media industries.* Routledge.

Kapoor, A., Singh, R., & Patel, M. (2023). Intelligent systems and the future of communication networks. *IEEE Communications Magazine, 61*(2), 45–52.

Knijnenburg, B. P., Willemsen, M. C., Kobijn, Z., & Cremonesi, P. (2020). A survey of recommender systems for news personalization. *ACM Transactions on Intelligent Systems and Technology (TIST), 11*(4), 1–66.

Latzer, M. (2013). *Convergence, co-evolution and complexity in European communications policy.* Working Paper of the Media Change & Innovation Division. Zurich, Switzerland.

Li, Y., Fang, S., & Zhu, X. (2023). A survey of artificial intelligence in multimedia content creation. *ACM Transactions on Multimedia Computing, Communications, and Applications (TOMM), 19*(1), 1–40.

Liu, Y., & Silverman, B. (2023). *Artificial Intelligence in media: Transformation of content production and distribution.* Media Studies Press.

McLuhan, M. (1994). *Understanding media: The extensions of man.* MIT Press.

McQuail, D. (1987). *Mass communication theory: An introduction.* Sage Publications, Inc.

Meikle, G., & Young, S. (2011). *Media convergence: Networked digital media in everyday life.* Palgrave Macmillan.

Miller, V. (2011). *Understanding digital culture.* Sage.

Molina, M. (2020). What is an intelligent system? *arXiv preprint* arXiv:2009.09083.

Onyeator, I., & Okpara, N. (2019). Human communication in a digital age: Perspectives on interpersonal communication in the family. *New Media and Mass Communication, 78*(1), 35–45.

Peil, C., & Sparviero, S. (2017). Media convergence meets deconvergence. In *Media convergence and deconvergence* (pp. 3–30).

Ramanujam, R. C. (2019). *Mass communication and its digital transformation.* APH Publishing Corporation.

Schäfer, M. S., & van Es, K. (2017). *The datafied society: Studying culture through data.* Amsterdam University Press.

Schultz, D. E., Block, M. P., & Raman, K. (2012). Understanding consumer-created media synergy. *Journal of Marketing Communications, 18*(3), 173–187.

Smith, J., & Gomez, F. (2023). Environmental communication in the age of media convergence: Opportunities and challenges. *Journal of Environmental Communication, 17*(1), 33–47.

Tandoc Jr., E. C., Lim, D., & Ling, R. (2018). Diffusion of disinformation: How social media users respond to fake news and why. *Journalism, 19*(11), 1370–1387. https://doi.org/10.1177/1464884918758715

Watkins, C. (2012). Digital divide: Navigating the digital edge. https://moody.utexas.edu/sites/default/files/watkins_1.pdf

Chapter 16

Recent advancements in the security of conventional networks with software-defined network

Vinay Gautam and Jyoti Rani

16.1 INTRODUCTION

In recent years, the network environment has gone through various phases of innovation. The software-defined network (SDN) is one of these, which was introduced in 2010 as a new development in computer networks [1, 2]. SDN is described as a centralized programmable network decoupled from the data plane. SDN can be referred to as a network where hardware devices are embedded with software services. These software services act as controllers of the network devices. Conventionally, the data and control planes in the networking devices were coupled together, meaning that both the operating system and hardware, such as routers, switches, and firewalls, were kept in a single device. These devices were intelligent enough to handle network traffic. These devices intelligently manage and distribute traffic in the network. In some devices, the Command Line Interface mode is fixed for the device to become configured as per network requirements, which makes this process slow and prone to errors. Open-source interconnection (OSI) is a well-defined industry standard used as a model in the data plane [3, 4]. This makes it easier to develop and upgrade the functions without considering the other layers. This type of layering architecture is also used in other fields. The literature confirms that a huge amount of research and development has happened in recent years. After this study, it was observed that network control and management planes were not available within the network. Hence, SDN, like the innovative architecture introduced here, decouples control from the data plane. The layering structure of SDN is elaborated in the next section.

16.1.1 SDN architecture layers

SDN is a three-layer architecture with an application layer, control layer, and infrastructure layer. These are collectively used to manage and control networks centrally and intelligently. These innovations in SDN enhance the possibility of network programmability and help the industry in initiating innovations in the development process. Furthermore, these features accelerate innovation and introduce new features and services within the network [5, 6]. Layered architecture is depicted in Figure 16.1.

DOI: 10.1201/9781003541363-16

The centralization of SDN allows performance and policy management. Thus, it is a scalable, reliable, and flexible network for user services and provides an abstraction layer between hardware and software. A detailed description of the layers is provided below:

A. *Infrastructure layer*

This layer provides a service to connect all the hardware that exists in the system. The southbound Application Programmable Interface (API) acts as a control plane between software and hardware [7]. The Application Programmable Interface (API) provides services that can be used to interact with the upper control layer. The Application Programmable Interface (API) is used to establish communication between these two layers.

B. *Control layer*

The next layer after the infrastructure layer is the control layer. This is a prime layer of the SDN architecture that is used to establish any type of communication between network devices and keep track of the topology in the network. The northbound API is used by the control layer to translate all the communication from the control layer to the upper layer [8].

C. *Application layer*

The next layer of SDN architecture is the application layer. This plays an important role in the network by defining all the policies, features, and

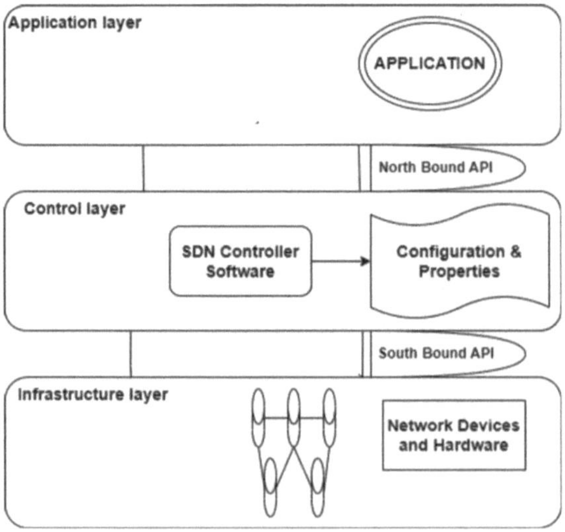

Figure 16.1 SDN architecture stack.

services. This layer clearly instructs all the network devices to follow the rules and regulations defined. Here, the applications request information regarding the devices and topology within the network. Whenever changes happen within the SDN architecture, the applications adapt their features and behave according to the changes. Whenever any changes happen within the policies, features, and services, the application adapts those and behaves accordingly. The northbound API is offered by the controller and the applications. Southbound API provides the necessary announcement between the controller and the network devices.

16.1.2 SDN architecture with virtualization

The SDN layer architecture is more adaptive to the data center scheme whenever the logical network slice has been used in the virtualization environment. Logical slicing is implemented with a hypervisor layer in the SDN stack, as depicted in Figure 16.2. The hypervisor layer acts as a controller of controllers and works on top of controllers. Here, it separates each controller to control its own physical and virtual hosts. This layer makes complete implementation independent of other controllers. The SDN network has its pros [9], but there have been recent challenges as well. Here, this chapter focuses on security, which is the prime challenge of the network.

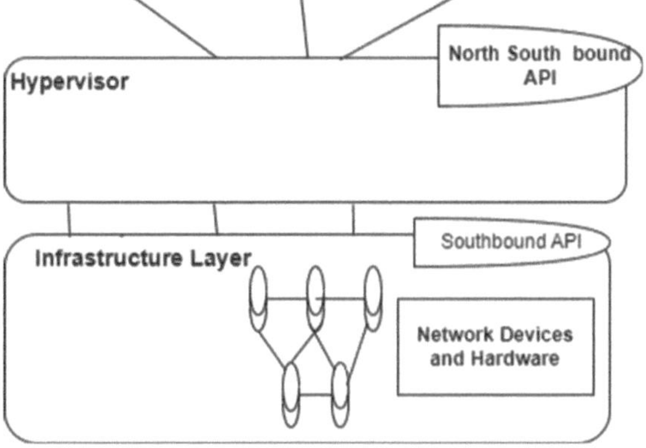

Figure 16.2 SDN stack with hypervisor.

Here, the SDN controls the complete network and has more capabilities to tackle security issues, allowing SDN to gain widespread popularity in the network market.

16.1.3 Centralization in SDN

Centralized control is the only way to view the resources worldwide and is maintained with a central control plane within the application layer. The SDN uses several networks that collaborate with the controller, leading to a centralized control structure [10–13]. The advantages of centralized control are that it:

a) Improves network management tasks.
b) Increases agility within the network.
c) Increases autonomy within the network.
d) Automatically detects and mitigates DDOS attacks.
e) Makes it easy to share resources because of the programmable network.
f) Creates a virtual network easily.

16.1.4 SDN benefits

In any network, data and control planes are important, and in SDN, the independent working of these two makes the network self-controlled through programming [14, 15]. The decoupling of these two planes enables the network to:

a) Enhance configurations
b) Improve performance
c) Encourage innovations and operations within the network
d) Encourage the design of other networks because of its convenience
e) Provide an isolated virtual network

The SDN control plane includes packet forwarding, link tuning, and avoiding the obstruction of layering [16, 17]. The instant network status and user-defined policies within the network permit real-time control. The list of features leads to network optimization and improved performance of the network.

i Configuration enhancement

In the case of network scaling, new nodes need to be inserted within the existing network, which requires network configuration [18]. The configuration function is used to achieve the normal function of the network. Manual configuration procedures have been utilized by manufacturers to maintain the heterogeneity within the network. Manual processing is not an easy task and may be the reason for errors within the network. That's why lots of effort is required to troubleshoot the network from errors, which is accepted as a big challenge. SDN also provides solutions for such issues within the network. Network management acts as a remedy for such issues within the network. The SDN uni-function to control all the network devices provides a better solution for scaling-up issues within the network [19]. The complete network is automatically configured and optimized using programming.

ii Performance improvement

The performance of the network is directly related to the network resources. The proper and appropriate utilization of network resources enhances the performance of the network. Huge amounts of technologies and stakeholders are related to a single network; that's why it is difficult to manage and optimize all the resources. Recent state-of-the-art techniques are used only to optimize or utilize a subset of a network. The prime reason for subset optimization is the local information of the network without considering cross-layer information. SDN provides solutions for the suboptimal performance-related issues of the network [20, 21]. The centralized control in the SDN provides a globalized view of the network and shares messages within layered architecture. The optimization issues can be resolved with the appropriate design of the centralized algorithm. SDN resolves most of the issues and improves the quality of service (QoS).

iii. Encouraging innovation:

The continuous development of network applications encourages the network to be more innovative to meet the demands of future applications. Rapid changes in the network cannot be implemented and deployed immediately. The real issue arises whenever new development happens in the conventional network, and it is really difficult to adjust the network to those developments [22]. In SDN, the network is reconfigurable and adjustable to all the changes. Innovative development in the applications is easily adaptable in the SDN. Table 16.1 compares SDN and conventional networks.

Table. 16.1 SDN vs conventional network

	Software-defined network	*Convention network*
Features	Component decoupling	Needs new protocol
Configuration	Automated	Manual with error
Performance	Excellent performance due to global control	Depending on the configurations
Innovation	Software-based easy upgradation	Difficult to implement new ideas

16.1.5 SDN challenges

Although SDN promises configuration enrichment, performance improvement, and advancement, it still faces lots of challenges. A few basic issues still remain unresolved in SDN:

a) Security, which is the most required feature of a network
b) Open flow drivers missing from SDN
c) Standard API
d) High-level programming

16.1.6 Motivation of research

SDN provides an innovative network for vendors to deploy their applications. It is very difficult to replace traditional networks with SDN. The prime concerns regarding SDN implementation are security, lack of experienced experts, and technical support. Elsewhere, various prototype implementations of SDN have been referred to as small test beds.

16.1.7 Contribution of the research

a. Detailed analysis of SDN compared to a conventional network
b. Detailed analysis of security perspectives of SDN network
c. Analysis of various components that contribute to the security perspective in SDN

16.2 RELATED WORK

Recent studies have evaluated relevant aspects, such as analysis of concepts, methods, effects, and challenges, in SDN development. Network technologies are the base technology for the underlying success of cloud computing [21]. However, the slow growth of the technology leads to various issues related to the implementation of the latest applications into underlying classical technologies. The SDN can act as a new-generation network and resolve all the issues faced with underlying conventional network technology. The objective of this chapter is to focus on the effects and challenges of SDN:

a. Implementation challenges of SDN
b. Security challenges of SDN

Figure 16.3 provides an overview of the literature survey, indicating the main challenges involved in implementing SDN. The literature review was based on four areas:

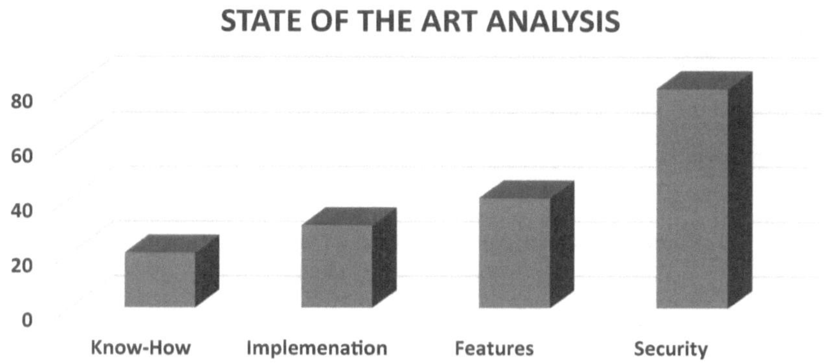

Figure 16.3 Overview of literature survey.

a) Know-how
b) Implementation
c) Features
d) Security

Several issues were discussed by researchers related to network technology. However, the major focus of the discussion was on network security issues and SDN-based solutions [22]. The other categories are implementation and know-how of existing network technology. This category is concerned about the infrastructure available to support SDN.

16.2.1 SDN security analyses

Secure communication depends upon five basic properties of security: confidentiality, integrity, availability, authentication, and nonrepudiation. A network that is not secure is exposed to various types of attacks and damages that need to be addressed with appropriate solutions. The security professional secures data, network assets, and communication transactions within the network, and SDN architecture has been used for this [23, 24]. In recent years, the complexity of network systems has increased rapidly, increasing the risk of bugs and errors within networks. This requires programmers to develop software solutions.

16.2.1.1 Security challenges

In the past, there were challenges in securing SDN. SDN allows for easy automation of the network with programs and dynamic policy flow. This may lead to security vulnerabilities. The SDN provides agility, simplified control, and real-time programmability [25]. However, security issues still need to be addressed, as security is an important concern to protect the

Table 16.2 Security challenges and resolutions

Layers	*Attacks on network*	*Reason of attack*	*Resolutions*
Data forwarding layer	Man-in-middle attack	TLS support is required for security	FlowChecker [26] ForNOX [27] VeriFlow [28] Controller replication [29]
	DoS attack	Restricted storage area and massive packet load in a short period	FlowVisor [30] Virtual source Address Validation Edge (VAVE) [31] Resonance [32]
Control layer	DoS/DDoS	Hard to distinguish right and wrong packets, and the low capacity of the controller	FloodGuard [33] DDoS Blocking Application [34] Content-Oriented Networking Architecture (CONA) [35]
	Related to distributed multi-controllers	Access control distribution, hard-to-implement multi-tenant controller, inappropriate multiple controller	DISCO [36] McNettle [37] HyperFlow [38]
	Applications	Open programmatic interface Malicious applications	SE-Floodlight [39] FRESCO [40]
Application layer	Unauthorized access	Controller software vulnerability	PermOF [41] NICE [42] Verificare [43] VeriCon [44]
	Security rules and configuration conflicts	Variety of application software Difference of access control and accountability for various application software	Flover [45] Anteater [46] NetPlumber [47]

network from malicious attacks. The following aspects need to be considered to protect the network:

1) Availability
2) Performance
3) Integrity and confidentiality

16.2.1.2 Process for SDN security

The aspects listed above need to be in place to secure the network.

a) *Authentication and authorization*

The SDN can be accessed by authorized and authenticated actors only. These two properties should be detailed enough to stop illegal access to the network.

b) *Resiliency*

The network should be able to recover automatically from any possible attack and start an alternate network whenever attacks happen.

c) Multi-domain isolation

The network must maintain isolation between tenants. The isolations are detailed as follows:

i. Resource isolation – the tenants must be in non-preemption mode and not be able to steal resources such as bandwidth.
ii. Traffic isolation – the network must not allow tenants to observe the traffic of others.

d) *Repudiation*

The actions of actors in the system must be logged and secure. The internal and external operations of actors within the system must be visible and recorded so that any issue can be resolved. The appropriate approach is to identify the issue and mitigate it.

e) *Resilient control plane*

SDN is composed of three different elements: SDN apps, the SDNC, and NES. The complete communication within the SDN is controlled and critical.

f) *Strong authentication and authorization*

This is used to recognize an unidentified source and its access privileges. This will protect the network from unauthorized excess and protect the network from these attacks:

i. Incorrect response
ii. Readjustment of the request path
iii. Redirecting unwanted traffic flow
iv. Acquiring access to some device or component

16.2.1.3 Challenges of network security implementation

In recent years, the growth in the area of IT infrastructure has led to the development of various computing technologies, such as cloud computing and fog computing, that have drastically changed data center technologies. These changes have also considerably changed the behavior of users. Complete operations were shifted to automation and virtualized. Therefore, the needs of every type of user must be considered, including security concerns regarding data centers and complete IT infrastructure. These concerns act as major barriers to IT infrastructure. Therefore, the protection of data and other services is of paramount importance for service providers. Although the risk is high, it can be mitigated with existing strategies and with some advanced solutions. In the current scenario, new advanced solutions should be developed to address the issues that existing solutions cannot. In these cases, the security solution struggles to provide complete protection. Some security issues are:

a) *Denial-of-service attacks*
b) *Vulnerability attacks*
c) *Connection flooding*
d) *Bandwidth flooding*

16.3 DISCUSSION OF SECURITY ENHANCEMENT

The data and control planes in the classical ethernet network are tightly coupled, so attackers can easily penetrate the network. The control and data planes in the SDN are decoupled. Consequently, the network is hard to penetrate. Although SDN infrastructure can tackle most network issues, still new issues will be identified that need to be tackled. In this part of the chapter, we will discuss the issues and their possible solutions. Before the attack, the attackers try to control the network infrastructure resources. With access to these resources, the attacker can employ various types of attacks. The motivations of attackers are:

a) To gain access to the private network topology and traffic
b) To control all network devices
c) Eavesdropping
d) Manipulation of network flow
e) To cause traffic jams over the network

A) *Illegal network access and expansion*

Illegal access to private networks comes under the category of threats. Unauthorized access to a network segment is referred to as an unauthorized join. In a network, switches are used to expand a network. If one switch is accessible to an unauthorized object, it may access the complete network, and he/she can expand the network with various access points (wired/wireless). This issue can be tackled by restricting the network device with extra security and location settings, and maintaining authorized access to the network. However, the task of disabling or enabling security configurations is tedious and needs an expert or professional. But in SDN, security configurations are easy to configure and take less time and effort. The control in the SDN is centralized so it is easy to maintain authorized access over the network.

B) *Network remote access*

The unauthorized object can access administrative and other services after illegal access to the network. The services are compromised by illegal access to the network. The illegal entry into the network is known as phishing. This attack can be overcome with the help of firewall implementation, but it is very expensive, and expert knowledge is required to configure it. SDN provides a perfect solution to such an issue.

16.4 CONCLUSION AND FUTURE WORK

This paper focused on recent research in SDN security and predicted the future of this area. The research focused on a new perspective of SDN, focused on security and vulnerabilities in the network. The advantages of SDN also introduce new challenges in security. Security within the network can be improved with SDN security. In the SDN, the centralized controller has overall control. The controller monitors overall processes, and recognizes and manages network hosts. Traffic is routed by the SDN with the help of the implementation of new features. An authentication mechanism has been implemented by the SDN controller, where the network can be restructured according to demand. As the SDN behaves as a network controller, it can remove protocols, and thereby their vulnerabilities, that are not required. In this chapter, a valid solution has been described to resolve most security issues. Future work should focus on vulnerabilities in SDN networks.

REFERENCES

1. Xia W, Wen Y, Foh CH, Niyato D, Xie H. A Survey on Software-Defined Networking. *IEEE Communication Surveys & Tutorials*, 17(1), 27–51, 2014.
2. Gautam V. MASSS—multi-agent-based steganography security system for VANET. In *Proceedings of 3rd International Conference on Computing Informatics and Networks*, 2021 (pp. 159–172). Springer Singapore.
3. Aqdus A, Amin R, Ramzan S, Alshamrani SS, Alshehri A, El-kenawy ESM. Detection Collision Flows in SDN Based 5G Using Machine Learning Algorithms. *Computers, Materials & Continua*, 75(1), 1413–1435, 2023.
4. Bhattacharya B, Das D. SDN based Architecture for QOS Enabled Services across Networks with Dynamic Service Level Agreement. In *2013 IEEE*

International conference on advanced networks and telecommunications systems (ANTS), 2013 (pp. 1–6). IEEE.
5. Zhang J, Liu Y, Li Z, Lu Y. Forecast-Assisted Service Function Chain Dynamic Deployment for SDN/NFV-Enabled Cloud Management Systems. *IEEE Systems Journal*, 17(3), 4371–4382, 2023.
6. Maleh Y, Qasmaoui Y, El Gholami K, Sadqi Y, Mounir S. A Comprehensive Survey on SDN Security: Threats, Mitigations, Future Directions. *Journal of Reliable Intelligent Environments*, 9(2), 201–239, 2023.
7. Li S, Liao W. Software Defined Networks. *IEEE Communication Magazine*, 51(2), 113, 2013.
8. Vaezpour E. Deep Learning-Driven Multi-Objective Dynamic Switch Migration in Software Defined Networking (SDN)/Network Function Virtualization (NFV)-Based 5G Networks. *Engineering Applications of Artificial Intelligence*, 125, 106714, 2023.
9. Kreutz D, Ramos FM, Verissimo P. Towards secure and dependable software-defined networks. In *Proceedings of the second ACM SIGCOMM workshop on Hot topics in software defined networking*, 2013 (pp. 55–60). ACM.
10. Bakshi K. Considerations for software defined networking (SDN): Approaches and use cases. In *2013 IEEE Aerospace Conference*, 2013 (pp. 1–9). IEEE.
11. Mehdi SA, Khalid J, Khayam SA. Revisiting Traffic Anomaly Detection using Software Defined Networking. In *Recent Advances in Intrusion Detection: 14th International Symposium, RAID 2011* (pp. 161–180). Springer, Berlin.
12. Raghavan B, Koponen T, Ghodsi A. Software defined networking: A new paradigm for virtual, dynamic, flexible networking. *Proceedings of the IEEE*, 103(1), 14–76, 2014.
13. Raghavan B, Casado M, Koponen T, Ratnasamy S, Ghodsi A, Shenker S. Software-defined internet architecture: decoupling architecture from infrastructure. In *Proceedings of the 11th ACM Workshop on Hot Topics in Networks*, 2012 (pp. 43–48), ACM.
14. Kobayashi M, Seetharaman S, Parulkar G, Appenzeller G, Little J, van Reijendam J, Weissmann P, McKeown N. Maturing of Open Flow and Software-defined Networking through deployments. *Computer Networks*, 61, 151–75, 2014.
15. Galis A, Clayman S, Mamatas L, Rubio Loyola J, Manzalini A, Kuklinski S, Serrat J, Zahariadis T. Softwarization of Future Networks and Services - Programmable Enabled Networks as Next Generation Software Defined Networks. In *IEEE SDN for Future Networks and Services (SDN4FNS)*, 2013 (pp. 1–7). IEEE.
16. Caraguay ALV, Lopez LIB, Villalba LJG. Evolution and Challenges of Software Defined Networking. In *IEEE SDN for Future Networks and Services (SDN4FNS)*, 2013 (pp. 1–7). IEEE.
17. Cahn A, Hoyos J, Hulse M, Keller E. Software-Defined Energy Communication Networks: From Substation Automation to Future Smart Grids. In *IEEE International conference on smart grid communications (SmartGridComm)*, 2013 (pp. 558–563). IEEE.
18. Lilhore UK, Simaiya S, Pandey H, Gautam V, Garg A, Ghosh P. Breast cancer detection in the IoT cloud-based healthcare environment using fuzzy

cluster segmentation and SVM classifier. In *Ambient Communications and Computer Systems: Proceedings of RACCCS 2021*, 2022 (pp. 165–179). Singapore: Springer Nature Singapore.
19. Wang J, Wang Y, Hu H, Sun Q, Shi H, Zeng L. Towards a Security-Enhanced Firewall Application for OpenFlow Networks. In *Cyberspace Safety and Security: 5th International Symposium, CSS*, 2013 (pp. 92–103). Springer, Cham.
20. Dixit A, Hao F, Mukherjee S, Lakshman TV, Kompella R. Towards an elastic distributed SDN controller. *ACM SIGCOMM Computer Communication Review*, 43(4), 7–12, 2013.
21. Gautam V. Analysis and application of vehicular ad hoc network as intelligent transportation system. In *Mobile Radio Communications and 5G Networks: Proceedings of MRCN 2020*, 2021 (pp. 1–17). Springer Singapore.
22. Gupta M, Sommers J, Barford P. Fast, accurate simulation for SDN prototyping. In *Proceedings of the second ACM SIGCOMM workshop on Hot topics in software defined networking*, 2013 (pp. 31–36). ACM.
23. Jarraya Y, Madi T, Debbabi M. A Survey and a Layered Taxonomy of Software-Defined Networking. *IEEE Communication Surveys & Tutorials*, 16(4), 1955–1980, 2014.
24. Kumar G, Kumar R. A survey on planar ultra-wideband antennas with band notch characteristics: Principle, design, and applications. *AEU-International Journal of Electronics and Communications*, 109, 76–98, 2019.
25. Dogra R, Rani S, Sharma B. A review to forest fires and its detection techniques using wireless sensor network. In *Advances in Communication and Computational Technology: Select Proceedings of ICACCT 2019, 2021* (pp. 1339–1350). Springer Singapore.
26. Al-Shaer E, Al-Haj S. FlowChecker: configuration analysis and verification of federated OpenFlow infrastructures. In *Proceedings of the 3rd ACM Workshop on Assurable and Usable Security Configuration*, 2010 (pp. 37–44). ACM.
27. Porras P, Shin S, Yegneswaran V, Fong M, Tyson M, Gu G. A security enforcement kernel for OpenFlow networks. In *Proceedings of the First Workshop on Hot Topics in Software Defined Networks*, 2012 (pp. 121–126). ACM
28. Khurshid A, Zhou W, Caesar M, Godfrey P. "Veriflow: verifying network-wide invariants in real time", *ACM SIGCOMM Computer Communication Review*, 42(4), 467–472, 2012.
29. Fonseca P, Bennesby R, Mota E, Passito A. A replication component for resilient OpenFlow-based networking. In: IEEE Network Operations and Management Symposium (NOMS), pp. 933–939, 2012.
30. Sherwood R, Gibb G, Yap K K, Appenzeller G, Casado M, McKeown N, Parulkar G. "Flowvisor: a network virtualization layer", *OpenFlow Switch Consortium, Tech. Rep*, 2009.
31. Yao G, Bi J, Xiao P. Source address validation solution with OpenFlow/NOX architecture. In: 19th IEEE International Conference on Network Protocols (ICNP), pp. 7–12, 2011.
32. Braga R, Mota E, Passito A. Lightweight DDoS flooding attack detection using NOX/OpenFlow. In: IEEE 35th Conference on Local Computer Networks (LCN), pp. 408–415, 2010.

33. Wang H, Xu L, Gu G. FloodGuard: a dos attack prevention extension in software-defined networks. In: 45th Annual IEEE/IFIP International Conference on Dependable Systems and Networks (DSN), pp. 239–250, 2015.
34. Lim S, Ha J I, Kim H, Kim Y, Yang S. A SDN-oriented DDoS blocking scheme for botnet-based attacks. In: IEEE Sixth International Conference on Ubiquitous and Future Networks (ICUFN), pp. 63–68, 2014.
35. IETF Locator/ID Separation Protocol (LISP) [Online]. Available: http://datatracker.ietf.org/wg/lisp/
36. Phemius K, Bouet M, Leguay J. Disco: distributed multidomain sdn controllers. In: IEEE Network Operations and Management Symposium (NOMS), pp. 1–4, 2014.
37. Voellmy A, Wang J. Scalable software defined network controllers. In: Proceedings of the ACM SIGCOMM 2012 Conference on Applications, Technologies, Architectures, and Protocols for Computer Communication, pp. 289–290, 2012.
38. Tootoonchian A, Ganjali Y. HyperFlow:a distributed control plane for OpenFlow. In: Proceedings of the 2010 Internet Network Management Conference on Research on Enterprise Networking. USENIX Association, pp. 3–3, 2010.
39. Security-enhanced floodlight. [Online]. Available: http://www. sdncentral.com/education/toward-secure-sdn-controllayer/2013/10/
40. Shin S, Porras P, Yegneswaran V, Fong M, Gu G, Tyson M. FRESCO: Modular Composable Security Services for SoftwareDefined Networks. In: Proceedings of Network and Distributed Security Symposium, pp. 1–16, 2013.
41. Wen X, Chen Y, Hu C, Shi C, Wang Y. Towards a secure controller platform for openflow applications. In: Proceedings of the Second ACM SIGCOMM Workshop on Hot Topics inSoftware Defined Networking, pp. 171–172, 2013.
42. Canini M, Venzano D, Peresini P, Kostic D, Rexford J. A NICE wayto test OpenFlow applications. In: Proceedings of the 9th USENIX Conference on Networked Systems Design and Implementation, 2012.
43. Skowyra R, Lapets A, Bestavros A, Kfoury A. Verifiablysafe software-defined networks for CPS. In: Proceedings of the 2nd ACM International Conference on High Confidence Networked Systems, pp. 101–110, 2013.
44. Kujur P, Biswal SP, Patel S. "Security challenges and analysis for SDN-based networks", *Software Defined Networks: Architecture and Applications*, 321–346, 2022.
45. Bhatia J, Dave R, Bhayani H, Tanwar S, Nayyar A. SDN-based real-time urban traffic analysis in VANET environment. *Computer Communications*, 149, 162–175, 2020.
46. Singh P, Gupta P, Jyoti K, Nayyar A. "Research on auto-scaling of web applications in cloud: survey, trends and future directions", *Scalable Computing: Practice and Experience*, 20(2), 399–432, 2019.
47. Kumar A, Krishnamurthi R, Nayyar A, Luhach AK, Khan MS, Singh A. "A novel Software-Defined Drone Network (SDDN)-based collision avoidance strategies for on-road traffic monitoring and management", *Vehicular Communications*, 28, 100313, 2021.

Chapter 17

Analyzing security and privacy in social communication networks

Meet Kapoor, Rishabh Aggarwal, Shivani G Sabharwal, and Suman Madan

17.1 INTRODUCTION

The adoption of the internet has led to the massive promotion of online social networks (OSNs), and making OSNs an essential component of the everyday routine of individuals. Social media and networks greatly help individuals from different parts of the world to get connected, and this has made social networking a key area of use for internet users. There are 4.9 billion active social media users, which represents 59.9% of the global population and about 92.7% of internet users. According to a study [1], the number of new users who joined social media between April 2022 and 2023 exceeded 140 million, which means that there was an increase of around 3.2% yearly. The average time spent by an individual on social media sites like Facebook, Instagram, and LinkedIn is 2 hours and 24 minutes individually. Collectively, this reaches up to 11.5 billion hours daily [1]. One of the main issues that we must focus on is the security and privacy of the data on social networks worldwide, as not only individuals but also different organizations use the data present on OSNs to help them find different hidden growth opportunities [2]. The government also uses the data for framing and designing policies, making service delivery easier. The data collected from the OSN is known as social networking data. It is rich in the personal information of a user, for instance, his/her profile, credentials, interests, email addresses, and contact number. Thus, it is very important to protect the sensitive information available on the OSN, because if it is easily accessible to anyone other than the owner, it may significantly harm the user if the information is used for unethical purposes [3].

To maintain the privacy of the data on the OSN, many tools have been developed. Two of them are:

i. **Privacy-preserving data mining (PPDM)**: In order to maintain data privacy mining, algorithms have been developed, such as Secure Multipart Computation, Trust Third Party, and Data Obscuration, which modify the data so that the real values are not disclosed [4].

DOI: 10.1201/9781003541363-17

ii. **Privacy-preserving data publishing (PPDP):** This process involves two phases, data collection and data publishing, and involves information holders, publishers, and information recipients. At the time of data collection, data publishers collect the data from the owners of the data, and during the data publishing phase, the processed data is delivered to the data recipients. Figure 17.1 represents the workings of PPDP [5].

Thus, to precisely maintain privacy in OSN there are some basic requirements namely:

i. **Confidentiality:** This indicates that the data is secured from any type of unauthorized intervention. A crucial part of confidentiality is the user's identification and authentication.
ii. **Integrity:** This refers to the state in which data is in complete form and has not been altered or modified by any unauthorized user.
iii. **Availability:** This means the data should be readily accessible to the user at the time it is needed; otherwise it is useless. It involves the proper maintenance of hardware and the operating system of the system in which the data is stored [6]. Figure 17.2 represents the basic requirements of OSN privacy.

17.1.1 Motivation

Social communication networks are increasingly being analyzed for their security and privacy. This is due to the rise in cyber threats, such as data breaches, identity theft, and cyber espionage, which require robust defenses. The security of sensitive personal and private information is crucial to protect users from privacy violations and potential misuse of their

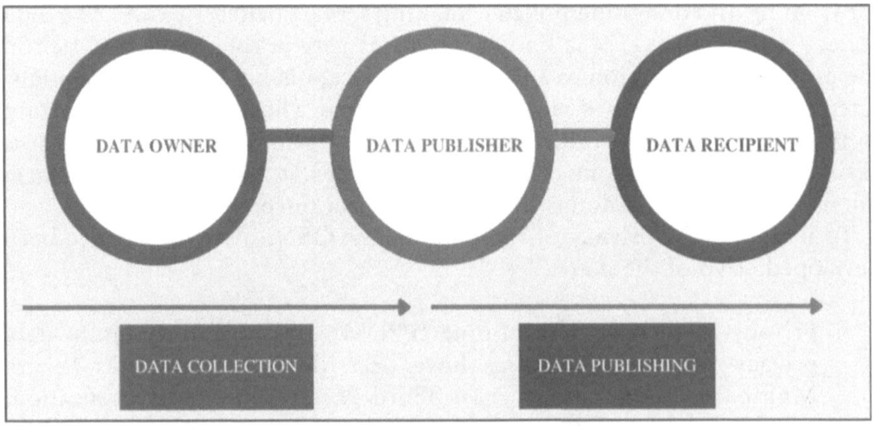

Figure 17.1 Workflow in privacy-preserving data publishing.

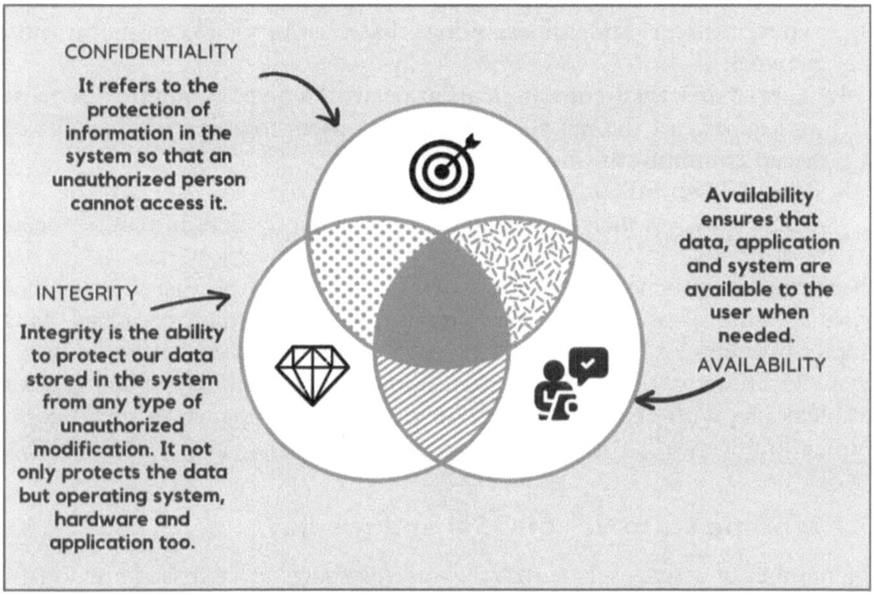

Figure 17.2 Security and privacy concerns in online social networks.

data. Maintaining user trust is vital for the success and sustainability of social communication platforms, and enhancing security measures helps in building and preserving trust. Regulatory compliance is also essential, as it ensures adherence to data protection and privacy regulations. As technology evolves, new communication channels and platforms emerge, presenting new security threats. Social engineering attacks are another threat, and analyzing security helps identify and mitigate vulnerabilities. Social communication networks are also crucial for corporate and national security, as breaches can expose sensitive information or compromise national security. Technological innovation is also essential to stay ahead of evolving threats, driving the development of new technologies and methodologies. Thus, analyzing security and privacy in social communication networks is crucial for safeguarding user data, maintaining trust, complying with regulations, and ensuring network resilience in the face of evolving cybersecurity threats. Security is a vital consideration for the following reasons:

1. It helps in protecting user's personal information and activities from cyberattacks and unauthorized access.
2. Social communication networks allow their users to share various types of content but also expose them to various risks related to security and privacy.

3. It is essential to develop effective solutions by analyzing different types of threats and vulnerabilities that exist in social communication networks.
4. Security in social communication networks helps maintain the trust and quality of the network and allows users to enjoy the benefits of social communication without fear.
5. Privacy in social communication networks gives the user control over who can access their personal information, activities, and preferences.

There are many benefits to using social media, but the major issue that arises are that of privacy and the threats relating the data breaches. This chapter provides an in-depth look at social communication networks, security in social networks, and how to minimize the probability of any type of data leakage, with details of recent incidents that happened in social communication networks that led to data breaches.

17.1.2 Some statistics of OSN and media

The number of active users across the internet space, as discussed previously, is approximately 4.9 billion. Of those who use OSNs, about 353.9 million individuals use Twitter monthly, 3.03 billion people use Facebook monthly, and 465 million people use Pinterest monthly. Many others are using other social media networks [7]. Figure 17.3 shows the number of users who are active on various social media platforms. According to many reports, there

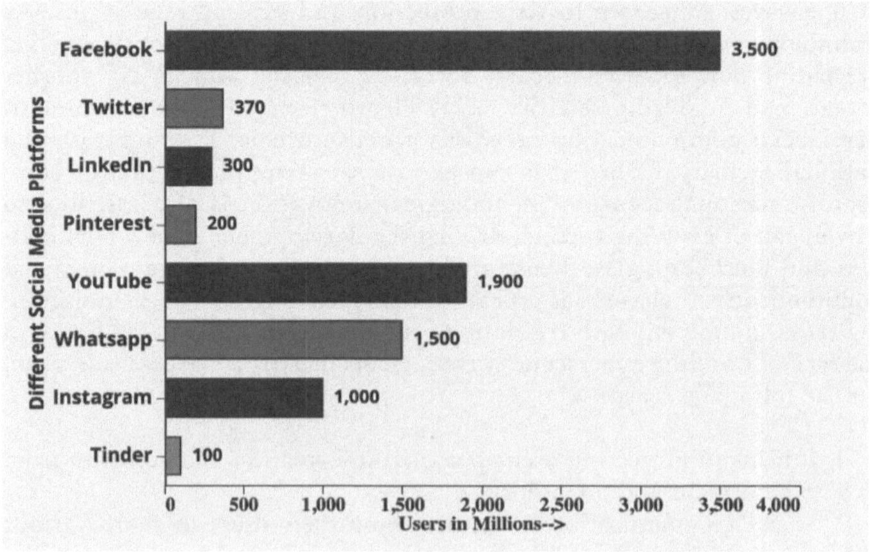

Figure 17.3 Number of active users on different social media networks.

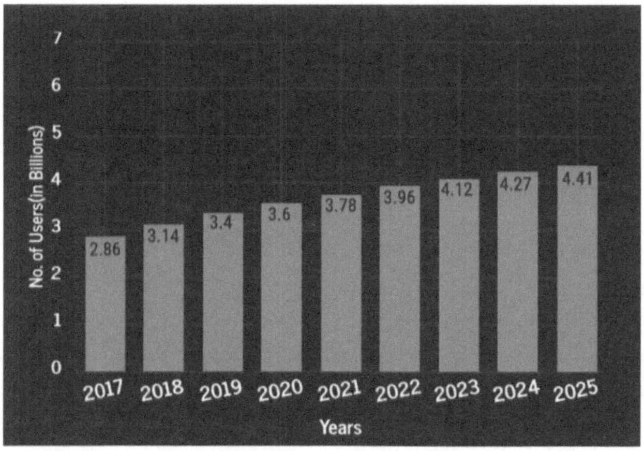

Figure 17.4 Number of users on online social media networks per year.

was a uniform increase of 16% of active users annually. It is also observed that about seven new profiles are established every second on social media. On average, users upload over 350,000 posts every day, and 298,000 statuses are updated. Because an enormous volume of data is being uploaded to social media sites, security threats may occur with high probability. Anyone can inject malicious statements into hidden social media data or inside uniform resource locators (URLs). In fact, about 83 million fake profiles and over 100,000 websites are attacked daily in data security breaches. Figure 17.4 illustrates the number of users on online social media networks annually. Increases in information and data in the OSN may lead to different cyber and online community threats [8].

17.1.3 Online community threats

The majority of people using online social platforms are unaware of the different types of threats. There are mainly four types of categories of risks and threats that may affect users and their data, as shown in Figure 17.5 [9–11]:

i. **Classical threats:** Normally, classical threats directly target the user's data, credentials, profiles, and accounts. Some of the most frequently experienced threats are phishing emails, malware, internet frauds, spammers, and cross-site scripting.
ii. **Modern threats:** In these types of threats, the user's data or information is hacked by prompting them to click on a link that takes them to a fake website from where their information can be accessed easily. This is known as click-jacking. In addition, there are many other types

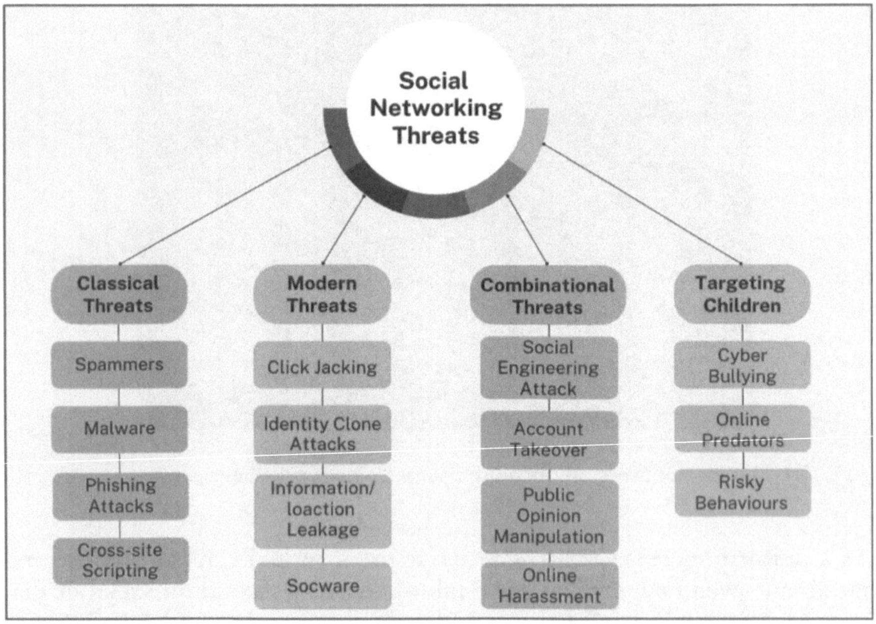

Figure 17.5 Major threats in social networks per category.

of modern threats. Some of these are identity clone attacks, information leakage, socware, and location leakage.

iii. **Combinational threats:** These types of threats are a combination of classical and modern threats. This suggests that these are more effective and have a greater probability of getting access to data.

iv. **Targeting children:** Children are a vulnerable population, and nowadays many are on OSN, where they sometimes communicate with strangers and share their personal information, thus leading to various threats and attacks. Table 17.1 illustrates these threats with their respective inclusive attacks.

17.2 RELATED WORK

With the increasing popularity and usage of OSNs in the early 2010s, their weaknesses or shortcomings became gradually apparent, which have become a topic of concern for researchers and cyber-experts. Suppose you share something very private with your very close friend, but your account experiences an OSN data breach and now your private information may be exposed to billions of users. This might hurt the user deeply and could affect the emotional state of the individual going forward [12, 13].

Table 17.1 Classical threats of social networking

Social networking threats	*Attack name*	*Description*
Classical threats	Malware	External malicious software can be employed to search and extract sensitive information from the victim's network.
	Phishing attacks	The attacker steals the user's identity and posts messages on behalf of the user.
	Cross-site scripting	Malicious scripts are injected inside the original scripts of a secure site to get access to the user's data.
	Internet fraud	An unauthorized person hijacks a genuine user's network or account.
	Spammers	A person sends a user random messages and emails that promote false claims.
Modern threats	Click-jacking	User's data is hacked by prompting them to click on a link that takes them to a fake website.
	Identity clone attacks	Usually, the attacker mimics or clones someone's identity, often with malicious intent.
	Information leakage	The data of any organization or user is leaked by any unauthorized person.
	Location leakage	Many users these days reveal their location when posting on OSNs, which may lead to location access.
	Socware	Malicious URLs are embedded inside posts and messages to perform unethical activities.
Combinational threats	Social engineering attacks	When a malicious actor gains access to a user's account mainly by stealing the user's credentials.
	Account takeover	In this kind of attack, the point of view, perception, and behavior of the public are influenced on the online platform, making them vulnerable.
	Public opinion manipulation	This usually refers to playing with the mind of the user and deliberately threatening or causing emotional damage to an individual or a group of people via an online network.
	Online harassment	This type of attack directly manipulates vulnerable human psychology and behavior to extract delicate information.
Targeting children	Cyberbullying	This is a type of online harassment where an individual is engaged in an intentionally hostile interaction by sending hurtful messages, spreading rumors, etc.
	Online predators	This is an action in which the attacker mainly targets a minor for illegal or harmful purposes.
	Risky behaviors	It refers to the behavior of users themselves that puts their online security and privacy at risk.

Thus, to prevent attacks, the following solutions are suggested:

1. **Identity exposure protection**: According to experts concerned with addressing privacy breaches, there are two major ways to achieve the goal of preventing attacks: modifying the social network data to minimize the chances of identifying the user and anonymizing the attributes.
   a) **Anonymizing OSN data:** One of the methods to maintain privacy is by modifying the whole structure of the social network itself. This can be undertaken by adjusting the connections between different individuals, which would make the work of hackers harder in identifying user information from the network.
   b) **Anonymizing user's attributes**: Anonymization can be applied to a user's many attributes in addition to social network structure. This means hiding sensitive user credentials such as location, name, or any type of personal details [14, 15]. The following are some of the techniques to achieve this:
      i) **Connection generalization:** This involves making the connections or the edges between the users less specific, so it is harder to link a user to their respective network.
      ii) **Randomization:** This involves randomizing the data to bring randomness to the network, making it difficult to identify the real identities of users from the data.
      iii) **Modification:** This involves changing the attributes and aspects of the network to protect the privacy of the user.
      iv) **Private mechanism:** This mechanism ensures that the data leaked from the network does not reveal specific information about a user, even if the attacker has some knowledge of the data.
2. **Access policy and control**: An individual's security assurance in OSN includes significantly different levels or complex factors. Existing privacy controls for OSN are barely utilized, as there is no awareness among users [16]. The statistics show that about 36% of the unmodified data according to the default privacy controls is shared and exposed to other users, which indeed should not happen.
   a) **Privacy fatigue:** This refers to a concept when the user becomes overwhelmed and shares or discloses more information over a period due to increases in complexity and less user-friendly privacy control. This can further compromise the privacy of the user [17]. The following are some of the efforts to protect the privacy of the user:
      i) **Ontology-based mechanism:** This mechanism uses ontologies to protect user privacy. It uses and supports semi-automation generation for a user's profile, making it easier for the user to control who can get access to their data [18].
      ii) **Social networking wizard:** This is a prototype that enables active learning. It gives privilege to friends and known

individuals of the user to access their information, based on their respective profiles according to the user's feedback [19].

iii) **Taxonomy-guided learning model:** This model warns the user when it predicts any kind of personal information will be leaked or is revealed by the user's post itself. It further provides him/her with the proper guidelines for preventing privacy leaks [20].

iv) **Privacy in photography:** Reports privacy breaches in photos involving anyone other than the user who posted [21].

v) **Profile control framework:** This is a type of framework for adjusting the privacy settings on a personal level based on the risk of privacy leakage involved.

3. **Tools and metrics:** In the majority of cases, the information revealed on the OSN is the outcome of voluntary activity. Thus, users should take care of their privacy while sharing information with other users on the OSN. This measures and improves the user's ability to identify the risk with their online activity. Some examples of tools and metrics are as follows [22–24]:
   a. **Online game:** Online gaming was developed for Facebook users wherein the personal information of users is taken and visible; thus, advice is offered for improving the privacy settings.
   b. **Risk qualification:** This is a concept that was proposed to measure how risky it would be to share personal details with friends, in terms of disclosure of personal data.
   c. **Inference probability:** This privacy tool was introduced to detect the probability or chances of deducing sensitive attributes from friendship links. Actions to reduce information leaks are recommended.
   d. **Attribute inference:** It was demonstrated that the personal attributes of the user can be inferred from social networks. Thus, a tool was developed to identify unintentional information leakage in OSN with the help of the risk associated with friend relationships.
   e. **Privacy metrics:** These are used to measure the amount of privacy a user has in a system and the level of privacy supplied by privacy-enhancing technologies. A framework was introduced to calculate the privacy score to indicate the possible privacy risk of the user.
   f. **Privacy score for education:** A game that simulates social networks was programmed to make the students aware of privacy. Privacy scores were introduced to reward the players for taking effective privacy steps.

17.3 ANONYMIZATION OPERATIONS AND RELATIONAL OSN DATA

The initial user's data or information is stored in an exclusive table that contains different rows (tuples) and columns. Each row has a unique ID, and each tuple has four different sorts of characteristics. Table 17.2 lists several user attribute types and provides unique examples for each, as well as solutions through an anonymization process [25–30].

Three privacy threats may affect the analysis of the published data. They are:

a. **Attribute disclosure (private data disclosure):** This type of threat mostly happens when the individual reveals information about his/her sensitive data or attributes, for instance, a person revealing his salary. These types of threats can be executed very conveniently in an imbalanced dataset.
b. **Membership disclosure:** This is a privacy threat in which the attacker can accurately determine that the individual's information is included or is absent in a published dataset.
c. **Identity disclosure:** This is the most common threat in PPDP. It usually occurs when the advisory can link a user within a privacy-protected database to their real identities. This is possible because the attacker extracts information from external sources, e.g. voter registration lists and online repositories.

Using any of the following anonymization operations on the original user data in relational form, the data owners can try to safeguard the data against any of the vulnerabilities mentioned above in the released dataset [27]:

Table 17.2 User's attributes and anonymization processes

User attributes	*Description*	*Attribute examples*	*Anonymization process*
Direct identifiers	They help in uniquely and directly identifying the user.	Social Security Number, name, phone number, email ID, etc.	It is ignored or removed before the process of anonymization.
Quasi-identifiers	They can be associated with supplementary information that can reveal someone's identity.	Zip code of the area, age, gender, date of birth, sex, etc.	Generalized, or we could say, suppressed, during the process of anonymization to protect the privacy of the user.
Sensitive attributes	Data/information/attributes that a user wants to hide from others.	Medication, disease, salary, religious views, etc.	This type of information is important for data analysis. Thus, it is retained.
Non-sensitive attributes	Attributes other than the above.	Interests, height, hobbies, adjectives, hair color, eye color, etc.	In most cases, it is not collected from the user but if done so it is posted or published as it is.

i. **Generalization:** In data privacy, protecting the sensitive data of the user while sharing the information involves the concept of replacing detailed values with more generalized or abstracted terms in the dataset itself. This helps in reducing the level or scope of details in the dataset. For example, in a dataset containing the age of an individual, say 35, the generalization operation will convert it and change it into an age range like 30–40.
ii. **Suppression:** This operation involves deliberately removing or replacing the quasi-identifier with specific values or symbols such as "*." It hides specific information or parts of the dataset that may lead to the identification of an individual, making it more challenging for malicious actors and attackers to track the user. For example, the contact number of the user 7658243561 will be converted to something like 765824****.
iii. **Anatomization:** This operation does not edit or modify the original data of the user, but it separates the sensitive attributes and quasi-attributes, breaking the association between the two. The data is then released in separate tables.
iv. **Perturbation:** The initial values of the attributes are changed with some artificially generated values. However, no statistical information in any of the datasets is altered by these replacements.
v. **Permutation:** This operation involves attributes and records that are split into multiple groups, with each group's sensitive-attribute values being shuffled at random. Therefore, due to the de-association of the quasi-attributes and sensitive-attribute boundaries in certain groups, erroneous data analysis may result, but the secrecy of the user is greatly protected.

Generally, more than one anonymization technique is applied to anonymize the individual's information. Moreover, the frequency of data release might vary based on the requirements and customer data.

17.4 SOCIAL NETWORKS AND GRAPH DATA

The data provided by the user can be structured into a graph, "G," which is further made up of three aspects. A user's set can be represented by "V" and social connections can be represented by "E." We can say $E \subseteq V \times V$. User's attributes are represented by "A." "A" consists of the quasi-identifiers and sensitive attributes of the social media networks. This can be expressed in the form of $A = \{Q, S\}$. Figure 17.6 shows a general example of graph data in social networks.

Theoretically, V stands for nodes, E for edges, and A for labels. Social network data is highly beneficial for analysis. Its utilities can be segregated into three levels [28]:

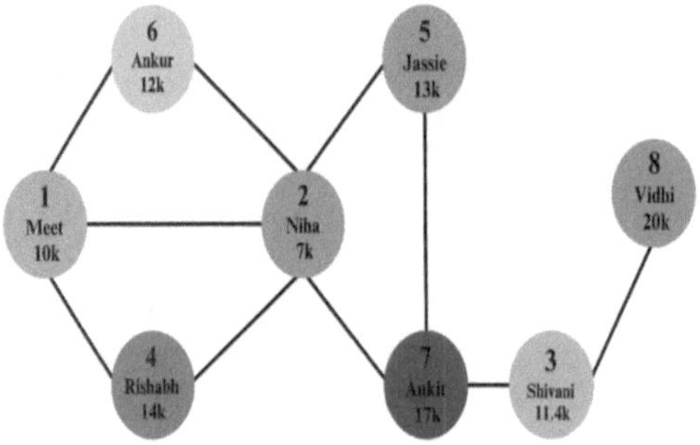

Nodes	1	2	3	4	5	6	7	8
1	-	1	0	1	0	1	0	0
2	1	-	0	1	1	1	1	0
3	0	0	-	0	0	0	1	1
4	1	1	0	-	0	0	0	0
5	0	1	0	0	-	0	1	0
6	1	1	0	0	0	-	0	0
7	0	1	1	0	1	0	-	0
8	0	0	1	0	0	0	0	-

Figure 17.6 Graph data between friends and family on SN and associated matrix.

a. **Level 1:** According to this level, the information providers, let's say the social network service provider, only publish the graph structure without any type of profile and label, so data analysts or data miners cannot get any type of solid information about any user on the SN. Thus, we can say that the only thing visible is the graph structure.
b. **Level 2:** In this level, the SN service providers make user profiles public but withhold the graph's structure. Data is published for data analyst processes after being saved in matrix form. Thus, all that is visible are node profiles.
c. **Level 3:** At this level of exposure, the owners of data disclose both the graph structure and the profiles of the user but only after applying some modification in the graph set and their respective profiles. Thus, this level exposes both the node profiles as well as the graph structure.

We know that the SN data and the user's information are very important and beneficial for data analysis. However, publishing the data on the OSN can lead to many privacy threats. There are mainly four types of privacy threats that can occur after publishing the graph data or set. They are listed as follows:

a. **Node re-identification:** This threat occurs when the privacy-preserved graph accurately discloses the person's identity. For instance, node 6, i.e., Ankur, can be identified by the opposer, but it is a privacy-preserved graph set, therefore it should not lead to any type of privacy breach.
b. **Edge disclosure:** This is another name for the revealing of a relationship or connection. This threat reveals the relationship between users connected to each other on an OSN graph set. For example, communication between the CEO of the company and the COO is confidential,

as it may involve important decisions about the company's growth, Communications may also comprise details of someone's private life, which, if leaked, could cause issues in the company and in individuals' lives.

c. **Vertex disclosure:** This occurs when the label of the linked vertex or the edge from a graph set is exposed to others, leading directly to an individual, which could cause a disclosure of identity.
d. **Affiliation link disclosure:** This happens when the relation between the user (V) and its affiliation (H) reveals a link and can directly lead to V. The affiliation group H could be involved in any type of unconventional activity (drugs/sexual preferences).

Before publishing the graph set, data owners can employ several anonymization strategies to safeguard the user in OSN data publication.

a. **Graph generalization:** This operation makes no conceivable changes to the graph's structure, but the nodes and edges are grouped together into super nodes and edges. This process functions sequentially in iteration. Each iteration keeps on making the super nodes' and edges' connections. The clustering of elements in super nodes and edges enhances the privacy of the graph's topology.
b. **Graph modification:** This operation helps in altering the graph set's nodes and edges. The specific criteria behind adding and removing edges and nodes are determined by the data publishing objective and the level of privacy protection by the data owners.
c. **Graph computation:** This involves processing data-miner queries without releasing the actual graph (G) to maintain privacy. Instead, computed graph properties such as degrees, centralities, size, and other relevant information are disclosed to data miners. This approach allows data miners to obtain essential insights and results from the graph while safeguarding sensitive data and preserving privacy.
d. **Hybrid operation:** A hybrid strategy combines multiple strategies for anonymization to improve the data's privacy-protection structure. For instance, it might require the usage of both graph modification and graph generalization to anonymize the graph G and generate a transformed graph G' [29].

17.5 PPDP AND ITS FUNDAMENTAL STAGES

The PPDP is made up of frameworks, techniques, tools, and solutions for exchanging important data with analysts and researchers while protecting users' privacy. The literature has extensively examined PPDP's ability to protect various facets of users' private information through public data analytics. Sharing information with analysts or researchers enables them to extract valuable insights from it. Even after anonymization, attackers can still extract any individual's identity or private information with the help of information acquired from external sources. To ensure the privacy of users, it is suggested that privacy-protection measures be implemented across the whole information processing cycle, which includes gathering, storing, processing, releasing, and destroying/archiving data [30]. Figure 17.7 represents a brief outline of the PPDP process that helps to preserve privacy. In total, there are five stages in the process of PPDP:

a. Gathering personal data from people.
b. Storage, interpretation, and anonymization of collected data.
c. Data anonymization for users.
d. Data publication or release under anonymity.
e. Analysis of published data to uncover embedded knowledge.

Stage 1: Gathering personal data from people:

Individuals' personal data is collected during this first stage of PPDP. With major technological advancements in today's world, there is a huge volume of data produced by various networks, online banking, the internet, the healthcare sector, social media platforms, third-party apps with SN integration, and several more offline industries. This record/information is gathered either directly from people or through smart gadgets. For example, when a person visits a hospital for treatment, they ask him/her for their personal information so that they can serve them better and understand them properly. Later, their data is preserved in the database of the hospital for any further needs.

Stage 2: Storage, interpretation, and anonymization of collected data:

In the second stage of PPDP, the collected data is stored, interpreted, and prepared for further operations such as anonymization. Large-scale databases are used by businesses nowadays to store user information and data that has been gathered. Facebook stores and manages user information, including likes, shares, and comments, using a MySQL database. Generally, the collected information about users can contain false values, incomplete attributes, or tuples. Thus, the data collected needs to be treated before the anonymization process.

Stage 3: Data anonymization for users:

The anonymization of data is the method employed most often to safeguard user privacy in data dissemination. In relational format, to protect user privacy, data anonymization modifies or encrypts the QI's real values to make information less visible. However, anonymization modifies the graph's structure if the data is provided in a graphical style to safeguard users' privacy and any sensitive attributes linked to them. Anonymization can be modified according to the user's privacy requirements and goals for the publication of the data.

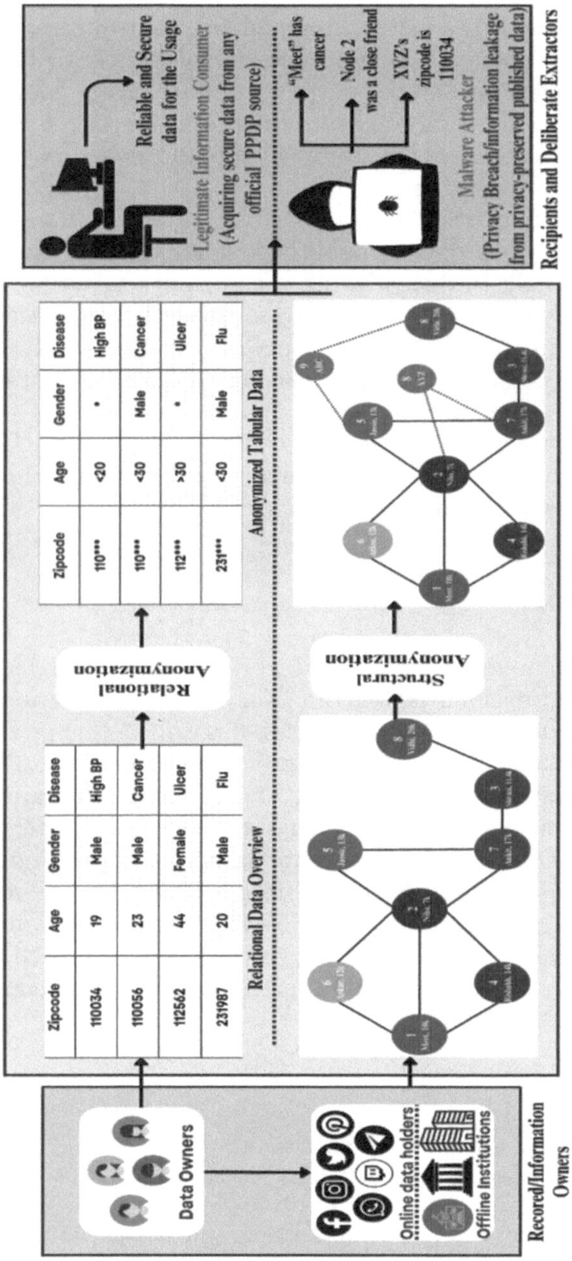

Figure 17.7 Working of the PPDP process.

Stage 4: Data publication or release under anonymity:

The outcome of the anonymization process depends upon the representation/style of the original user's data. The data that is anonymized can be used by analysts, researchers, data miners, analytics companies, and applications from outside sources. Prior to making anonymized data for analytics purposes publicly accessible, the data owner supervises the data to ensure the protection of the user's privacy. After detailed checks are performed by the owners of the data, they plan to release anonymous data.

Stage 5: Analysis of published data to uncover embedded knowledge:

When anonymous data is published, it is collected by intended recipients for analysis. Recently, many companies have been extracting social network data at a large scale to fulfill their scientific and business objectives. Data publication can put user privacy at risk since it can be copied by enemies who then utilize the vast quantity of information gleaned from outside sources to re-identify individuals. Therefore, users' privacy can be compromised by attackers having access to information, sophisticated programming abilities, tool knowledge, and knowledge of anonymization techniques.

17.6 CONCLUSION

This chapter has explored the complicated world of online social networks (OSNs), covering multifaceted aspects of privacy, statistics, relational data, and some privacy-related fundamental phases. It is evident that privacy concerns are paramount in the OSN landscape, necessitating careful consideration and robust safeguards to protect users' personal information. The utilization of statistics and graph data plays a vital role in understanding user behavior and network dynamics, enabling informed decision-making and user engagement strategies. Furthermore, the discussion of fundamental phases highlighted the evolution of OSNs from their inception to their current state, emphasizing the role of anonymization and the need for continuous adaptation and innovation in this ever-evolving digital realm.

REFERENCES

1. https://www.searchenginejournal.com/social-media-statistics/480507/.
2. Al-Turjman, Fadi, and Salama, Ramiz. (2021). Security in social networks; In book: Security in IoT Social Networks; doi: 10.1016/B978-0-12-821599-9.00001-7.
3. Madan, Suman, Bhardwaj, Kirti, and Gupta, Shubhangi; (2021). Critical Analysis of big data privacy preservation techniques and challenges, Springer's Int Conf on Innovatice computing and communications, Vol. 3, No. 1394, pp. 267–278. https://doi.org/10.1007/978-981-16-3071-2_23

4. Madan, S. (2020). A Literature Analysis on Privacy Preservation Techniques. In: *Advances in Computing and Intelligent Systems.* Springer. pp. 223–230. https://doi.org/10.1007/978-981-15-0222-4_19.
5. Madan, S., and Goswami, P. (2020). A privacy preservation model for big data in map-reduced framework based on k-anonymisation and swarm-based algorithms. *International Journal of Intelligent Engineering Informatics*, Vol. 8, pp. 38–53. https://doi.org/10.1504/IJIEI.2020.105433.
6. Gangarde, Rupali, Sharma, Amit, and Pawar, Ambika. (2020). Research opportunities in privacy of online social network data publishing. *International Journal of Advanced Science and Technology*, Vol. 29, No. 5, pp. 5095–5101.
7. https://datareportal.com/social-media-users
8. https://www.searchenginejournal.com/social-media-statistics/480507/
9. Jain, A.K., Sahoo, S.R., and Kaubiyal, J. (2021). Online social networks security and privacy: comprehensive review and analysis. *Complex & Intelligent Systems*, Vol. 7, pp. 2157–2177. https://doi.org/10.1007/s40747-021-00409-7.
10. Sathiya Devi, S., and Indhumathi, R. (2019). A Study on Privacy-Preserving Approaches in Online Social Network for Data Publishing. In: Balas, V. E. et al. (eds.), *Data Management, Analytics and Innovation, Springer's Advances in Intelligent Systems and Computing* 808, pp. 99–115. https://doi.org/10.1007/978-981-13-1402-5_8.
11. Cano, S., Rusu, C., and Quiñones, D. (2020). Evaluation of Customer eXperience and Behaviour: A Literature Review. In: Meiselwitz, G. (eds.), *Social Computing and Social Media. Participation, User Experience, Consumer Experience, and Applications of Social Computing. HCII 2020. Lecture Notes in Computer Science()*, Vol. 12195. Springer, Cham. https://doi.org/10.1007/978-3-030-49576-3_13.
12. Pensa, R.G., Di Blasi, G., and Bioglio, L. (2019). Network-aware privacy risk estimation in online social networks. *Social Network Analysis and Mining*, Vol. 9, No. 15. https://doi.org/10.1007/s13278-019-0558-x.
13. Zheleva, E., and Getoor, L. (2011). Privacy in Social Networks: A Survey. In: Aggarwal, C. (eds.), *Social Network Data Analytics.* Springer, Boston, MA. https://doi.org/10.1007/978-1-4419-8462-3_10.
14. Suman, M., and Goswami, P. (2017). Privacy preserving data publishing and data anonymization approaches: A review. https://doi.org/10.1109/CCAA.2017.8229787.
15. Huang, K., Hu, H., Zhou, S., Guan, J., Ye, Q., and Zhou, X. (2022). Privacy and efficiency guaranteed social subgraph matching. *The VLDB Journal — The International Journal on Very Large Data Bases*, Vol. 31, No. 3. pp. 581–602. https://doi.org/10.1007/s00778-021-00706-0.
16. Misra, G., and Such, J. M. (2016). How Socially Aware Are Social Media Privacy Controls? *Computer*, Vol. 49, No. 3, pp. 96–99. https://doi.org/10.1109/MC.2016.83.
17. Choi, Hanbyul, Park, Jonghwa, and Jung, Yoonhyuk. (2018). The role of privacy fatigue in online privacy behaviour. *Computers in Human Behavior*, Vol. 81, pp. 42–51. https://doi.org/10.1016/j.chb.2017.12.001.
18. Yadav, S., and Tiwari, N. (2023). Privacy preserving data sharing method for social media platforms. *PLoS One*, Vol. 18, No. 1, p. e0280182. https://doi.org/10.1371/journal.pone.0280182.

19. Livraga, Giovanni, Motta, Alessandro, and Viviani, Marco. (2022). Assessing User Privacy on Social Media: The Twitter Case Study. In *Proceedings of the 2022 Workshop on Open Challenges in Online Social Networks (OASIS '22)*. Association for Computing Machinery, New York, NY, USA, 1–9. https://doi.org/10.1145/3524010.3539502.
20. Song, X., Wang, X., Nie, L., He, X., Chen, Z., and Liu, W. (2018). A personal privacy preserving framework: I let you know who can see what. In: Proceedings of ACM SIGIR 2018, https://doi.org/10.1145/3209978.3209995.
21. Yu, J., Zhang, B., Kuang, Z., Lin, D., and Fan, J. (2017) iprivacy: image privacy protection by identifying sensitive objects via deep multi-task learning. *IEEE Transactions on Information Forensics and Security*, Vol. 12, No. 5. https://doi.org/10.1109/TIFS.2016.2636090.
22. Jain, A.K., Sahoo, S.R., and Kaubiyal, J. (2021). Online social networks security and privacy: comprehensive review and analysis. *Complex & Intelligent Systems*, Vol. 7, pp. 2157–2177. https://doi.org/10.1007/s40747-021-00409-7.
23. Safi, Seyyed Mohammad, Movaghar, Ali, and Ghorbani, Mohammad. (2022). Privacy protection scheme for mobile social network. *Journal of King Saud University - Computer and Information Sciences*, Vol. 34, No. 7, pp. 4062–4074. https://doi.org/10.1016/j.jksuci.2022.05.011.
24. Bioglio, L., Capecchi, S., Peiretti, F., Sayed, D., Torasso, A., and Pensa, R.G. (2019). A social network simulation game to raise awareness of privacy among school children. *IEEE Transactions on Learning Technologies*, Vol. 12, No. 4, pp. 456–469, https://doi.org/10.1109/TLT.2018.2881193.
25. Madan, S., and Goswami, P. (2021). Adaptive privacy preservation approach for big data publishing in cloud using k –anonymization. *Recent Advances in Computer Science and Communications*, Vol. 14, No. 8, pp. 2678–2688. https://doi.org/10.2174/2666255813999200630114256
26. Madan, S. (2021). Privacy-Preserved Access Control in E-Health Cloud-Based System. In *Disruptive Technologies for Society 5.0*; CRC Press: Boca Raton, FL, USA, 2021; pp. 145–162. https://doi.org/10.2174/2666255813999200630114256
27. He, Jing, and Sun, Yu. (2022). *Information security countermeasures for big data platforms based on cloud computing. Mobile Information Systems*, vol. 2022, Article ID 3981775, p. 11. https://doi.org/10.1155/2022/3981775.
28. Jain, A.K., Sahoo, S.R., and Kaubiyal, J. (2021). Online social networks security and privacy: comprehensive review and analysis. *Complex & Intelligent Systems*, Vol. 7, pp. 2157–2177. https://doi.org/10.1007/s40747-021-00409-7.
29. Kiabod, M., Dehkordi, M. N., and Barekatain, B. (2019). Tsram: Atime-saving k-degree anonymization method in social network. *Expert Systems with Applications*, Vol. 125, pp. 378–396.
30. Goswami, P., and Madan, S. (2022). Utilizing Deep Belief Network for Ensuring Privacy-Preserved Access Control of Data. In: Bashir, A.K., Fortino, G., Khanna, A., and Gupta, D. (eds), *Proceedings of International Conference on Computing and Communication Networks. Lecture Notes in Networks and Systems*, Vol. 394. Springer, Singapore. https://doi.org/10.1007/978-981-19-0604-6_45.

Chapter 18

Performance analysis of PEGASIS in 3D WSNs using variants of genetic algorithm applying energy-competent variants of genetic algorithm

Ramkrishna Ghosh, Suneeta Mohanty, and Prasant Kumar Pattnaik

18.1 INTRODUCTION

A wireless sensor network (WSN) is a group of hundreds or thousands of sensor nodes (SNs) that can sense and set up wireless communication between each other and perform computing and preparing actions [1]. The WSN, a dedicated network, comprises two major segments, SNs and the base station (BS). The SNs examine different environmental circumstances (heat, pressure, noise) and distribute the information to either the BS or to other SNs. WSN is predicted to be suitable for numerous usages in the areas of security, corporation observation, good fitness monitoring, etc. Exact data-oriented protocols are the requirement of WSNs. Protocols must be capable of gathering information and maximizing power utilization. The SNs are arranged to assess and gather information from adjacent nodes and convey this information to the BS. Various communication technologies make this possible. The expansion of minute, inexpensive, low-energy, multiple-operational intelligent SNs has been made possible due to engineering improvements in the areas of WSN, wireless communication systems, and micro-electromechanical systems (MEMS).

Every WSN comprises a radio transceiver with an inner antenna/link to the outer antenna, a microcontroller, an electronic circuit for interfacing with sensors, and a power connection, which is typically a battery. Consequently, since a power source is a significant feature in WSN, we require a way to increase the lifespan of the batteries of SNs [2]. The characteristics of WSNs are:

- Various network sizes: The range of a WSN may differ in area (1–100 SNs).
- Economical: SNs must be economical so that they may be utilized in huge quantities.
- Lengthy lifespan of the network: Resourceful protocols have to be outlined and employed so the WSN can remain as long as possible.

DOI: 10.1201/9781003541363-18

- Self-regulation: SNs must be capable of monitoring a WSN routinely without any exterior arrangement.
- Cooperation/data aggregation: SNs must be competent to aggregate data, improving the competence of the WSN.

18.2 RELATED WORK

Recently, many researchers have focused on power-proficient protocols for routing or algorithms. LEACH is one of the utmost popular hierarchical protocols that were suggested around 20 years ago [3]. There are two kinds of SNs: cluster heads (CHs) and ordinary nodes (ONs). Every ON assembles information from the region of interest and conveys the combination of data to a neighboring CH. Every CH controls the gathered information and passes it to the BS. CHs avoid prolonged remote transmission between ON and BS, and consequently, a large amount of power can be stored. Nonetheless, the choice of CHs is random. CHs can be inadequately allocated inside the entire WSN, leading to a worsening performance. CHs send information to the BS, which uses power. PEGASIS is a progression-built protocol for WSNs routing [4]. In this protocol, every SN is required to pass on data to an adjacent node that is nearer to the BS. Numerous chains might be built according to the algorithm based on the greedy approach. The leader of every link has the responsibility to relocate the information to the BS. Owing to the great responsibility of the leaders of chains, every SN develops into a leader to normalize power utilization. Due to multi-hop propagation, elongated distance communication between SN and BS is avoided and a lot of energy is stored. Concurrently, on account of multi-hop propagation, delays in the WSN can be severe, which would be unsuitable for time-sensitive usages. Table 18.1 shows details of a literature review.

Authors in [24–29] have shown interest in this field.

18.3 MOTIVATION

Nowadays, researchers suggest several protocols dependent upon chain such as PEGASIS, EEPB (energy-efficient PEGASIS-based protocol) [10], and IEEPB (improved energy-efficient PEGASIS-based protocol). Network lifespan is thought to be a particular benchmark for competency, but the reduction of data dispatch slowdown in PEGASIS is also a significant advancement. We examine the difficulties of a lengthy chain for competent power utilization and delay in data distribution examining BS movability. We focus on the concept of specified route movability and restrict the BS to restricted possible applications. We similarly reduce the load on one CL by beginning the concept of multi-head links. In our research work, we recommend PEGASIS, an excellent chain-built protocol for prolonging

Table 18.1 Literature review

Sl. No.	*Year of publication*	*Author*	*Proposed work*
1	2016	Xie, G. and Pan, F. [5]	The authors suggested a power-competent protocol for routine, applying a mobile sink dependent upon clustering. This is appropriate for WSNs with impediments. Mobile BS travels through the CHs and gathers information by means of single-hop communication. Competent scheduling dependent upon spanning graphs is introduced to locate a small route for mobile BS to evade obstruction. The outcome from the simulation demonstrates that the lifespan of the WSN is long-lasting and the problems of the WSN decrease.
2	2015	Zhu, C., Wu, S., Han, G., Shu, L. and Wu, H. [6]	The authors recommended a blueprint for information-gathering, applying mobile sink using tree clustering. This blueprint comprises three stages: i) construction of tree, ii) decomposition of tree, and iii) sub rendezvous points (SRP) election and data accumulation stages. Mobile BS travels near rendezvous points (RP) and SRP to gather data in single-hop transmission. After information-gathering in every round, SNs reassign RP and SRP. Simulation outcome shows that the recommended algorithm outperforms with regard to network lifespan, and the moving route of the mobile BS is small.
3	2010	Linping, W., Wu, B., Zhen, C., and Zufeng, W. [7]	The authors introduced a novel means of enhancing the competence of PEGASIS by selecting two CHs and separating the WSN into various information volumes.
4	2015	Mishra, A.K., Rahman, R.U., Bharadwaj, R., and Sharma, R. [8]	A well-organized chain formation system was worked out by authors to progress the WSN lifespan of PEGASIS. The system chooses SNs that are closest to the sink as possible leaders. The selection of CL in PEGASIS is random and therefore, decreases communication remoteness from CL to BS. In addition, a numerous overlaid chain configuration method is used for the communication of messages to the sink.
5	2018	Wang, J., Gao, Y., Yin, X., Li, F., and Kim, H-J.[9]	The authors recommended an enhanced PEGASIS (EPEGASIS) algorithm to lower the hotspot difficulty from four features. At first, the best possible transmission distance is decided to decrease power utilization in communication. Secondly, ten threshold values are put to limit the deceased SNs, and the mobile BS method is utilized to stabilize the power.

the lifespan of WSNs. In our research, we computed the lifespan of a 3D WSN BS inside and BS outside, using PEGASIS, PEG-GA, and variants of PEG-GA.

18.4 PEGASIS

Enhancements of the LEACH protocol were recommended. The protocol is known as PEGASIS, which is a close optimum chain-dependent protocol for enhancing the lifespan of WSNs. PEGASIS [11] is the familiar chain-dependent routing protocol for data collection. Stationary SNs are disseminated at random in the WSN, as displayed in Figure 18.1. The BS is located remotely from SNs and is predetermined. Every SN using a Global Positioning System (GPS) recognizes its own position and its adjacent SNs. Every SN is homogeneous and power-constrained with equal preliminary energy. These SNs can manage their energy and communicate with other SNs or the BS directly. The SNs are located at random in the network area and arrange themselves in the shape of progression. Nevertheless, the base station, or sink, calculates this progression and transmits this to all the SNs. PEGASIS selects a leader from the chain in each round with regard to remaining power, which gathers information from the adjacent node to be sent to BS. As a consequence, the average power used by each SN every round decreases.

Dissimilar to LEACH, PEGASIS eliminates cluster creation and employs simply one SN in a chain to transmit to the BS instead of many SNs. This methodology decreases the overhead and lessens the bandwidth necessities from the sink. Figure 18.3 illustrates that simply one CH leader SN

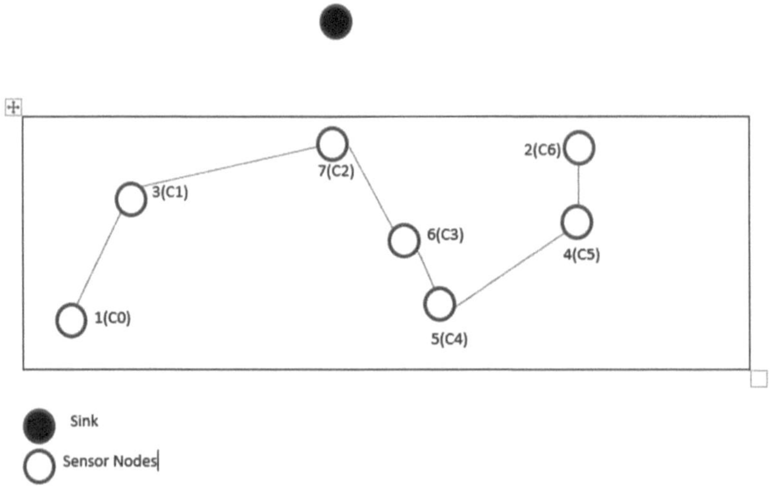

Figure 18.1 Chain formation of PEGASIS.

forwards the data to the sink. The foremost purpose of PEGASIS is for every SN to obtain information from adjacent nodes and to accept changes, for example, a change in the leader that will communicate with the sink. This method allocates power capacity equally among SNs in the WSN. Here, SNs are primarily positioned at random in the network area, and consequently, the ith SN is at a random position. These SNs are arranged to create a chain, which may be attained by the SNs themselves applying a greedy algorithm, beginning from a few SNs. On the other hand, the sink can calculate the chain and communicate this to every SN.

18.4.1 Formation of chain

Prior to the first round, a chain applying a greedy algorithm beginning from the outermost SN away from the BS is created. By using the greedy algorithm throughout the process for the formation of a chain, SNs choose the adjacent SN as the subsequent SN. For instance, when SNs are disseminated at random, as shown in Figure 18.2, the chain index of the SNs is different from the identification number of each SN. Every SN has a unique index in the chain. The arrangement of a chain in PEGASIS begins from the furthest SN away from the BS and employs the greedy algorithm. In Figure 18.2, an SN furthest away from the BS is sensor node 1, and the creation of the chain begins at sensor node 1. Consequently, the chain index of SN 1 turns into 0. According to the chain formation procedure using the greedy algorithm, the subsequent SN 1 (C0) joins to the adjacent SN3.

The index of sensor node 3 is increased by 1 for the chain index of sensor node 1. In PEGASIS, SNs in the chain may not be selected as the subsequent SN to avoid a loop. Therefore, they (i.e., sensor node 1 and sensor node 3)

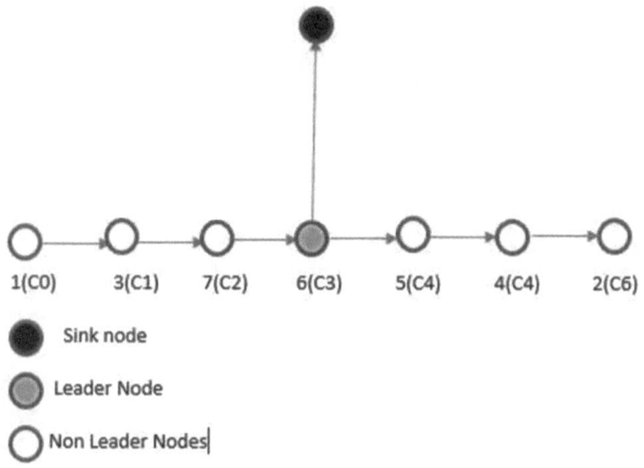

Figure 18.2 Data collection of PEGASIS.

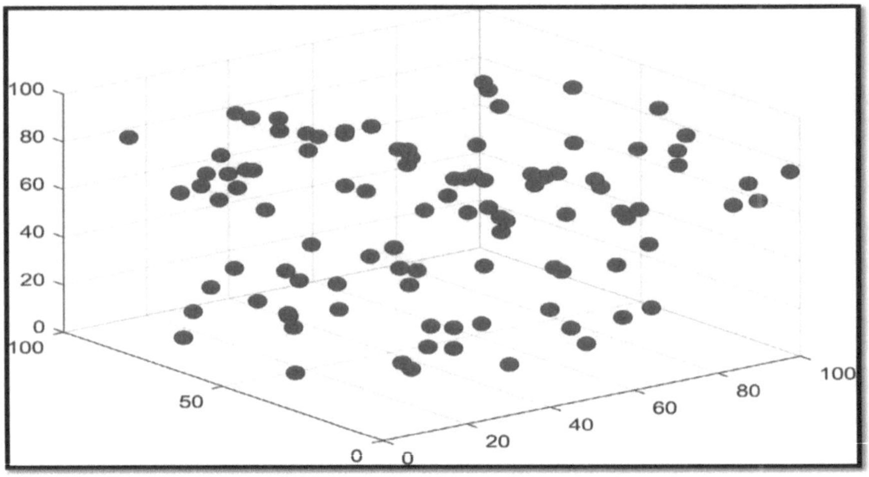

Figure 18.3 Network model of 100 SNs in 1 100m×100m×100m 3D space.

are debarred from the record of the subsequent applicant's sensor node 3. Until all the SNs are incorporated into the chain, the arrangement procedure of a chain is replicated.

18.4.2 Data collection

When chain creation is finished, one SN is elected at random as the leader to convey combined data to the sink in each round. The leader SN conveys a minute token to the ending SNs of the chain for data collection. By token passing, the data held by every SN is broadcast near the leader SN over the chain. SNs, except ending SNs, fuse their own data and data that was collected from adjacent nodes, and convey the integrated data to the BS, which is positioned remotely away from SNs. The adjacent nodes transmit the combined data to the other adjacent SNs through the chain, as displayed in Figure 18.2. The chief SN collects information from both adjacent nodes, and conveys the integrated data to the BS, which is positioned remotely away from SNs.

18.5 VARIANTS OF PEGASIS

PEGASIS is a conventional protocol. PEGASIS has multiple variants to minimize restrictions.

18.5.1 PEG-ANT

Guo et al. [12] proposed PEG-Ant, which employs Ant Colony Optimization (ACO) rather than a greedy algorithm to construct a smaller overall chain to decrease communication distance and consequently save power. It assists in overall optimization. The procedure of constructing this chain is fundamentally explaining the traveling salesman problem (TSP). The chain that builds the path is even more distributed. This balances the power utilization among SNs. Based on the present energy of every SN, the leader is selected, which accurately transmits to the sink in every round of communication. This protocol enhances the WSN lifespan.

18.5.2 EEPB

This is an extended PEGASIS algorithm. The algorithm is based on span and is utilized to outline the link. This may affect transmission remoteness between two SNs, which is lengthy in PEGASIS. Therefore, SNs expire sooner, as a lot of energy is utilized by them for the communication of data. Threshold distance i.e. the typical distance of the chain, is considered by an SN during the chaining procedure. If the span from the adjacent SN and crucial SN is higher than the threshold distance, the adjacent SN is the remote SN. If the adjoining SN links the chain, this turns into an elongated chain. EEPB prevents this occurrence by applying a distance threshold. This performs power saving on the threshold and balances the power utilization of all SNs.

18.5.3 IEEPB

This protocol controls the lack of EEPB. Whenever this constructs a chain, the threshold accepted is not certain and difficult to decide. This creates a long chain. A leader is selected that oversees SN power. The distance between the sink and the SN determines the selection of the leader. Based on this, the IEEPB evaluates the distance between two SNs twice and computes the shortest route to connect two nearby SNs. In choosing the leader, IEEPB considers the SN's power and the span between the sink and the SN, standardizes the two aspects, and allocates various weight coefficients to SNs. Lastly, the SN with the least weight becomes the leader.

This protocol has superior power competence and consequently extended WSN lifespan.

18.5.4 PEGASIS with double cluster head (PDCH)

This protocol stabilizes the weight of each SN and increases the WSN lifespan. Usually, this protocol makes use of one CH which transmits to the sink. At this point, instead of one, double CHs are utilized in one chain in a hierarchical arrangement such that lengthy chaining is prevented. PDCH performs better than PEGASIS by removing active cluster creation, reducing the distance between SNs, decreasing the number of messages conveyed to and from SNs, and requiring one communication to the sink in each round. As the weight of energy is disseminated among SNs, the WSN lifespan increases and consequently improves its performance.

18.5.5 PEG-ACO vertical clustering

The variants of PEGASIS demonstrated earlier aspire to lessen the boundaries of PEGASIS. PEG-ACO vertical clustering utilizes clustering in accordance with the x coordinates of SNs. By means of ACO, a chain is constructed in each cluster, as displayed in Figure 18.4. Multi-hop communication is taken on by CHs to convey the information to the sink. Every CH is chosen based on a weighted outcome between power and range. The accomplishment of this protocol has been evaluated, and developments have been found in lifespan, remaining power, and detainment. A different kind of clustering technique, horizontal clustering, continuously selects the CH accountable for communication to the sink in the topmost cluster, has also been considered. It is well known that the investigation is performed merely for identical WSNs with a fixed sink positioned outside the WSN. Although a sink inside the WSN increases its lifespan, it is not always achievable depending on the purpose. Both cases will be dealt with individually.

18.6 3D WSN

Most protocols for routing initiated in WSN consider protocols for routing in 2D. The reality is that we reside in a three-dimensional (3D) universe. Roy and Mukherjee [13] remarked that a 3D WSN strategy is needed for

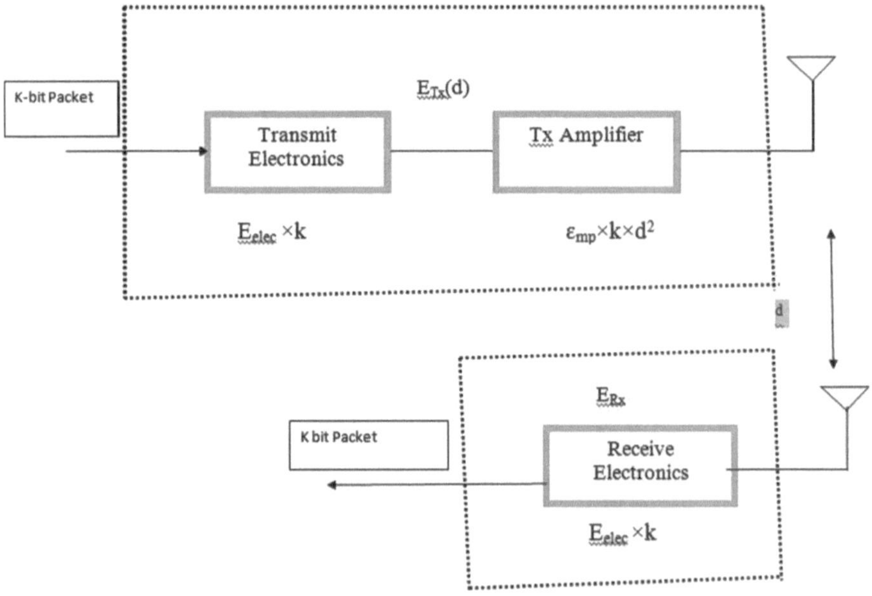

Figure 18.4 First-Order Radio Model.

many usages instead of a 2D WSN design. An example of such a design is the scheme of underwater WSNs [14]. Further possible usages involve organizing WSNs on trees with changeable heights, track subways, and underground sources. WSNs organized in unbalanced surroundings comprise this group. Literature provides information on the difficulty of coverage in 3D. Investigations into protocols for routing are comparatively lacking. Kohli et al. [15] and Baghouri et al. [16] proposed expanding LEACH to 3D. Consequently, the addition of PEGASIS to 3D will be a valuable investigation prospect to improve power competence in 3D usage.

18.7 METHODOLOGY

18.7.1 Network model

N quantity of SNs are scattered at random in a square area whose distance is L, displayed in Figure 18.3, and with a fixed BS. SNs are thought to be organized once. Each SN is thought to have a predetermined quantity of energy.

18.7.2 Radio model

The radio model of first-order is shown in Figure 18.4. It is used to work out the power usage of transmitting SNs. It is easy to employ and extensively utilized in WSN works [16]. Therefore, evaluations of preceding protocols show they are simpler. The circuits of the transmitter and receiver require power to be operated. Next is a task based on the quantity of bits (k) [18]. In the radio model, the radio scatters Eelec=50 nJ/bit to pass the sender or recipient electric circuit and εamp=100 pJ/bit/m^2 for the sender amplifying device. The wireless network has an energy controller and may expend the lowest possible power to spread the planned receivers. This may be switched off to eliminate unintentional communications. An r^2 power waste is utilized because of channel communication.

$$ETx/Rx = Eelec \times k \tag{18.1}$$

where E_{elec} is the power exhausted for every bit to operate the sender or recipient.

$$E_{Tx-amp} = \{\varepsilon_{fs} \times k \times d^2 if d < d_0 \varepsilon mp \times k \times d^4 if d > d_0 \tag{18.2}$$

The threshold d0 is utilized to decide which model has been given multichannel and is written as

$$d_0 = \sqrt{\frac{Efs}{Emp}} \tag{18.3}$$

where ε_{fs} and ε_{mp} are two constants that are utilized for the model-free space or multipath.

Therefore, the whole power needed for communication is:

$$E_{Tx}=\{E_{elec}\times k+\varepsilon_{fs}\times k\times d^2 ifd<d_0\ E_{elec}\times k+\varepsilon mp\times k\times d^4 ifd>d_0 \quad (18.4)$$

The power needed for reception comprises of power needed to pass through the network.

$$ERx = Eelec\times L \quad (18.5)$$

The value of power for data accumulation is [19]

$$EDA\text{-}tot = s\times k\times Eda \quad (18.6)$$

where E_{da} is the power of each bit exhausted in collection, and s is the number of signals being combined.

Reception is a high-power procedure; consequently, the number of collections and communications must be lowered. LEACH and PEGASIS utilize identical constants (Eelec, εamp, and k) for computing energy charges; consequently, PEGASIS preserves power by decreasing d and the number of communications and received information for every node. Consequently, for a d^4 model, PEGASIS will be superior in preserving power in comparison with LEACH.

18.7.3 CH election

PEGASIS selects its CH depending on the strategy of rotation. Although this policy assists in load balancing, this is confirmed to be insufficient. Therefore, many other techniques have been created for selecting a more appropriate CH. Ramluckun and Bassoo [17] introduced a technique that works out a weighted score based on i) the distance to BS and ii) the residual power of the SN. The range of nearby CHs can be considered if the cluster is a secondary cluster for protocols with inter-cluster communication. The primary CH is identified as the CH that sends out data to the BS. The CHs that do not send out data to the sink are characterized as secondary CHs. Ramluckun and Bassoo [17] utilized identical weights to select a chief and subordinate CH. Furthermore, the weights allotted were equivalent to 0.5 to provide the remaining power and span in the selection procedure of CHs. Instinctively, this would suggest that having identical weights (0.5) for both power and range would lead to the best outcomes. This necessitates allocating dissimilar weights to decide prime and subordinate CHs. Moreover, the weights would not be equivalent to 0.5.

$$\text{Outcome}(\text{primary}) = w_1 \times E_{res}(\text{Primary CH}) + \frac{w_2}{\alpha_{Bs}} \quad (18.7)$$

$$\text{Outcome}(\text{Secondary}) = w_{3\times} E_{\text{res}}(\text{Secondary CH}) + \frac{w_4}{d_{nextCH}} \tag{18.8}$$

The CHs of adjacent clusters are selected one by one after the designation of the main CH. This is carried out by shifting to the right and left of the main cluster (vertical clustering) or inside and outside (spherical clustering). Logically, when the main CH is created at a boundary cluster, the selection of the CH is maintained in a single route. A boundary cluster will be either the leftmost or rightmost cluster in vertical clustering. In the event of spherical clustering, this will be either the nearest or furthest cluster. Therefore, d_{nextCH} is in the range of the adjacent CH that has previously been selected. The weights w1, w2, w3, and w4 are decided after preliminary simulations. The weights w1 and w3 differ from 0.1 to 1, and the grouping of weights that leads to First Node Death (FND) is preferred.

18.7.4 Genetic algorithm

A genetic algorithm can be utilized. Therefore, a close optimum chain span is attained for solving the traveling salesman problem (TSP) in PEGASIS. This algorithm is simple to carry out and similarly less costly in comparison to methods like Ant Colony Optimization (ACO) [19]. The genetic algorithm is simulated from nature and tries to duplicate the procedures that happen in Darwinian progression. A genetic code is utilized to denote the explanation, and a population that consists of many genetic codes is employed. A genetic code comprises chromosomes, and with regard to TSP, a chromosome is an indicator signifying a particular SN. A resolution is the listing of SNs in a particular fashion. Adjoining SNs in the record are cities that should be traveled consecutively or, in PEGASIS, SNs that are connected at one time in the link. The benefits of a finding are decided by a fitness function. In this situation, this is merely the distance covered by visiting SNs in the direction stated by the genetic code. In our work, GA would be useful to solve TSP in three measurements. Similarly, considering protocols that utilize clustering, the GA could be utilized for every unconnected cluster to detect the small linkspan. Thereby, the complete chain length would be reduced, leading to power-competent 3D WSNs. This is an iterative procedure. For innovative results that can give an improved finding, GA operators are utilized on genetic codes at every repetition. The elementary GA operators are:

a. **Selection:** The operator includes supporting the choice of results having improved fitness.
b. **Crossover:** The operator is identical to the reproduction procedure that occurs in progression. This includes accepting more than one finding and joining them to yield a fresh finding.

c. **Mutation:** The operator involves changing a section of the result, and this aids in guaranteeing divergence in a population [20]. It comprises of the following:
   i. Swap procedure: Two SNs are chosen randomly in the solution, and their places are interchanged. For example, regarding positions 2 and 5:

Before	*After*
p →q →r →s →t →u	p →t →r →s →q → u

   ii. Flip procedure: A portion of the solution is flipped at random after flipping at positions 2 and 5 as follows:

Before	*After*
p → t →s →r →q →u	
p →q →r →s →t →u	

The GA algorithm utilizes selection and mutation. The stages in the GA algorithm are as follows.

1. Begin the preliminary population. The preliminary population comprises many findings that are produced randomly. Variable globalMin specifies the shortest chain span that has been so far identified, and this is allocated to ∞.
2. Compute the fitness of every representative of the population. populationMin is allotted to ∞. populationMin is a variable that directs the top fitness value (minimum general chain length) in the present population.
3. Document finding having the finest fitness in bestPopSolution. bestPopSolution is a variable that archives the finest finding in the present population. populationMin is allocated the value of finest fitness.
4. If populationMin<globalMin, a fresh value globalMin is documented, e.g., population Min and the globalBestSolution is documented as bestPopSolution. Variable global Best Solution designates the finest finding that has been accomplished so far.
5. The direction of findings in the population is randomized. For example, finding 1 may turn into finding 30, etc.
6. For every four representatives of the population (solutions 1–4, 5–8, etc), the finest finding is found. The stages to generate four new ways are as follows:
   (a) Generate haphazard way insertion points. These comprise two points that specify wherever a finding could be improved.
   (b) The finest finding of the preliminary four is drawn as fresh path1.
   (c) The invert process is utilized with the finest finding to generate fresh path2.

(d) The interchange process is employed with the finest finding to generate fresh path3.
(e) The shift process is used with the finest finding to generate fresh path4.

7. Thereafter, a fresh population is designed comprising fresh findings.
8. Stages 2 to 7 are repeated until all repetitions are finalized.

18.8 PROPOSED CLUSTERED PROTOCOLS

18.8.1 PEGASIS and PEG-GA

PEGASIS is extensive in 3D spaces by taking into consideration the third measurement for the purpose of constructing chains with the greedy algorithm. If not, the selection of CHs and communication stay the same. PEG-GA is identical to that of PEGASIS, except that GA is used to make the chain.

18.8.2 Vertical clustering

This is used in 2D networks extending to 3D, leading to clusters with the appearance of cuboids rather than rectangles. It works by applying clustering considering the value of the x coordinates of SNs. The selection of CH and data communication takes place in the same way in the two dimensions. The mechanism of CH selection introduced in Section 18.7.3 is applied. A chain is made in every cluster by applying the genetic algorithm, and information is directed to every CH related to PEGASIS. After that, multi-hop inter-cluster communication happens left to right and right to left until the main CH is attained. It conveys information to the base station. The threshold for the Cth cluster is provided by:

$$\frac{(C-1)R}{C_{total}} < x < \frac{CM}{C_{total}} \tag{18.9}$$

where C is the Cth cluster, C_{total} is the whole quantity of clusters, and x is the x coordinate of SNs.

18.8.3 Spherical clustering

This is achieved as stated in the range of SNs from the midpoint of the WSN. We think about an unreal sphere of radius equivalent to the range of the furthermost SN from the midpoint of the WSN. The sphere has a midpoint (M/2, M/2, M/2) and is split radically. SNs are allocated a cluster in consonance with the radii of the sphere they are in. Selection of CH and data communication takes place the same way as vertical clustering, except that multi-hop transmissions arise inside or outside before the main CH has been chosen.

The threshold for the Cth cluster is provided by:

$$\frac{(C-1)R}{C_{total}} < r < \frac{CM}{C_{total}} \tag{18.10}$$

where R is the range from the midpoint of the furthermost SN, and r is the range of SNs from the midpoint.

18.8.4 Quantity of cluster selection

The major purpose of clustering is to decrease the huge delay formed by a great number of hops in a lengthy link. Growing, the number of clusters in a 3D WSN boosts the amount of inter-cluster transmissions. As a result, the power utilized per round is also boosted. For that reason, a trade-off between delay and power depletion may occur. The average power exhausted per SN each round for the first 1000 rounds and the extreme delay are calculated. The whole power exhausted is:

$$\begin{aligned} E_{exhausted} &= \sum_{0}^{N} Initial\ Energy - \sum_{0}^{N} Residual\ Energy \\ &= \mathrm{N} \times \mathrm{E}_0 - \sum_{0}^{N} Residual\ Energy \end{aligned} \tag{18.11}$$

The typical power exhausted by each node each round for the first 1000 rounds is set by:

$$\mathrm{E}_{\mathrm{avg}} - 1000 = \frac{Eexhausted\ at\ 1000\ rounds}{N \times 1000} \tag{18.12}$$

18.8.5 FLOW DIAGRAM OF RECOMMENDED PROTOCOL

Figure 18.5 represents the important phases occurring in the recommended clustered protocol.

Where C is the Cth cluster, C_{total} is the whole quantity of clusters, and x is the x coordinate of SNs.

18.9 METRICS

18.9.1 Lifetime

WSN lifetime is a significant metric for design factors in WSNs that is related to the number of rounds at which a definite quantity of SNs still

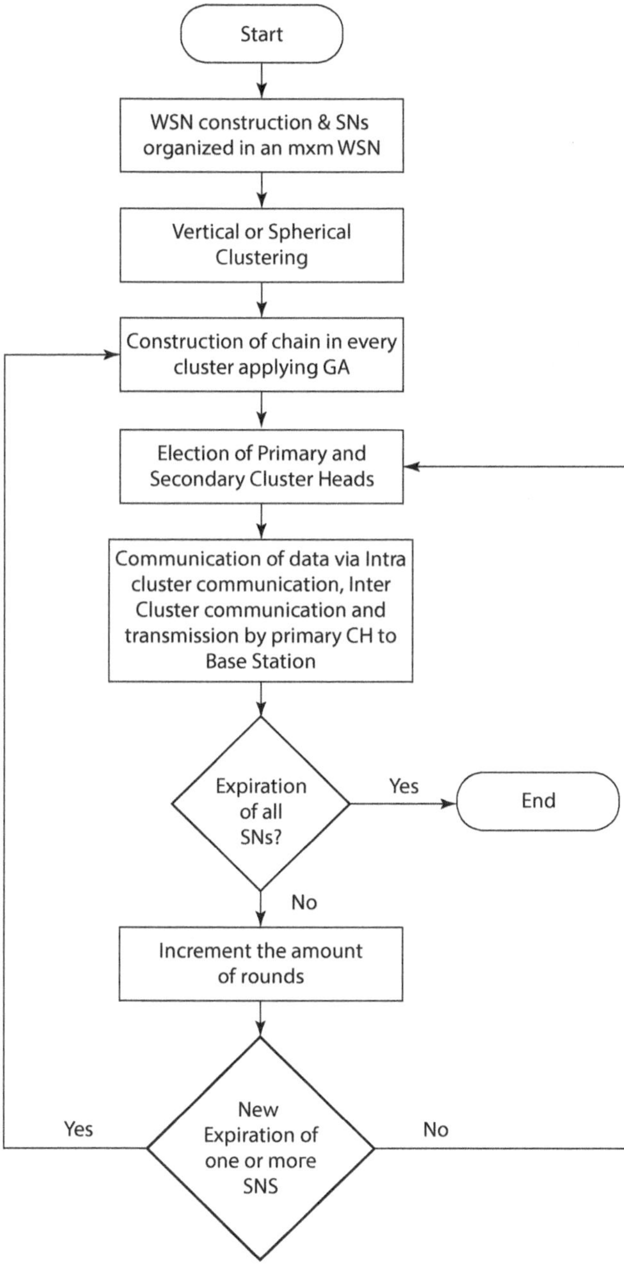

Figure 18.5 Key phases in the proposed clustered protocol.

have energy remaining and stay operational, therefore sustaining the application. There are a number of definitions for this.

i. The first time an SN completely depletes its energy, which is referred to as survival period or First Node Death (FND).
ii. The time that a proportion of SNs stay functioning in the WSN. If this occurs at 50%, this is called Death of Half N.
iii. The time at which every SN has expired is generally known as Last Node Death (LND).

18.9.2 Residual energy

Residual energy indicates the whole quantity of power that stays within the entire WSN.

18.9.3 Load balancing

This is related to how well the WSN routing protocol can allocate power utilization uniformly between the SNs. This can be evaluated by means of the proportion of remaining power of the entire WSN at FND. A lower proportion of remaining power at 1st node death indicates improved load balancing.

18.10 RESULTS AND DISCUSSION

The simulations were accomplished by using MATLAB 18a. Network and GA metrics are shown in Table 18.2.

Table 18.2 Population and GA metrics

Specifications	*Values*
Quantity of SNs (N)	100
Size of the network (M)	100
BS inside coordinates	(50,50,50)
BS outside coordinates	(50,50,300)
Packet size (K)	2000 bits
Eelec	50 nj/bit
εfs	10 pj/m^2
εmp	0.0013 pj/bit/m^4
Eda	5 nj/bit/signal
Preliminary power (E_0)	0.5
C	1ms
Popular size	100
Number of iterations	10000

Two cluster sizes from every protocol were selected for two purposes: i) reducing delay and ii) decreasing the average power utilized per SN per round. This is shown in Table 18.3.

The lifetime for a three-dimensional base station (inside) based on PEGASIS is displayed in Table 18.4.

From Table 18.4, it can be seen that PEGASIS performs the worst in relation to 1st node death. PEG-GA vertical clustering with size 3 cluster is superior to PEGASIS by 824% with regard to stability period. This outstandingly high percentage of enhancement attained for the greedy algorithm is lower in three-dimensional networks than in two-dimensional ones.

Variants of PEG-GA vertical clustering perform better than their corresponding PEG-GA spherical clustering. The lifetime of a 3D base station inside for variants of PEGASIS is shown in Figure 18.6.

Table 18.5 shows the extremely reduced performance of PEGASIS in a three-dimensional setting. The GA protocols demonstrate a much better performance in relation to 1st node death. Compared to the BS inside situation, PEG-GA is unaltered, acting upon protocols with regard to 1st node death. This is due to the technique of CH selection that is entirely power and distance unaware. The CH is a great communication-power liability and will therefore use significantly more power than in the BS inside case.

Table 18.3 Quantity of cluster selection

	Size of the cluster	
Protocol	*Base station inside*	*Base station outside*
Spherical clustering for PEG-GA	3 & 8	3 & 8
Vertical clustering for PEG-GA	3 & 9	3 & 9

Table 18.4 Lifespan for 3D base station inner side

Routing protocol	*First Node Death (FND)*	*50% of the node expiry*	*Last Node Death (LND)*
PEGASIS	234	1750	2155
PEG-GA	2130	2295	2930
Spherical clustering of PEG-GA having size of cluster three	1940	2265	2340
Spherical clustering of PEG-GA having size of cluster eight	1600	2170	2235
Vertical clustering of PEG-GA having size cluster three	2162	2262	2322
Vertical clustering of PEG-GA having size of cluster nine	1752	2125	2225

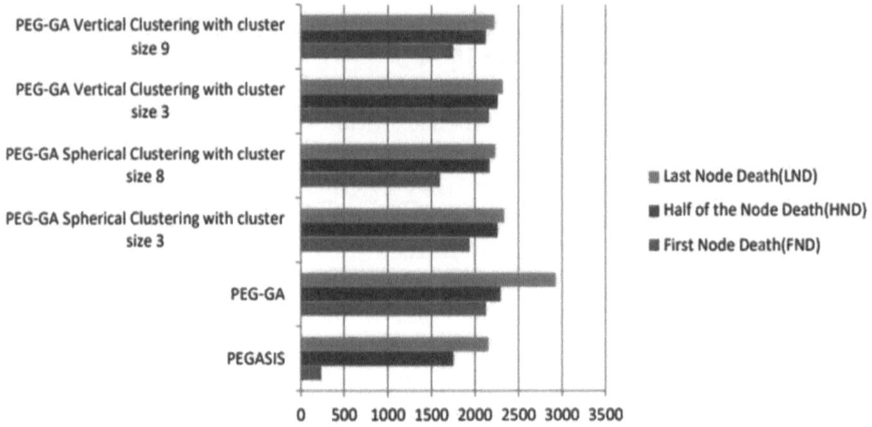

Figure 18.6 Lifetime for 3D base station inside for variants of PEGASIS.

Table 18.5 Residual energy (%) at death of first node for three-dimensional base station inside

Routing protocol	*First Node Death (FND)*	*50% of the node expiry*	*Last Node Death (LND)*
PEGASIS	305	1160	1360
PEG-GA	1115	1578	1771
Spherical clustering PEG-GA having size of cluster three	1562	1623	1673
Spherical clustering PEG-GA having size of cluster eight	1464	1537	1587
Vertical clustering PEG-GA having size of cluster three	1593	1646	1693
Vertical clustering PEG-GA having size cluster nine	1542	1595	1657

If a CH's efficiency is small, this great power expenditure could cause it to expire. Vertical clustering of PEG-GA with size 3 cluster performs better in more protocols for 1st node death, with a 422.2% and a 42.8% enhancement over PEGASIS and PEG-GA, respectively.

The lifetime for a 3D base station outside for variants of PEGASIS is displayed in Figure 18.7.

Table 18.6 shows that the vertical clustering of PEG-GA with a size 3 cluster has outstanding distributing load characteristics. Spherical clustering protocols are not superior to their corresponding vertical clustering protocols because of the unequal dissemination of SNs in clusters. This leads to identical SNs often additionally utilized for inter-cluster communication

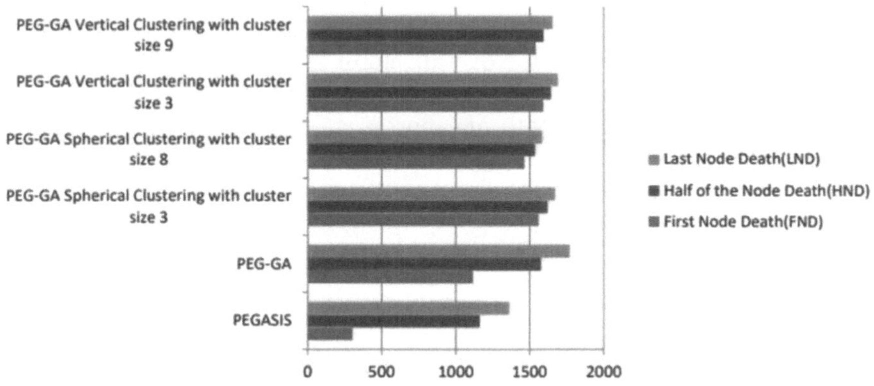

Figure 18.7 Lifetime for 3D base station outside for variants of PEGASIS.

Table 18.6 Residual energy (%) at death of first node for three-dimensional base station outside

Protocol	*Residual energy (%)*
PEGASIS	72.2
PEG-GA	25.5
Spherical clustering of PEG-GA having size of the cluster three	1.3
Spherical clustering of PEG-GA having size of the cluster eight	2.4
Vertical clustering of PEG-GA having size of the cluster three	1.4
Vertical clustering of PEG-GA having size of the cluster nine	1.5

in a number of clusters. The outcome is further seen in a huge quantity of clusters with a greater variation of remaining energy at First Node Death, assessed in spherical clustering of PEG-GA with cluster size 8, and vertical clustering of PEG-GA having cluster size 9.

18.11 CONCLUSION AND FUTURE WORK

In this chapter, an innovative technique of incorporating PEGASIS, clustering, and GA to decrease communication span was planned. A changeable weight CH selection dependent upon range and power for improved distribution of a load was approved. A new type of PEGASIS was modified for 3D WSNs, and variations of this were introduced. Simulations in MATLAB 2018a revealed the advantage and competence of our planned protocols with regard to FND. Our proposed protocols show superiority with regard to remaining power, distributing load, and detainment achievement. Our future research work will develop clustering to be further efficient.

After clustering, instead of creating every chain independently, the difficulty can be seen in the numerous TSP. After that, a meta-heuristic can be used to resolve it to reduce overall chain length. By trial and error, weights for the CH selection were decided. An algorithm such as particle swarm optimization (PSO) with an appropriate fitness function could be implemented to assist weight selection. Heterogeneous wireless sensor networks (HWSN) and mobility of SNs and BS are also regions that could be investigated.

REFERENCES

1. Khamforoos, K. and Khamforoush, H. (2009). *A new routing Algorithm for Energy Reduction in Wireless Sensor Networks*, IEEE.
2. Yu, Y. and Song, Y. (2010). An Energy Efficient Chain-Based Routing Protocol in Wireless Sensor Network. 2010 International Conference on Computer Application and System Modeling (ICCASM 2010), IEEE, 486–489.
3. Heinzelman, W. R., Chandrakasan, A. and Balakrishnan, H. (2000). Energy-Efficient Communication Protocol for Wireless Microsensor Networks, *IEEE Computer Society*, 18, 8020.
4. Lindsey, S. and Raghavendra, C. S. (2003) PEGASIS: power-efficient gathering in sensor information systems," in Proceedings of the IEEE Aerospace Conference, 3.
5. Xie, G. and Pan, F. (2016) Cluster-based routing for the mobile sink in wireless sensor networks with obstacles, *IEEE Access*, 4, 2019–2028.
6. Zhu, C., Wu, S., Han, G., Shu, L. and Wu, H. (2015) A tree-cluster-based data-gathering algo0rithm for industrial WSNs with a mobile sink, *IEEE Access*, 3(4), 381–396.
7. Linping, W., Wu, B., Zhen, C. and Zufeng, W. (2010) Improved algorithm of PEGASIS protocol introducing double cluster heads in wireless sensor network, Comput, *Mechatronics Control Electronic Engineering (CMCE)*, 1, 148–151.
8. Mishra, A.K., Rahman, R.U., Bharadwaj, R. and Sharma, R. (2015) An enhancementof pegasis protocol with improved network lifetime for wireless sensor networks, in: Power, Communication and Information Technology Conference (PCITC), 142–147.
9. Wang, J., Gao, Y., Yin, X., Li, F., and Kim, H-J.(2018) An enhanced PEGASIS algorithm with mobile sink support for wireless sensor networks, *Wireless Communications and Mobile Computing*, 1–9. https://doi.org/10.1155/2018/9472075.
10. Yueyang, L., Hong, J., and Guangxin, Y. (2006). An energy-efficient pegasisbased enhanced algorithm in wireless sensor networks, *China Communications*, 91–97.
11. Lindsey, S. and Raghavendra, C. S. (2002). PEGASIS: power-efficient gathering in sensor information systems, *Proceedings of the IEEE Aerospace Conference*, 3, 1125–1130.
12. Guo, W., Zhang, W. and Lu, G. (2010). PEGASIS protocol in wireless sensor network based on an improved ant colony algorithm. *Second International Workshop on Education Technology and Computer Science*, 3, 64–67.

13. Roy, S. and Mukherjee, N. (2012). Topology construction of 3d wireless sensor network. in: Advances in Computing and Information Technology, Springer, 533–542.
14. Heidemann, J., Stojanovic, M., and Zorzi, M. (2012). Underwater sensor networks: applications, advances and challenges, *Philosophical Transactions of the Royal Society A*, 370, 158–175.
15. Kohli, S., Bhattacharya, P.P., and Jha, M.K. (2016). Implementation of homogeneous leach protocol in three dimensional wireless sensor networks, *International Journal of Sensors, Wireless Communications Control*, 6, 4–11.
16. Baghouri, M., Hajraoui, A. and Chakkor, S. (2015). Three-dimensional application-specific protocol architecture for wireless sensor, *Networks,* 15, 352–360.
17. Ramluckun, N. and Bassoo, V. (2018). *Energy-efficient chain-cluster based intelligent routing technique for Wireless Sensor Networks, Applied Computing and Informatics*, Elsevier. https://doi.org/10.1016/j.aci.2018.02.004.
18. Heinzelman, W., Chandrakasan, A., and Balakrishna, H. (2000). Energy-Efficient Communication Protocol for Wireless Microsensor Networks. In Proceedings of the Hawaii Conference on System Sciences.
19. Mukhairez, H.H. and Maghari, A.Y. (2015). Performance comparison of simulated annealing, GA and ACO applied to TSP, *International Journal of Intelligent Computing Research*, 6(4), 647–654.
20. Gupta, D. and Ghafir, S. (2012). An overview of methods maintaining diversity in genetic algorithms, *International Journal of Emerging Technology and Advanced Engineering*, 2, 56–60.
21. Warrier, M.M. and Kumar, A. (2016). An energy efficient approach for routing in wireless sensor networks, *Procedia Technology*, 25, 520–527.
22. Min, B.G., Prak, J., Kim, H.G., and Shon, J. G. (2019). *Improvement of Multi-Chain PEGASIS Using Relative Distance, Advanced Multimedia and Ubiquitous Engineering*, Springer.
23. Hussein, A.A. and Khalid, R. (2019). Improvements of PEGASIS routing protocol in WSN, *International Advance Journal of Engineering Research*, 2(11), 1–14.
24. Ghosh, R., Mohanty, S., Pattnaik, P. K., and Pramanik, S. (2021). A novel performance evaluation of resourceful energy saving protocols of heterogeneous WSN to maximize network stability and lifetime, *International Journal of Interdisciplinary Telecommunications and Networking (IJITN)*, 13(2), 72–88.
25. Ghosh, R., Mohanty, S., Pattnaik, P.K., and Pramanik, S. (2020). A performance assessment of power-efficient variants of distributed energy-efficient clustering protocols in WSN. *International Journal of Interactive Communication Systems and Technologies (IJICST)*, 10(2), 1–14.
26. Ghosh, R., Mohanty, S., and Pattnaik, P. K. (2021). An evolving α -dependent mobility model for a fleet of unmanned aerial vehicles in wireless sensor networks. *International Journal of Communication Systems*, 34(14), e4926.
27. Chugh, P., Meenu, G., S. Indu, Chaudhary, G., Khari M., and Shanmuganathan, V. (2023). Advanced energy efficient pegasis based routing protocol for IoT applications, *Microprocessors and Microsystems*, 103(1), 104727. https://doi.org/10.1016/j.micpro.2022.104727.

28. Ali, K.H., and Shah, A. F. M. S. (2023). Performance Comparison of PEGASIS, HEED and LEACH Protocols in Wireless Sensor Networks, *Celal Bayar University Journal of Science*, 19(1), 11–18, https://doi.org/10.18466/cbayarfbe.1165816
29. Mote, T. S., and Jagtap, S. K. (2023). Adaptive threshold residual energy-based efficient sensor network protocol (ATREEN), *Journal of Sensors*, https://doi.org/10.1155/2023/1629438.

Chapter 19

Sustainable communication network architecture in agriculture

Pawan Whig, Jhansi Bharathi Madavarapu, Rattan Sharma, and Venugopal Reddy Modhugu

19.1 INTRODUCTION

In a world where agriculture is fundamental for sustenance and economic progress, there's an urgent need for innovation. As the global population grows, agriculture confronts the challenge of meeting escalating demands for food, fiber, and fuel while tackling environmental issues, resource scarcity, and the call for sustainable practices [1, 2]. This chapter explores the "Sustainable Communication Network Architecture in Agriculture," a crucial element in modern farming practices. It delves into the interplay between advanced communication technologies and agriculture, suggesting their potential for significant transformation [3].

Traditionally, agriculture has been associated with timeless landscapes [4]. However, behind this facade, a revolution is underway. This revolution doesn't involve just tractors and plows but data streams and digital connectivity. It's driven by the necessity to feed a growing global population while lessening agriculture's environmental impact. The chapter examines the foundations and applications of sustainable communication network architecture in agriculture, highlighting the synergy between modern telecommunications and farming practices [5–7]. It unravels the technologies, strategies, and innovations fueling this transformation. From precision agriculture and smart farming to real-time data analytics and Internet of Things (IoT) applications, it reveals tools empowering farmers in the 21st century. Additionally, it showcases real-world case studies and practical implementations to illuminate the tangible benefits for farmers, agribusinesses, and ecosystems. The goal is to showcase how the convergence of agriculture and advanced communication technologies enhances productivity, resource efficiency, and sustainability, leading to a future of food security and environmental preservation [8].

The chapter is strategically structured with clear objectives, aiming to offer a comprehensive understanding of sustainable communication network architecture's significance in agriculture. It introduces sustainable agriculture's relevance and delves into communication network architecture, IoT integration, precision agriculture, and data-driven decision-making. It also

DOI: 10.1201/9781003541363-19

Table 19.1 Literature review

Reference	*Title*	*Key findings*
[1] Khan et al. (2021)	Current progress and future prospects of agriculture technology: Gateway to sustainable agriculture	• Agriculture technology contributes to sustainable agriculture. • Highlights the potential of technology in enhancing agricultural practices.
[2] Yang et al. (2022)	Exploring the spatial network structure of agricultural water use efficiency in China: A social network perspective	• Examines agricultural water use efficiency from a social network perspective in China.
[3] Nandyala & Kim (2016)	Green IoT agriculture and healthcare application (GAHA)	• Introduces the concept of green IoT for agriculture and healthcare applications.
[4] Pathmudi et al. (2023)	A systematic review of IoT technologies and their constituents for smart and sustainable agriculture applications	• Provides a systematic review of IoT technologies for sustainable agriculture.
[5] Dumitrache et al. (2017)	A cyber-physical systems approach for agricultural enterprise and sustainable agriculture	• Proposes a cyber-physical systems approach for sustainable agriculture.
[6] Islam et al. (2021)	A review of applications and communication technologies for Internet of Things (IoT) and UAV-based sustainable smart farming	• Reviews IoT and UAV-based technologies for smart farming.
[7] Srbinovska et al. (2015)	Environmental parameters monitoring in precision agriculture using wireless sensor networks	• Discusses environmental monitoring in precision agriculture using wireless sensor networks.
[8] Li (2012)	Application of the Internet of Things technology in precision agriculture irrigation systems	• Focuses on the application of IoT in precision agriculture, particularly in irrigation systems.
[9] Wongchai et al. (2022)	Artificial intelligence-enabled soft sensor and Internet of Things for sustainable agriculture using ensemble deep learning architecture	• Presents AI-enabled soft sensors and IoT for sustainable agriculture.
[10] Lu et al. (2020)	Energy efficiency optimization in SWIPT enabled WSNs for smart agriculture	• Addresses energy efficiency optimization in smart agriculture using SWIPT-enabled WSNs.
[11] Mekala & Viswanathan (2017)	A Survey: Smart agriculture IoT with cloud computing	• Surveys smart agriculture IoT with a focus on cloud computing.
[12] Sanjeevi et al. (2020)	Precision agriculture and farming using Internet of Things based on wireless sensor network	• Discusses precision agriculture and farming using IoT and wireless sensor networks.

addresses environmental sustainability, climate-resilient farming, improved market access, and economic viability for farmers through enhanced communication infrastructure. Real-world case studies are incorporated to provide practical insights. Lastly, the chapter explores future trends and research opportunities, catering to stakeholders interested in technology, sustainability, and agriculture.

19.2 LITERATURE REVIEW

The progression of agricultural technology has played a crucial role in advancing sustainable agriculture, a theme extensively discussed in various scholarly works. Khan et al. (2021) and Dumitrache et al. (2017) highlight technology's significance, focusing respectively on cyber-physical systems and IoT to improve agricultural practices and promote sustainability. Studies such as Yang et al. (2022) and Srbinovska et al. (2015) explore networks and wireless sensor systems used to monitor agricultural water use efficiency and environmental factors in precision agriculture. The integration of IoT, as seen in Pathmudi et al. (2023) and Li (2012), illustrates its importance in optimizing irrigation systems and fostering smart farming initiatives. Additionally, the combination of artificial intelligence, cloud computing, and deep learning, outlined in research by Wongchai et al. (2022) and Mekala & Viswanathan (2017), enhances the potential for sustainable agriculture through sophisticated sensor systems and intelligent data analysis. Collectively, these studies paint a comprehensive picture where technological innovations, particularly IoT, AI, and wireless networks, hold significant promise in reshaping agricultural practices toward sustainability as shown in Table 19.1.

19.3 SCOPE AND IMPORTANCE OF SUSTAINABLE COMMUNICATION IN AGRICULTURE

The scope and importance of sustainable communication in agriculture extend far beyond the confines of traditional farming practices. In an era defined by global challenges such as climate change, resource scarcity, and a growing population, the role of communication networks in agriculture has taken center stage [9–12]. Below, we explore the vast scope and critical significance of sustainable communication in agriculture, as shown in Figure 19.1.

Sustainable communication networks play a multifaceted role in revolutionizing the landscape of agriculture [13]. These networks serve as the linchpin for data collection from a plethora of sources, such as sensors, drones, and satellites, ushering in the era of precision agriculture. By harnessing real-time data, they enable the fine-tuning of resource allocation, optimizing the use of water, fertilizers, and pesticides, ultimately leading

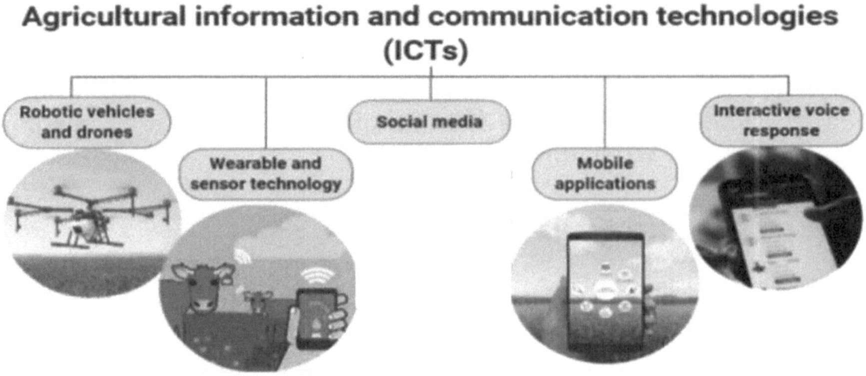

Figure 19.1 Sustainable communication practices in agriculture.

to enhanced efficiency, reduced wastage, and a diminished environmental footprint [14–22]. The precise provision of inputs for crop growth translates to increased yields and cost savings for farmers. Sustainable communication networks serve as the backbone of climate resilience, providing access to critical weather forecasts and early warning systems and aiding farmers in adapting to climate change [23–25]. These networks also bridge the gap between rural areas and markets, granting farmers access to buyers, financial services, and real-time market information [26–28]. Furthermore, they facilitate data-driven decision-making by collecting and analyzing extensive agricultural datasets, supporting sustainable practices and predictive analytics. Additionally, they contribute to environmental sustainability by enabling remote ecosystem monitoring and conservation efforts. These networks are instrumental in rural development, extending education and healthcare services to underserved regions. They promote collaboration and knowledge sharing within the agricultural community through online platforms and accelerate research and innovation by globally sharing agricultural data. Ultimately, sustainable communication networks form the foundation of a technologically advanced, sustainable, and resilient agricultural sector. Their extensive scope and significance empower farmers with the necessary tools and knowledge to navigate modern agriculture while fostering environmental responsibility, economic growth, and resilience amid global challenges. As agriculture progresses, sustainable communication networks will remain integral in shaping its future.

19.4 COMMUNICATION TECHNOLOGIES IN AGRICULTURE

A diverse array of communication technologies is revolutionizing agriculture, offering innovative solutions, and enhancing efficiency across the sector. These advancements span from the IoT, facilitating real-time data collection for precision farming, to the use of mobile applications that grant farmers access to crucial information on crop management, pest control, and market trends. Satellite technology, leveraging imagery and GNSS systems, provides accurate location data and enables remote sensing. Drones and UAVs play a pivotal role in crop monitoring, disease detection, and mapping through aerial surveillance tools. Big data analytics platforms process information from multiple sources, generating valuable insights. Additionally, weather forecasting services and climate data aid farmers with real-time updates on weather patterns and climatic conditions. Remote monitoring and control systems further enhance efficiency by allowing remote management of farm operations. Blockchain technology ensures transparency and traceability within supply chains, while communication platforms facilitate knowledge sharing among farmers, researchers, and experts. The advent of 5G connectivity has notably bolstered high-speed, low-latency networks, enabling real-time data exchange in agricultural operations. These diverse communication technologies collectively drive innovation, significantly benefiting farmers and addressing global challenges related to food security and sustainability.

19.4.1 Communication networks in agriculture

Communication networks play a crucial role in modern agriculture by facilitating efficient data exchange, remote monitoring, and connectivity among various stakeholders in the agricultural ecosystem. These networks are instrumental in enabling real-time data collection and monitoring in fields using IoT sensors, weather stations, and drones. This data includes essential information such as soil moisture, temperature, crop health, and weather conditions, all vital for informed decision-making. Additionally, these networks support seamless data transmission to cloud-based platforms, enabling farmers to collaborate with experts, researchers, and extension services. This collaboration facilitates data-driven precision agriculture, providing farmers with actionable insights to optimize planting, irrigation, fertilization, and pest control strategies based on thorough data analysis.

Moreover, communication networks provide farmers with crucial market information, prices, and demand trends, allowing them to make informed decisions about crop selection and harvest timing to maximize profitability. They also contribute to efficient supply chain management by offering real-time tracking capabilities for agricultural products, reducing waste, and ensuring timely delivery to consumers. These networks enable farm automation by facilitating remote monitoring and control of farming equipment and systems, enhancing efficiency while reducing labor costs.

Access to weather forecasts and historical climate data through these networks empowers farmers to effectively plan planting and harvesting schedules, mitigating the risk of weather-related crop losses. Furthermore, communication networks serve as a channel for research and extension services to disseminate valuable research findings, best practices, and advisory services to farmers, thereby enhancing agricultural productivity and sustainability. Overall, these networks are pivotal in modern agriculture, driving the sector toward greater productivity and sustainability through data-driven decision-making, precision practices, and improved connectivity with markets and experts.

19.4.2 Challenges

Communication networks play a crucial role in addressing the diverse challenges encountered by modern agriculture. These challenges span resource efficiency, climate change resilience, pest and disease management, market access, rural development, data-driven agriculture, and environmental sustainability. These networks facilitate data-driven resource management, aiding in the sustainable use of resources while minimizing waste and costs. They also provide timely weather data, enabling farmers to adapt to changing climate conditions. Additionally, through sensors and remote monitoring, these networks assist in early pest and disease detection. Farmers benefit from market information and expanded marketing opportunities, overcoming challenges related to accessing markets and obtaining fair prices for their produce. Rural connectivity promoted by these networks enhances access to essential services like education, healthcare, and financial resources. Advanced communication technologies handle vast agricultural data, extracting valuable insights for informed decision-making. Ultimately, these networks serve as a linchpin in addressing agricultural complexities, empowering farmers, and promoting sustainable practices to bolster productivity, resilience, and environmental conservation in agriculture.

19.5 SUSTAINABLE COMMUNICATION NETWORK ARCHITECTURE

Ensuring reliability for instant decision-making and secure data transmission is paramount in a sustainable communication network structure. Security forms the foundation, especially in today's landscape, of escalating cyber threats. To safeguard sensitive information from unauthorized access, data encryption is employed both during transit and while stored. Stringent access control mechanisms only grant network entry to authorized users and devices. Robust authentication and authorization protocols verify user and device identities before allowing access. Regular audits and monitoring promptly detect and mitigate vulnerabilities, upholding system integrity. Interoperability, a critical principle, ensures seamless integration with various devices and systems. The architecture adopts industry-standard communication protocols and data formats for compatibility, supported by well-documented APIs enabling third-party integration. Given agriculture's energy-intensive nature, the architecture emphasizes energy-efficient devices, particularly in remote areas with limited power resources. Devices feature power-saving modes to minimize energy consumption, promoting sustainability. Designs incorporate weatherproofing and lightning protection for resilience in harsh environmental conditions. Edge computing enhances data processing closer to sources, reducing latency and bandwidth requirements and benefiting time-sensitive farming tasks. Scalable cloud-based data storage accommodates expanding agricultural data volumes, ensuring efficient retrieval and analysis. Open data standards prevent vendor lock-in, fostering transparent and flexible data management. Embracing environmental sustainability, the architecture integrates renewable energy sources and selects energy-efficient hardware components to reduce the carbon footprint. Adherence to agriculture-specific data privacy regulations and responsible data handling are paramount, coupled with user-friendly interfaces for seamless monitoring and interaction. Community engagement involving local farmers ensures the architecture meets their specific needs, enhancing their livelihoods and fostering ownership.

19.6 RENEWABLE ENERGY SOURCES

Incorporating renewable energy sources into agricultural communication network architecture is a multifaceted and crucial undertaking that aims to enhance sustainability and resilience within the agricultural sector. This comprehensive approach involves assessing energy needs across critical infrastructure components like devices, routers, and servers. Solar power emerges as a prominent solution, utilizing strategically positioned solar panels to harness abundant sunlight. Battery storage systems complement solar energy by storing excess power, ensuring a continuous energy supply even during periods of low sunlight or at night. Remote stations powered by solar energy extend connectivity to remote agricultural areas where grid power is limited. Additionally, small wind turbines complement solar panels by offering consistent power in regions with reliable wind patterns. Hybrid systems seamlessly combine solar and wind energy sources, guaranteeing uninterrupted energy flow despite weather variations. Micro-hydroelectric systems, where feasible, offer eco-friendly energy solutions. Furthermore, exploring bioenergy options, such as converting agricultural waste into biogas, diversifies the renewable energy portfolio and reduces reliance on conventional power sources. Ensuring energy efficiency within agricultural communication network architecture involves investing in energy-efficient hardware components like low-power sensors, routers, and communication devices. Smart energy management systems automate device scheduling based on demand, optimizing energy consumption and maximizing savings. IoT sensors facilitate remote monitoring and maintenance, enabling real-time tracking of energy production and system performance, reducing downtime crucial for time-sensitive agricultural operations. Redundancy planning, incorporating multiple renewable energy sources and backup options like diesel generators, ensures continuous network operation. Educating farmers and network users on responsible energy usage fosters a culture of sustainability within the agricultural community. Environmental impact assessments are integral to ensure that renewable energy installations do not harm local ecosystems or wildlife, local regulations are complied with, and necessary permits are obtained. Evaluating the long-term economic viability of renewable energy integration involves assessing the return on investment (ROI) and total cost of ownership (TCO) over the system's lifecycle. Community involvement is crucial, garnering support for renewable energy projects and ensuring broader community benefits. By blending solar, wind, hydroelectric, and bioenergy solutions, integrating renewable energy into agricultural communication networks reduces carbon footprint and strengthens resilience by reducing dependence on centralized power grids. This sustainable approach paves the way for an environmentally conscious and resilient agricultural future, supporting the farming community while nurturing the planet for generations to come.

19.6.1 Importance of renewable energy in agricultural context

Farmers who invest in renewable energy systems gain a degree of energy independence. They become less vulnerable to power outages, fluctuations in energy prices, and disruptions caused by grid failures. This independence enhances the reliability and resilience of agricultural operations.

19.6.1.1 Rural electrification

Renewable energy projects can bring electricity to remote and off-grid farming communities. By extending the reach of electricity, these projects enhance the quality of life, support education, and enable economic development in rural areas.

19.6.1.2 Compliance with sustainability goals

Many countries and regions have set ambitious sustainability targets, which often include the promotion of renewable energy. Adopting renewable energy solutions aligns agriculture with these goals, positioning it as a responsible and forward-thinking industry.

19.6.2 Solar and wind energy integration for communication infrastructure

Solar power is a versatile and reliable energy source for agricultural communication infrastructure:

- **Solar panels:** Photovoltaic solar panels convert sunlight into electricity, providing a consistent source of power for devices, sensors, and communication equipment.
- **Off-grid installations:** Solar panels can be used in off-grid areas, making it feasible to establish communication networks in remote agricultural regions.
- **Battery backup:** Battery storage systems store excess solar energy for use during nighttime or cloudy periods, ensuring uninterrupted communication.

Wind energy complements solar power, especially in regions with consistent wind patterns:

- **Small wind turbines:** Small wind turbines can be integrated with solar panels to provide additional energy, especially during the night or cloudy days.
- **Hybrid systems:** Hybrid systems that combine both solar and wind power sources offer reliability by ensuring a continuous power supply.

19.7 AGRICULTURAL INSIGHTS

In the realm of modern agriculture, data stands as a crucial driver, shaping farming practices through its extensive collection, analysis, and interpretation. The application of data analytics transforms raw data into actionable insights, enabling precision agriculture tailored to specific field conditions. This precision leads to optimized resource allocation, increased crop yields, and predictive modeling aiding in informed decision-making regarding weather patterns, disease outbreaks, and crop growth. Moreover, data analytics enhances resource efficiency by optimizing irrigation, fertilization, and pesticide use based on soil conditions, moisture levels, and nutrient content. It also provides valuable market intelligence, supporting farmers in making informed decisions about crop selection, pricing, and distribution. By leveraging data analytics, modern agriculture tackles complex challenges while promoting sustainability, resilience, increased productivity, and contributing to global food security.

Exploring the socio-economic impacts of sustainable communication network architecture in agriculture reveals transformative effects on smallholder farmers and rural communities. These networks offer unprecedented access to real-time market information, online marketplaces, and digital payment systems, significantly enhancing income potential and breaking the cycle of poverty among rural farmers. Additionally, digital financial tools promote financial inclusion, enabling secure transactions, savings, and access to credit. Furthermore, improved agricultural training through online resources and expert consultations empowers farmers with previously inaccessible knowledge, leading to improved crop yields and sustainable practices, and bolstering broader food security. Ultimately, connectivity acts as a catalyst for economic and social upliftment, empowering smallholder farmers and fostering resilience in the face of agricultural challenges. Various methods of IoT for smart agriculture are shown in Figure 19.2.

The integration of communication networks into agriculture has revolutionized market access and transformed how the agricultural industry operates. One significant change is the increased transparency in supply chains, ensuring better traceability and quality control from the farm to the end consumer. This visibility has minimized post-harvest losses, ensuring consumers receive top-quality products while benefiting farmers.

Moreover, these networks have remarkably improved distribution efficiency. Real-time data on inventory and demand fluctuations allow for precise coordination in transportation, reducing waste and costs significantly. They've also empowered agribusinesses by linking processing and packaging facilities to markets, leading to increased profits and efficiency.

Beyond farming, these networks have had a profound impact on rural communities. They've encouraged economic diversity by opening up new markets and providing essential information, reducing reliance solely on traditional

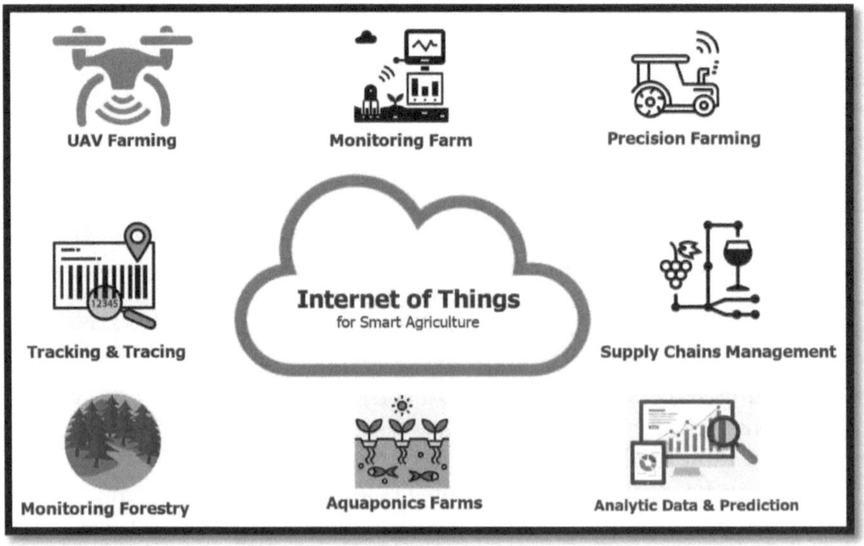

Figure 19.2 Various methods of IoT for smart agriculture.

farming. Healthcare and education have been transformed through remote access initiatives, positively impacting the well-being and skills of rural populations. Additionally, easier access to government services has improved the overall quality of life in these areas. However, challenges remain. Ensuring that farmers and rural communities have the necessary digital skills is crucial. Developing infrastructure in remote regions is complex and requires substantial investment. Furthermore, protecting personal and financial data on these networks is a significant concern. These networks have had a substantial impact on agriculture, empowering farmers, improving supply chains, and fostering sustainability. Overcoming challenges related to infrastructure, digital literacy, and data security will be key to maximizing these benefits. The ongoing development of communication networks holds the promise of a brighter future for agriculture and rural communities.

19.8 CASE STUDIES AND SUCCESS STORIES

The case studies highlight the pivotal role of sustainable communication networks in driving transformative change within agriculture. They vividly portray how technology, when applied thoughtfully and inclusively, can empower farmers and revolutionize farming practices. These real-world examples illuminate the capacity of communication networks to enhance productivity while promoting sustainability and resilience in agriculture. More importantly, they underscore how these networks are not just technological advancements but crucial assets in uplifting rural communities and ensuring the stability of the global food supply.

19.8.1 Smart agriculture network in India

In the heart of rural India, an ambitious initiative took shape, aiming to revolutionize age-old farming practices by introducing a visionary smart agriculture network. This transformative endeavor focused on efficiency and sustainability, leveraging renewable energy and a multitude of IoT devices. In an environment where farmers grappled with unpredictable weather, uncertain soil conditions, and volatile market dynamics, this innovative network brought about a seismic change. Equipped with smartphones, farmers gained access to a treasure trove of agricultural knowledge. Real-time weather forecasts became invaluable, aiding precise irrigation scheduling by providing critical data on rainfall and temperature. IoT sensors played a pivotal role by gathering soil quality data, offering insights into nutrient levels and moisture content, and revolutionizing fertilization and irrigation methods. More significantly, access to market prices empowered farmers to make informed decisions on crop selection and pricing strategies. This amalgamation of information led to unprecedented data-driven decision-making. The outcomes were astonishing, with crop yields skyrocketing by an impressive 20%, a remarkable boost in food production crucial for India's agrarian economy. Besides heightened productivity, farmers directly accessed consumers through online markets, bypassing intermediaries, resulting in fairer prices and increased income. Furthermore, the project's commitment to renewable energy sources ensured sustainability, significantly reducing the carbon footprint of agricultural operations.

In the agricultural heartland of the United States, a shift toward precision farming techniques resulted in a significant 20% reduction in fertilizer costs. This move aligned with sustainability objectives by curbing chemical runoff into waterways. Notably, these efficiency gains didn't compromise crop yields; many farmers actually maintained or increased their productivity. The success of precision farming had a ripple effect throughout the farming community. Nearby farmers swiftly embraced similar technologies, signaling a shift toward more sustainable and resource-efficient practices across the region. These pioneering farmers showcased that leveraging technology and data-driven decision-making could pave the path for sustainable agriculture, a lesson that holds relevance not only for the United States but also for agricultural regions worldwide.

In the diverse landscapes of sub-Saharan Africa, where agriculture is the backbone of livelihoods, an influential agribusiness conglomerate initiated a transformative journey. Central to this transformation was the implementation of a comprehensive communication network architecture, seamlessly linking farmers, distributors, and processing units.

19.9 DISCUSSION

The innovative sustainable communication network architecture designed for modern agriculture promises a transformative shift by enhancing connectivity and enabling real-time data exchange among stakeholders and machinery. It integrates advanced technologies like IoT and data analytics, empowering data-driven decision-making in farming practices. Prioritizing scalability, reliability, and security, it also emphasizes renewable energy sources and low-power protocols for sustainability, aiming to reduce the carbon footprint while optimizing energy usage. Despite its potential, challenges in infrastructure, technological adoption, and stakeholder awareness require collaborative efforts for successful implementation and usability in agricultural settings.

19.10 CONCLUSION

In summary, the integration of sustainable communication network architecture into agriculture has sparked transformative opportunities, reshaping farming approaches, rural development, and food security. This technology, driven by renewable energy, energy-efficient tools, and data analytics, empowers smallholder farmers and promotes environmentally conscious practices. Its socio-economic impact spans financial inclusion, improved healthcare, gender equality, and rural community growth. However, challenges like digital literacy and privacy concerns persist. The showcased case studies vividly demonstrate the tangible benefits, providing a glimpse of a future where these networks ensure a resilient and prosperous agricultural sector. To fully realize this potential, ongoing investment, innovation, and dedicated efforts to address challenges remain imperative for a sustainable agricultural future worldwide.

19.11 FUTURE SCOPE

The future of sustainable communication network architecture in agriculture is brimming with promising opportunities and evolving trends. Emerging technologies like advanced 5G systems, evolving edge computing, and the pervasive influence of artificial intelligence are set to empower farmers with unparalleled data-driven insights and autonomous decision-making capabilities. The expansion of sustainable energy solutions, potentially integrating hydrogen and biomass sources alongside advancements in energy storage and harvesting, will revolutionize agricultural practices. Precision agriculture will flourish, driven by enhanced satellite and drone technologies, robotics, and more sophisticated IoT sensors. Continued advancements in data analytics and AI will offer deeper insights and predictive analytics for climate resilience. Efforts toward digital inclusion will bridge rural-urban divides in technology access, ensuring effective utilization. Simultaneously, a strong focus on sustainability, meeting ESG goals, and responding to consumer demand for eco-friendly farming practices will accelerate the adoption of sustainable communication networks in agriculture. These transformative trends herald a future where agriculture becomes more resilient, efficient, and sustainable, effectively addressing the pressing challenges of global food security for generations to come.

REFERENCES

1. Khan, N., Ray, R. L., Sargani, G. R., Ihtisham, M., Khayyam, M., & Ismail, S. (2021). Current progress and future prospects of agriculture technology: Gateway to sustainable agriculture. *Sustainability*, 13(9), 4883.
2. Yang, G., Gong, G., & Gui, Q. (2022). Exploring the spatial network structure of agricultural water use efficiency in China: A social network perspective. *Sustainability*, 14(5), 2668.
3. Nandyala, C. S., & Kim, H. K. (2016). Green IoT agriculture and healthcare application (GAHA). *International Journal of Smart Home*, 10(4), 289–300.

4. Pathmudi, V. R., Khatri, N., Kumar, S., Abdul-Qawy, A. S. H., & Vyas, A. K. (2023). A systematic review of IoT technologies and their constituents for smart and sustainable agriculture applications. *Scientific African*, e01577. https://doi.org/10.1016/j.sciaf.2023.e01577
5. Dumitrache, I., Caramihai, S. I., Sacala, I. S., & Moisescu, M. A. (2017, May). A cyber physical systems approach for agricultural enterprise and sustainable agriculture. In *2017 21st International Conference on Control Systems and Computer Science (CSCS)* (pp. 477–484). IEEE.
6. Islam, N., Rashid, M. M., Pasandideh, F., Ray, B., Moore, S., & Kadel, R. (2021). A review of applications and communication technologies for internet of things (IoT) and unmanned aerial vehicle (UAV) based sustainable smart farming. *Sustainability*, 13(4), 1821.
7. Srbinovska, M., Gavrovski, C., Dimcev, V., Krkoleva, A., & Borozan, V. (2015). Environmental parameters monitoring in precision agriculture using wireless sensor networks. *Journal of Cleaner Production*, 88, 297–307.
8. Li, S. (2012, August). Application of the internet of things technology in precision agriculture irrigation systems. In *2012 International Conference on Computer Science and Service System* (pp. 1009–1013). IEEE.
9. Wongchai, A., Shukla, S. K., Ahmed, M. A., Sakthi, U., & Jagdish, M. (2022). Artificial intelligence-enabled soft sensor and internet of things for sustainable agriculture using ensemble deep learning architecture. *Computers and Electrical Engineering*, 102, 108128.
10. Lu, W., Xu, X., Huang, G., Li, B., Wu, Y., Zhao, N., & Yu, F. R. (2020). Energy efficiency optimization in SWIPT enabled WSNs for smart agriculture. *IEEE Transactions on Industrial Informatics*, 17(6), 4335–4344.
11. Mekala, M. S., & Viswanathan, P. (2017, August). A Survey: Smart agriculture IoT with cloud computing. In *2017 international conference on microelectronic devices, circuits and systems (ICMDCS)* (pp. 1–7). IEEE.
12. Sanjeevi, P., Prasanna, S., Siva Kumar, B., Gunasekaran, G., Alagiri, I., & Vijay Anand, R. (2020). Precision agriculture and farming using Internet of Things based on wireless sensor network. *Transactions on Emerging Telecommunications Technologies*, 31(12), e3978.
13. Whig, P., Kouser, S., Velu, A., & Nadikattu, R. R. (2022). Fog-IoT-assisted-based smart agriculture application. In *Demystifying Federated Learning for Blockchain and Industrial Internet of Things* (pp. 74–93). IGI Global.
14. Whig, P., Nadikattu, R. R., & Velu, A. (2022). COVID-19 pandemic analysis using application of AI. *Healthcare Monitoring and Data Analysis Using IoT: Technologies and Applications*, 1. https://doi.org/10.1049/pbhe038e_ch1
15. Whig, P., Velu, A., & Bhatia, A. B. (2022). Protect Nature and Reduce the Carbon Footprint With an Application of Blockchain for IIoT. In *Demystifying Federated Learning for Blockchain and Industrial Internet of Things* (pp. 123–142). IGI Global.
16. Whig, P., Velu, A., & Naddikatu, R. R. (2022). The Economic Impact of AI-Enabled Blockchain in 6G-Based Industry. In *AI and Blockchain Technology in 6G Wireless Network* (pp. 205–224). Springer.
17. Whig, P., Velu, A., & Nadikattu, R. R. (2022). Blockchain Platform to Resolve Security Issues in IoT and Smart Networks. In *AI-Enabled Agile Internet of Things for Sustainable FinTech Ecosystems* (pp. 46–65). IGI Global.

18. Whig, P., Velu, A., & Ready, R. (2022). Demystifying Federated Learning in Artificial Intelligence with Human-Computer Interaction. In *Demystifying Federated Learning for Blockchain and Industrial Internet of Things* (pp. 94–122). IGI Global.
19. Whig, P., Velu, A., & Sharma, P. (2022). Demystifying Federated Learning for Blockchain: A Case Study. In *Demystifying Federated Learning for Blockchain and Industrial Internet of Things* (pp. 143–165). IGI Global.
20. Kumar, A., Gupta, S., Yathiraju, N., Bose Chakraborty, S., Dodda, S., & Verma, D. (2023). "The Novel E-Way of Identifying the Face Mask and To Ware the System in the Crowd Management System," 2023 3rd International Conference on Advance Computing and Innovative Technologies in Engineering (ICACITE), Greater Noida, India, pp. 999–1002, doi: 10.1109/ICACITE57410.2023.10182427.
21. Pandey, R., Saha, S., Yathiraju, N., Abdulrahman, I. S., Nittala, R., & Tripathi, V. (2023). "Integration of RFID and Image Processing for Surveillance ABased Security System," 2023 3rd International Conference on Advance Computing and Innovative Technologies in Engineering (ICACITE), Greater Noida, India, pp. 380–384, doi: 10.1109/ICACITE57410.2023.10182987.
22. Nagaraju Naik, M., Kaur, A., Yathiraju, N., Das, S., & Pant, K. (2023). "Improved and Accurate Face Mask Detection Using Machine Learning in the Crowded Places," 2023 3rd International Conference on Advance Computing and Innovative Technologies in Engineering (ICACITE), Greater Noida, India, pp. 572–576, doi: 10.1109/ICACITE57410.2023.10182567.
23. Yathiraju, N., Raman, P., Madala, R., Surgonda Patil, P., Kumar, A., & Ashwin, S. (2023). "Research and Innovation to Market Development: Artificial Intelligence in Business," 2023 Eighth International Conference on Science Technology Engineering and Mathematics (ICONSTEM), Chennai, India, pp. 1–6, doi: 10.1109/ICONSTEM56934.2023.10142715.
24. Yathiraju, N., Sankar, A., & Sandhiya, S. (2022). "Cardiac Disease Prediction for Heart Monitoring using Data Mining Techniques," 2022 International Interdisciplinary Humanitarian Conference for Sustainability (IIHC), Bengaluru, India, pp. 1282–1287, doi: 10.1109/IIHC55949.2022.10060047.
25. Yathiraju, N., & Mohapatra, A. (2023). The Implications of IoT in the Modern Healthcare Industry post COVID-19. *International Journal of Smart Sensor and Adhoc Network*, 3(4), Article 3.
26. Yathiraju, N., & Dash, B. (2023). Gamification Of E-Wallets With The Use Of Defi Technology-A Revisit To Digitization In Fintech. *International Journal of Engineering, Science*, 3(1), 2582–9734.
27. Yathiraju, N., & Dash, B. (2023). Big data and metaverse revolutionizing the futuristicfintech industry. *International Journal of Computer Science & Information Technology (IJCSIT)*, 15(1), 2023. doi:10.5121/ijcsit.2023.15101
28. Ahammad, D. S. H., & Yathiraju, D. (2021). maternity risk prediction using iotmodule with wearable sensor and deep learning based feature extraction and classification technique. *Research Journal of Computer Systems and Engineering*, 2(1), 40–45.

Chapter 20

Modernizing healthcare through IoT-based e-health systems

Nafees Akhter Farooqui, Iqra Ashraf, and Rupam Jaiswal

20.1 INTRODUCTION

The Internet of Things (IoT) is emerging in healthcare tracking devices to provide effective emergency services [1]. It is also utilized as e-fitness software with high-tech components. Smartphones with fitness sensors are vital to people's daily lives. [2]. The boundary surveillance system gathers data from wards and diagnostic equipment for effective and eco-friendly healthcare administration [3]. IoT healthcare devices improve patient care by tracking and evaluating the environment.

IoT-based healthcare systems use larger batteries to power patient wearables and sensors [4]. Frequent charging of these gadgets and mobile devices can exhaust users and necessitate assistance [5]. The cloud data center's excessive energy use raises cloud computing costs. Currently, health monitoring systems require low latency and energy-intensive cloud services. Cloud data centers' excessive energy use raises cloud computing costs [6].

Many frameworks have been developed to implement IoT in fitness services. [7]. A typical IoT layout and implementation technique is provided by this framework. It addresses all realistic implementation issues of the time, including conversation elements, technologies, hardware structure, statistics storage, flow, and access techniques. Based on Zhao et al. [8], a framework remotely monitored elderly people. They focused on IoT utility. Their version met elderly needs. Machine learning algorithms were used in their proposed version [9]. With Basanta et al.'s help, Help to You (H2U) was offered to screen elderly and impacted people [10].

This study was inspired by the growing use of IoT in healthcare infrastructures. Using wearable technology, environmental monitoring, diagnostic data collecting, and patient care management, the IoT has the potential to completely transform the healthcare industry.

DOI: 10.1201/9781003541363-20

20.2 LITERATURE REVIEW

There has been a significant boost in the use of IoT technology in healthcare due to the introduction of healthcare tracking devices, which are essential for providing quick emergency assistance to patients [11]. One area where the IoT has found new uses is in e-fitness software, which takes advantage of medical technological improvements [12]. The ubiquitous smartphone is now an integral component of the Healthcare IoT due to its ability to link to sensors and provide real-time information about patients' health statuses [20].

20.2.1 Management of healthcare through effective surveillance systems

Healthcare monitoring systems collecting data from diagnostic instruments and hospitals is a major step forward. Next, healthcare management that is both efficient and environmentally conscientious is made possible using the data that has been extracted [13].

20.2.2 Studying the environment for better patient care

IoT-enabled healthcare systems improve patient care by tracking and evaluating the surroundings. Monitoring and evaluating environmental elements may help us improve patient wellness [22].

20.2.3 Problems with power usage

Several challenges exist when trying to use IoT for healthcare. The need for a larger energy source is a practical challenge with wearables and implanted sensors [14]. The constant requirement to charge these devices and mobile devices is linked to both the user experience and the workload of healthcare providers and attendants [15].

20.2.4 Healthcare-related IoT frameworks

Several structures have been proposed to facilitate the integration of the IoT into healthcare services, considering the challenges involved [16]. The research consistently highlights three fundamental paradigms: utility, security, and performance. These paradigms exemplify the intricate characteristics of IoT technology in the healthcare field [17].

20.2.5 A balanced approach to overcoming implementation obstacles

Recent research emphasizes a framework that addresses security, performance, and uses of IoT technologies in healthcare. [7]. Overcoming real

IoT implementation challenges – communication entities, innovations, hardware architecture, data storage, data flow, and access mechanisms – must be achieved.

20.2.6 Frameworks centered on utilities

In the realm of practical applications, Zhao et al. put forth a system that considers the unique requirements of the elderly by means of machine learning methods for remote monitoring of their health state [18, 19].

20.3 H-IOT DEVELOPMENT

The provision of healthcare is essential to society. The use of IoT seeks to improve the efficacy of medication. People can access records relating to illnesses and treatments more effectively. Humans are now able to access real-time clinical or therapeutic data via the IoT. In this chapter, we will examine the advantages and disadvantages of implementing Healthcare IoT (H-IoT) in the healthcare system.

Healthcare IoT revolutionizes healthcare for patients and clinicians. Healthcare IoT focuses on remote patient monitoring. IoT will boost medication efficacy. Individuals can better access and share medical records, known in the field of e-health as Electronic Health Records (EHR). Figure 20.1 explains the structural design of the smart healthcare system.

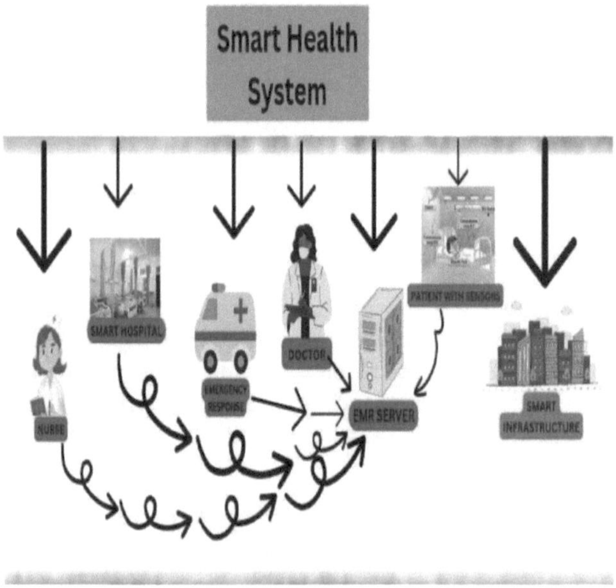

Figure 20.1 Smart healthcare system.

Monitoring patients in the early stages of disease can improve recovery. It will also help doctors respond rapidly to heart attacks and the severe symptoms of diabetes. The IoT makes EHRs digitally practical, thus telemonitoring systems can eliminate scientific errors. Digital EHR retrieval and access are easier and more user-friendly. Healthcare IoT employs Personal Area Networks (PANs). Smartwatches, wristbands, and approach sensors can monitor personal location networks.

Healthcare providers will be able to provide treatment, if necessary, thanks to remote access to this data. This aids in early disease identification, information sharing, and self-management of health concerns. Thus, IoT healthcare applications comprise various sensors and gadget monitoring. To share data, all units are parallel and linked [18, 19].

20.4 H-IOT APPLICATIONS

H-IoT is key in the management of modern devices. Today's science enables seamless connectivity between medical equipment and carers, which is advantageous to patients, doctors, and nurses. It enables the dialogue systems within these devices to communicate continuously, with remarkable performance and innovation, allowing the transmission of patient medical records in a timely manner. It can also update the patient's circumstances before their arrival.

A healthcare application for the IoT is remote patient monitoring, which comprises constant, comprehensive monitoring of patients remotely by means of sensors and cameras that track vitals and assess mental and physical states. Figure 20.2 explains health-based IoT equipment. The H-IoT is explained as follows.

20.4.1 The development and health of seniors

Community-based healthcare for the aging population must be prioritized. Senior welfare research is growing. Residential doctors hire scientists and nursing assistants from client organizations to oversee patient treatments from single centers using video and advanced telemedicine. Technology, clinical treatment, education, economic strategies, and marketing have made older people healthier and more active.

20.4.2 Pregnancy incubator

Doctors, pediatric nurses, and parents have criticized incubators, and some infant incubators have been rejected due to inadequate design. To ensure proper functionality, it is essential to regularly assess the temperature of an infant in an incubator. Clinical studies at local hospitals in Malawi involved monitoring the air temperature of infant incubators. [21, 24]. To save young

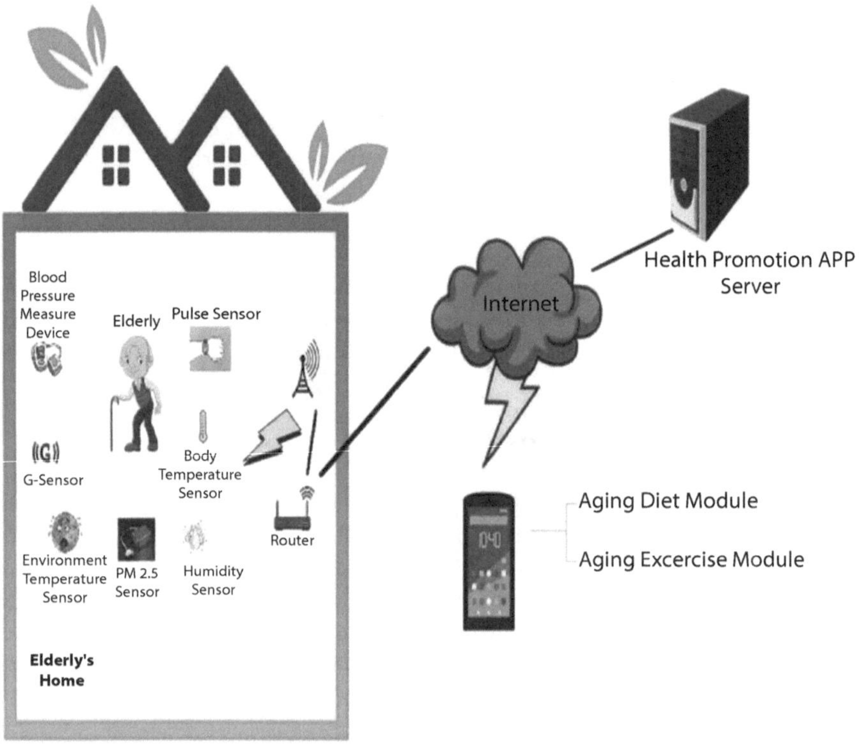

Figure 20.2 H-IoT (Healthcare IoT)-based applications.

lives, appropriate data is needed for a fast, cheap, precise solution. An inexpensive, lightweight infant incubator is recommended by the International Health Organization and Health Promotion Engineering technicians. An innovative incubator that monitors the baby's vital signs and makes autonomous environmental adjustments based on sensor data could be the result of extensive research [23].

20.5 IOT TOWARD COVID-19

The IoT network of mechanical, digital, and permanent computation objects can communicate any data without human intervention. Parts are coded separately. Proven IoT technology supports real-time analytics, computer-based research, and sensory products. Smart lighting, home security, and other technologies help people satisfy their real-world demands, making the IoT part of their daily lives [24]. COVID-19 may be controlled in numerous ways with the IoT. A few examples follow:

Virtual patient monitoring: IoT devices let clinicians monitor COVID-19 patients' symptoms and vital signs remotely. This frees up hospital beds for more serious cases and reduces healthcare worker exposure.

Contact tracing: IoT sensors can track people and identify those who had close contact with a COVID-19-positive person. Technology helps healthcare staff quickly identify and isolate viral transmitters.

Monitor interpersonal distancing: IoT sensors can be used to keep surveillance on crowds in public areas and notify people when they are not following social distancing rules. In regions with significant traffic, this may help slow viral infection spread.

Fever inspection: One of the main signs of COVID-19 is fever, which may be promptly and reliably screened for using IoT devices. Technology can assist in locating possible carriers of viruses and keeping individuals out of public areas.

IoT-enabled remote monitoring, interaction tracking, environmental monitoring, social distancing, and temperature evaluation could help minimize cases of COVID-19. IoT improves architecture and simplifies data integration and exchange for people who need assistance and wireless services [25].

Remote patient monitoring, in-home care for older adults, tracking rehabilitation and recovery, intelligent diagnostic equipment to automate and centrally manage operations, and smart hospital buildings to improve hospital care for medical staff and patients are all healthcare applications of IoT. Figure 20.3 depicts IoT applications in different domains.

All healthcare devices must be properly designed for doctors, patients, and governments. Medical device law compliance, constant communication, and future adaptation are needed. Healthcare devices must safely record and send vital data and stay connected. Medical devices must comply with IEC 62304, ISO 13485:2016, Medical Device Regulation 2017/745, GDPR, and NIST Special Publication 1800-30 privacy and cybersecurity regulations. These devices must disable remote applications or upgrade security to ensure performance.

20.6 H-IOT ARCHITECTURE

Healthcare IoT architecture combines IoT technologies and cloud computing with medicine. It allows sensors and clinical units to provide patient records to a healthcare network. An H-topology IoT connects the unique aspects of a healthcare system or network and their consistent causality. A typical H-IoT device has a publisher, broker, and subscriber (Figure 20.4) [27].

ECG, electromyogram, blood pressure, heart rate, temperature, oxygen saturation, etc. can be recorded [26]. This information can be continuously

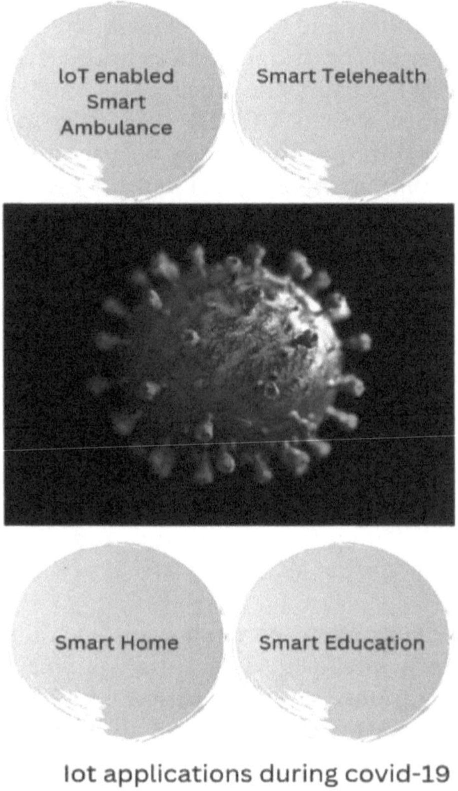

Figure 20.3 IoT applications during pandemic situations.

communicated to a distributor via a network. The H-IoT hybrid grid uses each IoT network and cloud device for a unique purpose. H-IoT architecture is hard to standardize because it relies on healthcare needs and applications. There are many different sorts of architectures found in H-IoT devices [28–30].

ECG monitoring: Electrocardiograms reveal atrium and ventricular depolarization and repolarization. An electrocardiogram (ECG) can detect many heart issues. Bansal and Gandhi suggest a microelectronic, big data, IoT-enabled ECG monitor for long-term, continuous monitoring [31–33].

Monitoring blood sugar: Chronically elevated blood glucose causes diabetes. Its widespread occurrence is a health risk. Type 1, type 2, and gestational diabetes are most common. The most common diabetes diagnosis method is "fingerpicking" followed by blood glucose testing. Recent IoT improvements have made blood glucose monitoring

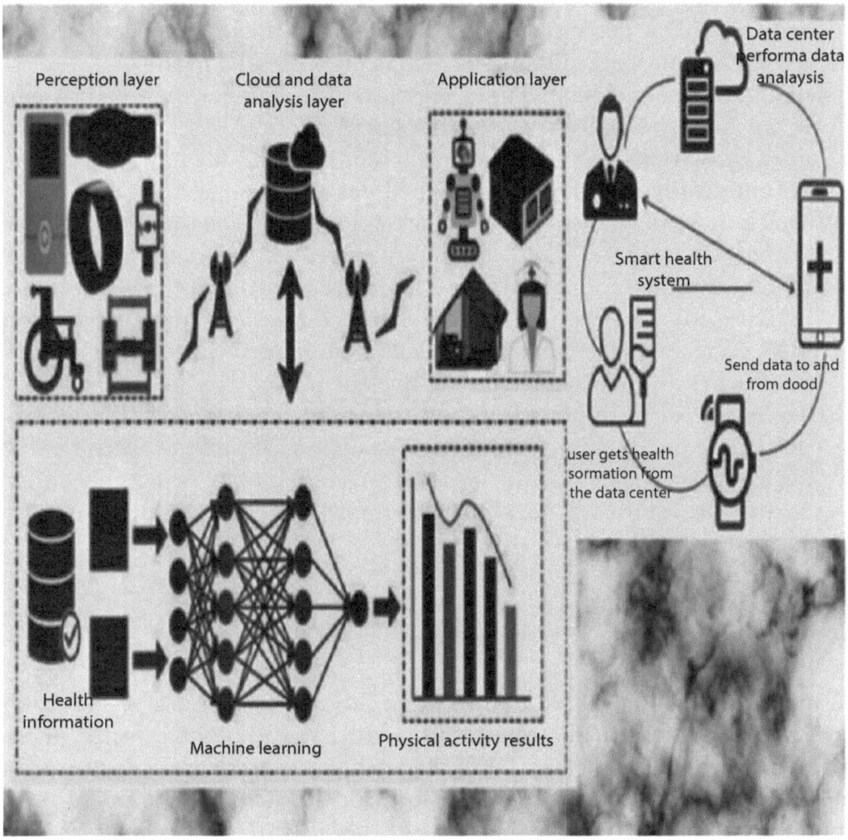

Figure 20.4 IoT-based monitoring system.

wearables noninvasive, enjoyable, convenient, and safe [34–36]. For continuous blood sugar monitoring, an M-IoT noninvasive glucometer was recommended [37].

Temperature monitoring: Several diagnostics assess homeostasis using core body temperature. Gunawan made an Arduino-Raspberry Pi IoT temperature monitor. A mobile-friendly webpage provided temperature data from a database [38, 39].

Blood pressure (BP) monitoring: Hospital patients need regular measurements of blood pressure. IoT-based BP monitors assess systolic and diastolic pressure. Cloud storage is possible. This device's accuracy and efficiency were confirmed by 100 people. Guntha's IoT-based BP system of measurement uses cloud and fog computing [40].

Oxygen saturation monitoring: Pulse oximeters detect oxygen saturation levels noninvasively, making them useful medical analytical tools. The microcontroller server allows pulse oximeter and WLAN router integration. Von Chong et al. [41] suggest a multispectral sensor to reduce LED's detrimental impacts. A low-power, low-cost remote patient monitoring system has also been presented [42].

Medication management: Fuzzy logic analyses temperature sensor data in an adaptive IoT-based smart pharmaceutical system. The device efficiently treats fever by constantly monitoring body temperature and automatically modifying drug schedules and dosages.

Wheelchair management: The IoT can track wheelchair usage, performance, and repair. This technology improves wheelchair maneuverability. Internet-connected wheelchairs track battery life, position, and activity. IoT can also detect wheelchair difficulties quickly, enabling fast service. IoT-enabled wheelchair management could improve healthcare and wheelchair users' lives.

IoT-based medical applications: As the importance of IoT devices was not immediately apparent, several academic disciplines were slow to accept them. These could be used in hemoglobin detection, cancer treatment, cellular abnormality monitoring, surgical procedures performed remotely, etc.

20.7 CONCLUSION

The Internet of Things-based electronic healthcare system is a game-changing technical innovation that is revolutionizing the healthcare sector. It has improved the quality, efficiency, and affordability of healthcare delivery dramatically. Healthcare practitioners can provide more individualized care and respond more quickly to patient needs when they use IoT devices and sensors to track patient vital signs in real time. Better diagnoses, treatments, and results have resulted from the system's improved medical data management, which has made it simpler for healthcare providers to obtain and analyze patient information. The IoT-based e-healthcare system offers considerable potential to improve healthcare delivery and the overall patient experience, despite some limitations such as data security and privacy concerns. This has changed the focus of healthcare from institutions to individuals. In this chapter, we described the features and developments in H-IoT gadgets. Problems with the H-IoT device's development, production, and deployment were detailed. Future progress and attention over the next few years will be shaped by these issues. This chapter serves as an up-to-date resource for readers interested in learning more about H-IoT devices and developing new uses for them.

REFERENCES

1. A.M. Rahmani, T.N. Gia, B. Negash, A. Anzanpour, I. Azimi, M. Jiang, and P. Liljeberg, "Exploiting Smart e-Health Gateways at the Edge of Healthcare

Internet-of-Things: A Fog Computing Approach," *Future Generation Computer Systems*, vol. 78, pp. 641–658, 2018.

2. Qingling, Liu, K. G. Mkongwa, and C. Zhang, "Performance Issues in Wireless Body Area Networks for the Healthcare Application: A Survey and Future Prospects," *SN Applied Sciences*, vol. 3, pp. 1–19, 2021.
3. Barbaros Preveze, Ahmed Alkhayyat, Firas Abedi, Aqeel Mahmood Jawad, and Ali S. Abosinnee, "SDN-Driven Internet of Health Things: A Novel Adaptive Switching Technique for Hospital Healthcare Monitoring System," *Wireless Communications and Mobile Computing*, vol. 2022, 2022.
4. V. Subramaniya Swamy, G. Manogaran, R. Logesh, V. Vijayakumar, N. Chilamkurti, D. Malathi, and N. Senthil Selvan, "An Ontology-Driven Personalized Food Recommendation in IoT-Based Healthcare System," *The Journal of Supercomputing*, vol. 75, pp. 3184–3216, 2018.
5. P. Verma, S. K. Sood, and S. Kalra, "Cloud-Centric IoT Based Student Healthcare Monitoring Framework," *Journal of Ambient Intelligence and Humanized Computing*, vol. 9, no. 5, pp. 1293–1309, 2018.
6. Nafees Akhter Farooqui, Ritika Mehra, and Samrah Butool Faridi, "An Intellectual Analysis of Structural Healthcare Systems in India Using Intelligence-Based Techniques." In *Security Implementation in Internet of Medical Things*, pp. 37–58. CRC Press, 2023.
7. V. Gupta, H. Singh Gill, P. Singh, and R. Kaur, "An Energy Efficient Fog-Cloud Based Architecture for Healthcare," *Journal of Statistics and Management Systems*, vol. 21, no. 4, pp. 529–537, 2018.
8. M. Elhoseny, G. Ramírez-González, O. M. Abu-Elnasr, S. A. Shawkat, N. Arunkumar, and A. Farouk, "Secure Medical Data Transmission Model for IoT-Based Healthcare Systems," *IEEE Access*, vol. 6, pp. 20596–20608, 2018.
9. Braimoh Ikharo, Adaorah Obiagwu, Chukwuemeka Obasi, Suleiman U. Hussein, and Precious Akah, "Security for Internet-of-Things Enabled E-Health Using Blockchain and Artificial Intelligence: A Novel Integration Framework." In 2021 1st International Conference on Multidisciplinary Engineering and Applied Science (ICMEAS), pp. 1–4. IEEE, 2021.
10. Chi, Hao Ran, Maria de Fátima Domingues, Hongxu Zhu, Chunguo Li, Kazuyuki Kojima, and Ayman Radwan. "Healthcare 5.0: In the Perspective of Consumer Internet-of-Things-Based Fog/Cloud Computing." *IEEE Transactions on Consumer Electronics*, 2023.
11. B. Oryema, "Design and implementation of an interoperable messaging system for IoT healthcare Services," in Proceedings of the 2017 14th IEEE Annual Consumer Communications & Networking Conference (CCNC), pp. 45–52, Las Vegas, NV, USA, January 2017.
12. A. Ahad, M. Tahir, and K.-L. A. Yau, "5G-Based Smart Healthcare Network: Architecture, Taxonomy, Challenges and Future Research Directions," *IEEE Access*, vol. 7, pp. 100747–100762, 2019.
13. M. N. Birje and S. S. Hanji, "Internet of Things Based Distributed Healthcare Systems: A Review," *Journal of Data, Information and Management*, vol. 2, 2020.
14. K. T. Kadhim, "An Overview of Patient's Health Status Monitoring System Based on Internet Of Things (IoT)," *Wireless Personal Communications*, vol. 114, pp. 1–28, 2020.

15. Y. Yuehong, "The Internet of Things in Healthcare: An Overview," *Journal of Industrial Information Integration*, vol. 1, pp. 3–13, 2016.
16. G. Shanmugasundaram and G. Sankarikaarguzhali, "An Investigation on IoT Healthcare Analytics," *International Journal of Information Engineering and Electronic Business*, vol. 9, no. 2, p. 11, 2017.
17. J.-Y. Lee and R. A. Scholtz, "Ranging in a Dense Multipath Environment Using an UWB Radio Link," *IEEE Journal on Selected Areas in Communications*, vol. 20, pp. 1677–1683, 2002.
18. H. Aftab, K. Gilani, J. Lee, L. Nkenyereye, S. Jeong, and J. Song, "Analysis of Identifiers in IoT Platforms," *Digital Communications and Networks*, vol. 6, no. 3, pp. 333–340, 2020.
19. G. Cerruela Garc´ıa, I. Luque Ruiz, and M. G´omez-Nieto, "State of the Art, Trends and Future of Bluetooth Low Energy, A. D. Dwivedi, L. Malina, P. Dzurenda, and G. Srivastava, "Optimized blockchain model for internet of things-based healthcare applications," 2019 42nd Int. Conf. Telecommun. Signal Process. TSP 2019, pp. 135–139, 2019, doi: 10.1109/TSP.2019.8769060.
20. M. S. Hossain and G. Muhammad, "Cloud-assisted Industrial Internet of Things (IIOT) Enabled framework for health monitoring," *Computer Networks*, vol. 101, pp. 192–202, 2016, doi: 10.1016/j.comnet.2016.01.009.
21. A. Darwish, A. E. Hassanien, M. Elhoseny, A. K. Sangaiah, and K. Muhammad, "The Impact of the Hybrid Platform of Internet of Things and Cloud Computing on Healthcare Systems: Opportunities, Challenges, and Open Problems," *Journal of Ambient Intelligence and Humanized Computing*, vol. 10, no. 10, pp. 4151–4166, 2019, doi: 10.1007/s12652-017-0659-1.
22. H. Ahmadi, G. Arji, L. Shahmoradi, R. Safdari, M. Nilashi, and M. Alizadeh, *The application of internet of things in healthcare: a systematic literature review and classification*, vol. 18, no. 4. Springer Berlin Heidelberg, 2019.
23. D. Pal, S. Funilkul, N. Charoenkitkarn, and P. Kanthamanon, "Internet-of-Things and Smart Homes for Elderly Healthcare: An End User Perspective," *IEEE Access*, vol. 6, pp. 10483–10496, 2018, doi: 10.1109/ACCESS.2018.2808472.
24. S. B. Baker, W. Xiang, and I. Atkinson, "Internet of Things for Smart Healthcare: Technologies, Challenges and Opportunities," *IEEE Access*, vol. 5, pp. 26521–26544, 2017, doi: 10.1109/ACCESS.2017.2775180.
25. G. Manogaran, R. Varatharajan, D. Lopez, P. M. Kumar, R. Sundarasekar, and C. Thota, "A New Architecture of Internet of Things and Big Data Ecosystem for Secured Smart Healthcare Monitoring and Alerting System," *Future Generation Computer Systems*, vol. 82, pp. 375–387, 2018, doi: 10.1016/j.future.2017.10.045.
26. L. M. Dang, M. J. Piran, D. Han, K. Min, and H. Moon, "A Survey on Internet of Things and Cloud Computing for Healthcare," *Electronics*, vol. 8, no. 7, p. 768, 2019.
27. B. Oryema, "Design and Implementation of an Interoperable Messaging System for IoT Healthcare Services," in Proceedings of the 2017 14th IEEE Annual Consumer Communications & Networking Conference (CCNC), pp. 45–52, Las Vegas, NV, USA, January 2017.

28. A. Ahad, M. Tahir, and K.-L. A. Yau, "5G-Based Smart Healthcare Network: Architecture, Taxonomy, Challenges and Future Research Directions," *IEEE Access*, vol. 7, pp. 100747–100762, 2019.
29. M.N Birje and S. S. Hanji, "Internet of Things Based Distributed Healthcare Systems: A Review," *Journal of Data, Information and Management*, vol. 2, 2020.
30. K. T. Kadhim, "An Overview of Patient's Health Status Monitoring System Based on Internet of Things (IoT)," *Wireless Personal Communications*, vol. 114, pp. 1–28, 2020.
31. M. Bansal and B. Gandhi, "IoT & Big Data in Smart Healthcare (ECG Monitoring)," in Proceedings of the 2019 International Conference on Machine Learning, Big Data, Cloud and Parallel Computing (COMITCon), pp. 390–396, Faridabad, India, February 2019.
32. H. Djelouat, M. Al Disi, I. Boukhenoufa et al., "Real-Time ECG Monitoring Using Compressive Sensing on a Heterogeneous Multicore Edge-Device," *Microprocessors and Microsystems*, vol. 72, p. 102839, 2020.
33. Al-Kababji, "IoT-Based Fall and ECG Monitoring System: Wireless Communication System-Based Firebase Realtime Database," in Proceedings of the 2019 IEEE Smart World, Ubiquitous Intelligence & Computing, Advanced & Trusted Computing, Scalable Computing & Communications, Cloud & Big Data Computing, Internet of People and Smart City Innovation (Smart World/SCALCOM/UIC/ATC/CBDCom/IOP/SCI), pp. 1480–1485, Leicester, UK, August 2019.
34. T. N. Gia et al., "Energy Efficient Fog-Assisted IoT System for Monitoring Diabetic Patients with Cardiovascular Disease," *Future Generation Computer Systems*, vol. 93, pp. 198–211, 2019.
35. T. N. Gia et al., "Energy Efficient Fog-Assisted IoT System for Monitoring Diabetic Patients with Cardiovascular Disease," *Future Generation Computer Systems*, vol. 93, pp. 198–211, 2019.
36. G. M. Bhat and N. G. Bhat, "A Novel IoT Based Framework for Blood Glucose Examination," in Proceedings of the 2017 International Conference on Electrical, Electronics, Communication, Computer, and Optimization Techniques (ICEECCOT), pp. 205–207, Mysuru, India, December 2017.
37. T. N. Gia, M. Ali, I. B. Dhaou et al., "IoT-Based Continuous Glucose Monitoring System: A Feasibility Study," *Procedia Computer Science*, vol. 109, pp. 327–334, 2017.
38. Sargunam and S. Anusha, "IoT Based Mobile Medical Application for Smart Insulin Regulation," in Proceedings of the 2019 IEEE International Conference on Electrical, Computer and Communication Technologies (ICECCT), pp. 1–5, Erode, India, October 2019.
39. R. S. Istepanian, "The Potential of Internet of m-health Things "m-IoT" for Non-Invasive Glucose Level Sensing," in Proceedings of the 2011 Annual International Conference of the IEEE Engineering in Medicine and Biology Society, pp. 5264–5266, Boston, MA, USA, March 2011.
40. Sunny and S. S. Kumar, "Optical Based Non-Invasive Glucometer with IoT," in Proceedings of the 2018 International Conference on Power, Signals, Control and Computation (EPSCICON), pp. 1–3, Thrissur, India, January 2018.

41. Von Chong, M. Terosiet, A. Histace, and O. Romain, "Towards a Novel Single-LED Pulse Oximeter Based on a Multispectral Sensor for IoT Applications," *Microelectronics Journal*, vol. 88, pp. 128–136, 2019.
42. E. C. Larson, "SpiroSmart: Using a Microphone to Measure Lung Function on a Mobile Phone," in Proceedings of the 2012 ACM Conference on Ubiquitous Computing, pp. 280–289, New York, NY, USA, September 2012.

Index